Bill Frohlich

Aquifer Testing

Design and Analysis of Pumping and Slug Tests

Karen J. Dawson
Jonathan D. Istok   Jack. rstok@orst.edu

Department of Civil Engineering 800-272-0082
Oregon State University
Corvallis, Oregon

LEWIS PUBLISHERS

Library of Congress Cataloging-in-Publication Data

Dawson, Karen J.
    Aquifer testing: design and analysis of pumping and slug tests/
Karen J. Dawson, Jonathan D. Istok.
        p.   cm.
    Includes bibliographic references and index.
    1. Groundwater flow--Measurement.   2. Aquifers--Measurement.
I. Istok, J. D.   II. Title.
GB1197.7.D39   1991
551.49--dc20   91-9392
ISBN 0-87371-501-2

LEWIS PUBLISHERS, INC.
121 South Main Street, Chelsea, Michigan 48118   313

PRINTED IN THE UNITED STATES OF AMERICA

To my husband Clay for making sure I smile every morning.

To my wife and friend Joan who makes life worth living and to my friends Ken, Ann, Kevin, Alan, and Lorrie who are always there when I need them.

Preface

Increased concern for the groundwater resource has arisen in the last few years. Stricter environmental regulations and growing competition for a limited resource have led to large increases in the number of groundwater investigations. These investigations almost always include the determination of aquifer properties (e.g., hydraulic conductivity and storativity) using aquifer tests, primarily pumping tests and slug tests. Because of the interdisciplinary nature of groundwater investigations, aquifer test design and data analysis are performed by people with a wide range of backgrounds including civil engineering, geology, water resources, and soil science. Very few of these people have received formal training in aquifer testing and this can sometimes lead to poor test design or incorrect interpretation of test data.

Although there is a large body of literature on this subject, most references explain only one or two methods. The use of inconsistent notation and advanced mathematics tend to obscure the simple principles (Darcy's Law, conservation of mass) on which all of these methods are based. Often the underlying assumptions required for method development are not clearly stated and this can make the selection of an appropriate method difficult. The application of a method may also require the user to perform complex calculations requiring specialized computer programs that are not widely available. Thus, there is a need for an easy-to-use reference where the available methods are described in sufficient detail and with the appropriate computer programs to permit routine application.

This manual was written to serve as a comprehensive reference for the design and analysis of pumping tests and slug tests for practicing groundwater hydrologists. The manual is divided into two parts. Part I contains general information about aquifer tests from preliminary test design to data analysis. Emphasis is placed on the proper selection of a conceptual model of the aquifer system. Part II contains a comprehensive description of nineteen conceptual models that may be used for test design and for the interpretation of test data. Each chapter is devoted to a different model and describes the hydrogeologic setting, a complete list of the assumptions used to derive the solutions for that model, and a step-by-step procedure for interpreting test data. Examples are presented to illustrate each step in the solution process. An extensive collection of computer programs (written in Basic) for use with each method is contained in Appendix B; these programs are also contained in the diskettes provided.

Acknowledgments

The authors would like to thank the many people who helped with the preparation of this book. The computer programs in Appendix B were written by Michael J. Stevens. Many of the drawings were prepared by Joan Istok. The first draft was typed by Elvina Lim. The students of my groundwater classes made numerous suggestions for improvement (and caught many blunders!). Bill Winkler and the members of the Faculty/Staff Fitness program kept our bodies healthy and our spirits strong!

Funding for this publication was provided by the Office of Research and Development, U.S. Environmental Protection Agency, under agreement R-815738-01 through the Western Region Hazardous Substance Research Center. The content of this publication does not necessarily represent the views of the agency.

This book would never have been completed without the tireless efforts of Ann Daggett. She served as chief editor, proofreader, typesetter, and cheerleader. Thanks Ann!

Karen Dawson is a member of the geotechnical staff of CH2M Hill in Bellevue, Washington. She received an M.S. in geotechnical engineering with an environmental engineering minor and focus on groundwater from Oregon State University. She also has B.S. degrees in civil and forest engineering from Oregon State University. Karen was formerly an employee of the Washington State Department of Transportation and involved with the design of the Seattle I-90 project.

Jonathan Istok is an Associate Professor in the Department of Civil Engineering at Oregon State University. He specializes in the development of statistical and mathematical models for groundwater flow and contaminant transport processes. He is the author of the American Geophysical Union Water Resources Monograph "Groundwater modeling by the finite element method."

Contents

Contents

Contents

Contents

Contents

Aquifer Testing

Design and Analysis of Pumping and Slug Tests

Chapter 1

INTRODUCTION

1.1   AQUIFER TESTS

The reader is assumed to have a basic understanding of groundwater hydraulics. This section briefly reviews terminology and a few fundamental concepts. Additional background material can be obtained from a number of introductory texts (e.g., Freeze and Cherry, 1979; de Marsily, 1986).

An *aquifer* is a geological formation or group of formations that yields significant quantities of water to wells. An *aquitard* is a formation through which water moves very slowly. An *aquiclude* is a formation which is for all practical purposes impermeable. It is possible for a soil or rock type to act as an aquifer or an aquitard in different geologic settings. For example, a layer of sandy silt may act as an aquifer when bounded by clay deposits, but as an aquitard when it occurs adjacent to a sand deposit. An *aquifer system* may consist of several aquifers, aquitards, and aquicludes.

In an aquifer system water levels and water pressure are expressed in terms of *piezometric head* (or simply *head*). Piezometric head is a measure of the energy per unit weight of groundwater relative to water in an arbitrary reference state. Since groundwater velocities are usually very small, kinetic energy is neglected and piezometric head at any point is defined as

$$h = z + \frac{p}{\gamma_w} \tag{1.1}$$

where

z = distance of the point above an arbitrarily located datum, L
p = water pressure at the point, $MT^{-2}L^{-1}$
γ_w = unit weight of water, $MT^{-2}L^{-2}$

z and p/γ_w are called the *elevation head* and *pressure head*, respectively (in this manual, combinations of the symbols M, L, and T are used to define the dimensions of a quantity, where M refers to mass, L refers to length, and T refers to time).

A *confined aquifer* is bounded above and below by an aquitard or aquiclude and the saturated thickness of the aquifer is determined by the elevations of the contacts (Figures 1.1, 1.2a). The *piezometric surface* defines the height to which groundwater will rise in a well that is screened over the entire aquifer saturated thickness. In a confined aquifer, the pressure head of the groundwater at the top of the aquifer is always greater than zero. If the piezometric surface is above the land surface at any point, groundwater in a well will flow to the surface without pumping (an artesian well). If the hydraulic conductivity of the upper or lower aquitard (or both) is large enough, the vertical flow of groundwater through these layers may be important; in this case the aquifer is sometimes called *partially-confined* and the aquitard through which water flows is called *leaky*. An *unconfined aquifer* (or *water table aquifer*) is bounded below by an aquitard or aquiclude and the saturated thickness of the aquifer is determined by the elevation on this contact and the position of the free surface (zero pressure head) called the *water table* (Figure 1.2b).

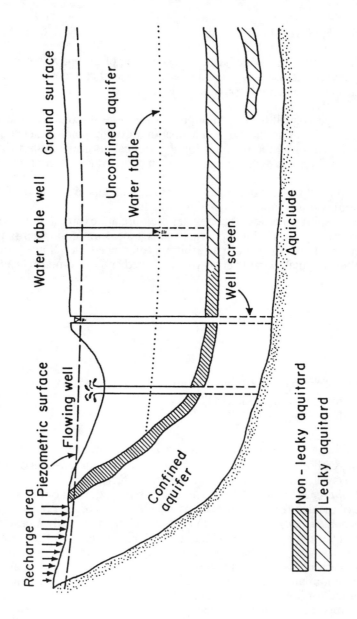

Figure 1.1   Confined and unconfined aquifers (after Hantush, 1964).

a

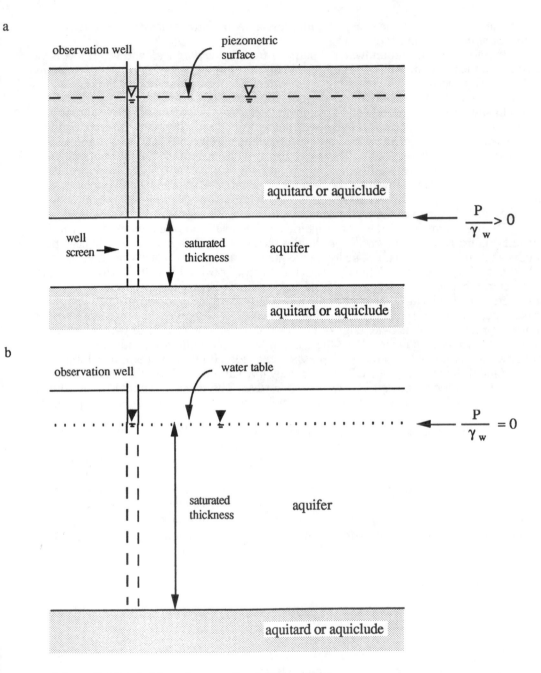

Figure 1.2   Confined (a) and unconfined (b) aquifers.

Aquifer tests are performed to determine the hydraulic properties of aquifer systems. The values of these properties for a specific aquifer are normally of interest, but the analyst will frequently need to determine these properties for aquitards that lie above or below the aquifer of interest. These properties may be needed, for example, to design municipal water supply well fields, to predict rates and directions of groundwater flow, and to design effective groundwater remediation systems (e.g., the "pump and treat" systems used at many hazardous waste sites). Aquifer tests may be further classified into two groups: *pumping tests* and *slug tests* .

In pumping tests, groundwater is pumped from a *pumping well* and water levels are measured in the pumping well and (perhaps) in one or more *observation wells*. A *constant discharge* test is performed by pumping at a constant rate for the duration of the test. In a *step discharge* test, the pumping rate is periodically increased during the test.

In slug tests, the groundwater level in an *injection or withdrawal well* is abruptly raised or lowered and water levels in the well are measured for a period of time following the initial change. *Slug injection* tests are performed by abruptly raising the water level in the well casing, usually by pouring water into the well casing from a container of known volume or by dropping a closed section of pipe into the well casing. *Slug withdrawal* tests are performed by abruptly lowering the water level in the well casing, usually by bailing water from the well, using compressed air to push water out of the well, or by removing a section of closed pipe from the well casing.

During a pumping test, the piezometric head in a surrounding *cone of depression* is lowered (Figure 1.3). The *hydraulic gradient* is the rate of change in piezometric head per unit distance along a flow path. The hydraulic gradient and groundwater velocities are highest near the pumping well and decrease radially away from the well. *Drawdown*, s, is the difference between the piezometric head at some time after the test begins and the piezometric head existing at the start of the test. Drawdown is computed from measurements of water levels in the pumping and observation wells. The distance from the pumping well center to the outer edge of the cone of depression (where drawdown is zero) is called the *radius of influence*, R, of the pumping well. As pumping continues, drawdown increases and the cone of depression expands. If pumping continues long enough (and if the pumping rate is constant) drawdown and the radius of influence become constant, a condition called *equilibrium* or *steady-state*. The *specific capacity* of the aquifer is the ratio of discharge to drawdown at the pumping well at a specified time since pumping began.

The same process occurs during a slug withdrawal test. During a slug injection test, however, the piezometric head in the surrounding aquifer is raised and the term *buildup* is used to refer to the difference between the piezometric heads before and during the test.

Pumping tests and slug tests are performed to determine several properties of an aquifer system. *Hydraulic conductivity* , K, (often referred to as *permeability* or the *coefficient of permeability*) is the rate at which groundwater flows through a unit area of aquifer or aquitard under a unit hydraulic gradient. It has dimensions of velocity (LT^{-1}) with typical units of ft/day, gal/(day $\cdot$ ft^2), m/sec, cm/sec, or m^3/(day $\cdot$ m^2). If the hydraulic conductivity and hydraulic gradient are known, the *apparent groundwater velocity* , v (LT^{-1}), can be computed using *Darcy's Law*

$$v = -K\frac{dh}{dx} \tag{1.2}$$

where dh/dx is the hydraulic gradient. The negative sign in Equation 1.2 is required because groundwater flows in the direction of decreasing head (dh/dx is negative).

Intrinsic permeability, k, is defined as $k = K\mu/\rho g$ where ρ is the fluid density, ML^{-3}, g is the gravitational acceleration, LT^{-2}, and μ is the fluid absolute viscosity, $ML^{-1}T^{-1}$. It is usually sufficient to describe groundwater flow in terms of hydraulic conductivity, K, since groundwater density and viscosity can be assumed to be constant in many aquifer systems. Important exceptions are aquifer tests conducted in geothermal aquifer systems, in very deep boreholes, in highly saline groundwater, or in the presence of nonaqueous phase liquids (e.g., petroleum).

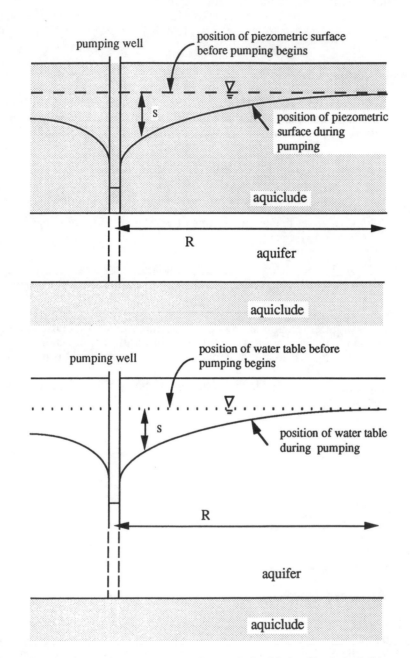

Figure 1.3 **Groundwater flow to a pumping well in (a) confined and (b) unconfined aquifers.**

Transmissivity (also called *transmissibility*), T, is the product of hydraulic conductivity and saturated thickness of the aquifer or aquitard

$$T = Km, \quad L^2T^{-1} \tag{1.3}$$

where

   m = saturated thickness, L

Storativity, S, (also called the *coefficient of storage*) is a dimensionless coefficient that represents the volume of water an aquifer or aquitard releases from storage per unit surface area per unit decrease in piezometric head. Storativity is made up of two terms:

$$S = S_y + S_s m \tag{1.4}$$

where

S_y = *specific yield*, dimensionless
    = the volume of water that a unit volume of aquifer or aquitard releases from storage by gravity drainage
S_s = *specific storage*, L^{-1}
    = the volume of water that a unit volume of aquifer or aquitard releases from storage by the expansion of water and compression of the soil or rock skeleton

In a confined aquifer, the aquifer remains saturated during pumping and specific yield is zero (Figure 1.4a). In an unconfined aquifer, a portion of the aquifer is dewatered by gravity drainage as the water table is lowered by pumping and specific yield is nonzero (Figure 1.4b).
   Specific storage can be expressed in terms of the compressibility of the aquifer or aquitard and the groundwater

$$S_s = \gamma_w \left(\frac{\theta}{E_w} + \frac{1}{E_s} \right) \tag{1.5}$$

where

γ_w = unit weight of water , $MT^{-2}L^{-2}$
θ = effective porosity , dimensionless
E_w = bulk modulus of elasticity of water, $M^{-1}T^2L$
E_s = modulus of elasticity of the aquifer or aquitard , $M^{-1}T^2L$

Porosity is the ratio of the volume of voids to the total volume. The voids include all liquid or air-filled spaces in the media, including water in immobile films surrounding soil particles and in pores not available for flow because of discontinuities. Conversely, the *effective porosity* includes only the portion of the volume of voids available for flow. In coarse-grained soils, the difference between porosity and effective porosity is very small, but the difference can be very large in fine-grained soils. The pore space is often expressed as the ratio of volume of voids to volume of solids, termed *void ratio*, e. Porosity, n, is related to the void ratio by

$$n = \frac{e}{1 + e} \tag{1.6}$$

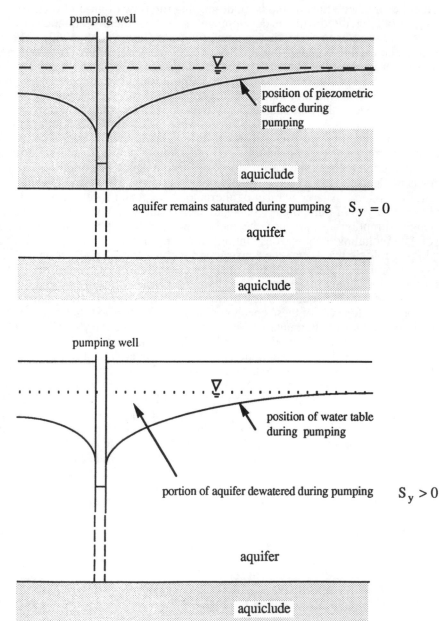

a

pumping well

∇

position of piezometric
surface during
pumping

aquiclude

aquifer remains saturated during pumping $S_y = 0$

aquifer

aquiclude

b

pumping well

∇

position of water table
during pumping

portion of aquifer dewatered during pumping $S_y > 0$

aquifer

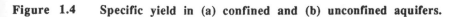

aquiclude

Figure 1.4   Specific yield in (a) confined and (b) unconfined aquifers.

1.2   ANALYSIS OF AQUIFER TEST DATA

The development of methods for interpreting aquifer test data has paralleled the development of mathematical methods for describing the flow of groundwater to wells. Thiem is generally credited with the development in 1906 of the first equations to describe the flow of groundwater to a well during a pumping test (Ferris et al., 1962). The Thiem equations can be used to compute transmissivity from measurements of drawdown in the pumping well or in observation wells. In the development of these equations Thiem was required to make a large number of simplifying assumptions about the geometry and hydraulic behavior of the aquifer including: the aquifer is isotropic and homogeneous and of infinite lateral extent, the pumping well is screened over the entire saturated thickness of the aquifer (the pumping well is said to *fully penetrate* the aquifer), the aquifer is bounded above by an aquiclude that is horizontal and of infinite lateral extent, the pumping rate is constant, and pumping has continued for a sufficient length of time to achieve equilibrium conditions (a complete list of the assumptions used in the development of the Thiem equations is in Chapter 6). Collectively these assumptions constitute a *conceptual model* for the aquifer system. A large number of alternate conceptual models have been proposed for use in the analysis of pumping test data and each of these is based on a different set of simplifying assumptions. In practice, the analyst selects a conceptual model based on a variety of data including: geologic descriptions of the aquifer materials obtained from drilling logs, geologic maps, etc., the construction records for the production well and (perhaps) one or more observation wells, the drawdown data collected during the test, and a set of assumptions about the geometry and hydraulic behavior of the aquifer system. Analytical solutions to the equations describing groundwater flow during pumping tests have been obtained for each of these conceptual models and methods have been developed for using these solutions to compute aquifer system properties from pumping test data.

A list of the conceptual models considered in this manual is in Table 1.1. Table 1.1 also includes a partial list of the assumptions used in the development of each model (complete lists of the assumptions are in Chapters 6 to 21). Methods have been developed for the analysis of drawdown data for equilibrium and transient flow conditions and for confined aquifers with and without aquitard leakage and storage, unconfined aquifers, confined aquifers undergoing conversion to an unconfined aquifer, fissure-block systems, and for "aquitard-aquifers." Some methods include the effects of pumping well storage, partial pumping and observation well penetration, and anisotropic aquifer properties.

Analytical solutions to the equations describing groundwater flow during slug tests have been obtained for several conceptual models and methods have been developed for using these solutions to compute aquifer system properties from drawdown or buildup data. A list of the conceptual models considered in this manual and a partial list of the assumptions for each method are in Table 1.2.

The values of aquifer properties computed by a particular method can only be considered correct if the assumptions included in the conceptual model on which the method is based are valid for the particular aquifer system being tested. Because the computed values of aquifer properties depend on the choice of conceptual model used to analyze the test data, the selection of an appropriate conceptual model is the single most important step in the analysis of aquifer test data.

Table 1.1 Selected conceptual models for interpreting pumping test data.

Model number	Flow condition	Aquifer type	Aquitard leakage	Aquitard storage	Well storage	Partial well penetration	Anisotropic properties	Reference
1	equilibrium	confined	no	no	no	no	no	Thiem (1906)
2	equilibrium	unconfined	-	-	no	no	no	Thiem (1906)
3	transient	confined	no	no	no	no	no	Theis (1935)
4	transient	confined	yes	no	no	no	no	Hantush and Jacob (1955)
5	transient	confined	yes	yes	no	no	no	Hantush (1964)
6	transient	confined	no	no	no	yes	yes	Hantush (1964)
7	transient	confined	yes	no	no	yes	yes	Hantush (1964)
8	transient	confined	no	no	yes	no	no	Papadopulos and Cooper (1967)
9	transient	confined	yes	no	yes	no	no	Lai and Su (1974)

Table 1.1   (Continued).

Model number	Flow condition	Aquifer type	Aquitard leakage	Aquitard storage	Well storage	Partial well penetration	Anisotropic properties	Reference
10	transient	confined (fissure-block system)	yes	yes	no	no	no	Boulton and Streltsova (1977)
11	transient	confined	no	no	no	no	yes	Papadopulos (1965)
12	transient	confined to unconfined aquifer conversion	no	no	no	no	no	Moench and Prickett (1972)
13	transient	unconfined	no	no	no	no	yes	Neuman (1972)
14	transient	unconfined	no	no	no	yes	yes	Neuman (1974)
15	transient	unconfined	no	no	yes	yes	yes	Boulton and Streltsova (1976)
16	transient	unconfined ("aquitard-aquifer")	yes	yes	no	yes	yes	Boulton and Streltsova (1975)

Table 1.2   **Selected conceptual models for interpreting slug test data.**

Model number	Flow condition	Aquifer type	Aquitard leakage	Aquitard storage	Partial well penetration	Anisotropic properties	Reference
17	transient	confined	no	no	yes	yes	Hvorslev (1951)
18	transient	unconfined or leaky confined	yes	no	yes	no	Bouwer and Rice (1976)
19	transient	confined	no	no	no	no	Cooper et al. (1967)

1.3   ORGANIZATION OF THIS MANUAL

The remaining four chapters in Part I contain information useful to the design and the analysis of an aquifer test. Step-by-step procedures for design are in Chapter 2. Chapter 3 describes methods for correcting test data for external conditions (e.g., changes in barometric pressure or surface loadings) that occur during a test. An overview of the methods used in aquifer test data analysis is in Chapter 4. Methods for analyzing pumping test data from step-discharge tests are described in Chapter 5.

The chapters in Part II contain descriptions of the methods for analyzing aquifer test data for each conceptual model in Tables 1.1 and 1.2. Each chapter contains a definition sketch, a list of assumptions, governing differential equations, boundary and initial conditions, an analytical solution to the governing equations, one or more methods for analyzing pumping test or slug test data to obtain aquifer system properties, and examples that illustrate the application of these methods.

Chapter 2

AQUIFER TEST DESIGN

Proper aquifer test design is essential to the successful determination of aquifer system properties. Without proper placement of observation wells, the selection of an appropriate pumping rate, injection or withdrawal volume, and a sufficient duration of measurement, collected data could be worthless. Careful test design may also allow the analysis of test data to be performed with simpler analytical methods. The steps that follow are presented with the assumption that a preliminary site investigation has been completed. The information collected during this investigation should include a description of the geologic materials, approximate thicknesses of formations, the identification of potential aquifers, aquitards, and aquicludes, and estimated water levels.

2.1 STEP 1. IDENTIFY SITE CONSTRAINTS

In most cases, conditions at the site impose constraints on aquifer test design. These should be recorded prior to the start of the design process. Some examples are:

1. Limitations on the placement of pumping, injection, or withdrawal wells and observation wells due to the locations of buildings, roads, property boundaries, and existing wells.

2. Limitations on pumping rate(s) for example, to prevent excessive drawdown in nearby wells or to limit the amount of pumped water to be disposed.

3. Limitations on test duration imposed by the project budget, or by difficulties in scheduling site access, equipment, and personnel.

4. Requirement that existing wells or pumps be used. This may place restrictions on the number and locations of observation wells and may introduce additional complexities into the analysis of test data including the effects of well storage, partial penetration of the aquifer by pumping and observation wells, and head losses due to flow through the pumping well screen and pump intake.

5. Limitations on the placement of pumping and observation wells due to the known or suspected presence of lateral discontinuities in aquifers and aquitards, the presence of aquifer recharge and discharge zones, etc. These features may also introduce complexities into the analysis of test data for example, the need to use image well theory (see Chapter 4).

2.2 STEP 2. LIST REQUIRED AQUIFER SYSTEM PROPERTIES

A list of required aquifer system properties is needed to select an appropriate conceptual model (see Section 2.4). Lists of the properties determined by each of the conceptual models considered in this manual are in Tables 2.1 and 2.2. Using the list of

Table 2.1   Pumping test design table.

Conceptual Model	Model Description	Aquifer System Properties Determined by Application of Model[1]	Simpler Alternative Model	Aquifer System Properties Determined by Alternative Model	Conditions on Drawdown Data to Apply Alternate Model
1	EQUILIBRIUM, CONFINED	• T	None		
2	EQUILIBRIUM, UNCONFINED	• K	None		
3	TRANSIENT, CONFINED	• T, S	1	• T	• equilibrium data • at least two observation wells or one well and an estimate of the radius of influence, R
4	TRANSIENT, CONFINED, LEAKY	• T, S always • K' if early data are available or if equilibrium data are available for at least three observation wells (equilibrium can be assumed when $t > \dfrac{8m'S}{K'}$)	3	• T, S	• $r < 0.01\sqrt{Tm'/K'}$

Table 2.1   (Continued).

Conceptual Model	Model Description	Aquifer System Properties Determined by Application of Model[1]	Simpler Alternative Model	Aquifer System Properties Determined by Alternative Model	Conditions on Drawdown Data to Apply Alternate Model
5	TRANSIENT, CONFINED, LEAKY, AQUITARD STORAGE	With early data only: • T, S always • (K'S'), (K"S") if the relation between (K'S') and (K"S") is known	4	• T, S, K'	• $r < 0.04 \mathrm{m}[KS_s/K'S_s']^{1/2}$
		With late data only: • T, S + S' for Case A • T, S, K' for Cases B and C if equilibrium data are available for at least three observation wells		• T, S • K' if equilibrium data are available for at least three observation wells	• $t > [(5\ S'_s m'^2)/K']$
		With early and late data: • T, S, K', S', K", S" for Case A if the relation between (K'S') and (K"S") is known, and for Cases B and C if equilibrium data are available for at least three observation wells	4		

Table 2.1 (Continued).

Conceptual Model	Model Description	Aquifer System Properties Determined by Application of Model[1]	Simpler Alternative Model	Aquifer System Properties Determined by Alternative Model	Conditions on Drawdown Data to Apply Alternate Model
6	TRANSIENT, CONFINED, PARTIAL PENETRATION, ANISOTROPIC	• K_r, S always • K_z as long as $r < 1.5m\sqrt{K_r/K_z}$, observation wells are not screened through the entire aquifer thickness, and a full time range of data is available	3	• T, S	• $r > 1.5m\sqrt{K_r/K_z}$, or $r > m\sqrt{K_r/K_z}$ and $t > (10Sm/4K_r)\sqrt{K_r/K_z}$ or the observation well is screened through the entire aquifer depth (Hantush, 1961)
7	TRANSIENT, CONFINED, LEAKY, PARTIAL PENETRATION, ANISOTROPIC	• K_r, K', S when a full range of data is available • K_z with a full range of times, but comparison with multiple sets of type curves is required	4	• K_r, K', S	• $r > 1.5m\sqrt{K_r/K_z}$ or $r > m\sqrt{K_r/K_z}$ and $t > 10(Sm/K_r)\sqrt{K_r/K_z}$
			6	• K_r, K', S	• $t > (mS)/(2K_z)$
8	TRANSIENT, CONFINED, WELL STORAGE	• S, T	3	• S, T	• $t > (2.5 \times 10^3 r_c^2)/T$ or $r > 300r_w$

Table 2.1　(Continued).

Conceptual Model	Model Description	Aquifer System Properties Determined by Application of Model[1]	Simpler Alternative Model	Aquifer System Properties Determined by Alternative Model	Conditions on Drawdown Data to Apply Alternate Model
9	TRANSIENT, CONFINED, LEAKY, WELL STORAGE	• S, T, K'	4	• S, T • K'	• $t > (2.5 \times 10^3\ r_c^2)/T$ • $t > (2.5 \times 10^3\ r_c^2)/T$ and equilibrium data are available for at least three observation wells
10	TRANSIENT, CONFINED, FISSURE-BLOCK SYSTEM	• $T_{fissure}$, T_{block}, $S_{fissure}$, S_{block}, average block size	3	• T, S	• when block size and fissure size become vanishingly small
11	TRANSIENT, CONFINED, ANISOTROPIC IN TWO HORIZONTAL DIRECTIONS	• T in the principal directions, the orientation of the principal axes, S (data from at least three observation wells are required)			
12	TRANSIENT, UNDERGOING CONVERSION FROM CONFINED TO UNCONFINED	• T, S, S_y, S_s (if the radius to water table conversion does not pass through an observation well, draw-down must also be measured in the pumping well)	3 3	• T, S_s • T, $S_y + S_s m$	• $t \le (R^2 S_s)/12T$ • $t > (5R^2 S_s)/T$

Table 2.1 (Continued).

Conceptual Model	Model Description	Aquifer System Properties Determined by Application of Model[1]	Simpler Alternative Model	Aquifer System Properties Determined by Alternative Model	Conditions on Drawdown Data to Apply Alternate Model
13	TRANSIENT, UNCONFINED, ANISOTROPIC	• K_r, K_z, S_y, S_s	2	• K_r	• equilibrium data available and Dupuit-Forchheimer assumptions are valid
			3	• K_r, S_y	• when delayed yield effects have dissipated (see Fig. 18.1 and Fig. 18.2)
14	TRANSIENT, UNCONFINED, PARTIAL PENETRATION, ANISOTROPIC	• K_r, K_z, S_y, S_s	3	• K_r, S_y	• $r > m\sqrt{K_r/K_z}$ and $t > (10S_y r^2)/T$
			3	• K_r, S_y	• $r > m\sqrt{K_r/K_z}$ and $t > (S_y r^2)/T$ if the observation well is fully screened over the saturated depth
			3	• K_r, S_s	• $r < 0.03m\sqrt{K_r/K_z}$ and $t < (S_s r^2)/T$ if the observation well is fully screened over the saturated depth

Table 2.1 (Continued).

Conceptual Model	Model Description	Aquifer System Properties Determined by Application of Model[1]	Simpler Alternative Model	Aquifer System Properties Determined by Alternative Model	Conditions on Drawdown Data to Apply Alternate Model
15	TRANSIENT, UNCONFINED, PARTIAL PENETRATION, ANISOTROPIC, WELL STORAGE	• K_r, K_z, S_y, S_s	3	• K_r, S_y	• at late times when the effects of storage and delayed yield have dissipated and at distances large enough to be unaffected by partial penetration (approximate time and distance depend upon many factors)
			8	• K_r, S_s	• at very early times before partial penetration and delayed yield affect drawdown (not recommended since the shape of the curves is similar)
			14	• K_r, K_z, S_y	• $t > 2.5 \times 10^3 \, r_c^2/T$

Table 2.1 (Continued).

Conceptual Model	Model Description	Aquifer System Properties Determined by Application of Model[1]	Simpler Alternative Model	Aquifer System Properties Determined by Alternative Model	Conditions on Drawdown Data to Apply Alternate Model
16	UNCONFINED, PARTIAL PENETRATION, AQUITARD-AQUIFER	• K_r, K_z, S_s, K', S_y'	7	• K_r, S_s, K'	• $r < 0.05m\sqrt{K_r m m'/K'}$

[1] K = hydraulic conductivity, LT^{-1}
K_r = horizontal hydraulic conductivity, LT^{-1}
K_z = vertical hydraulic conductivity, LT^{-1}
m = thickness, L
r = radial distance from pumping or injection well, L
r_c = well casing radius, L
r_w = effective well radius, L
S = storativity, dimensionless
S_s = specific storage, L^{-1}
S_y = specific yield, dimensionless
t = time since pumping began, T
T = transmissivity, L^2T^{-1}

Plain letters refer to properties of the aquifer, while single primed terms denote upper aquifer or single aquitard properties. Double primed terms refer to lower aquitard properties when the aquifer is bounded by two aquitards.

Table 2.2   Slug test design table.

Conceptual Model	Model Description	Aquifer System Properties Determined by Application of Model[1]	Conditions and Limitations
17	CONFINED, INFINITE OR SEMI-INFINITE DEPTH, INCOMPRESSIBLE AQUIFER, ANISOTROPIC	• T	• must estimate the degree of anisotropy for T • aquifer storativity is zero • solutions are generally limited to the case of either an infinite aquifer with the well screen in the middle, or a confined aquifer of infinite depth • finite well diameter
18	UNCONFINED OR LEAKY CONFINED, INCOMPRESSIBLE AQUIFER, PARTIAL PENETRATION	• T	• aquifer storativity is zero • finite well diameter
19	CONFINED, COMPRESSIBLE AQUIFER, WELL STORAGE	• T, S	• fully penetrating injection well • significant aquifer storage

[1] S = storativity, dimensionless
     T = transmissivity, L^2T^{-1}

required aquifer properties, one or more potentially useful conceptual models of the aquifer system can be selected. For example, consider the design of a pumping test being performed in a confined aquifer bounded below by an aquiclude and above by an aquitard and an unconfined aquifer (a "source bed") (Figure 2.1). During pumping, groundwater flows vertically from the source bed, through the aquitard and into the aquifer where it then flows laterally to the pumping well. Depending on the purpose of the investigation, the pumping test could be designed to determine a number of properties of the aquifer/aquitard/source bed system. If the horizontal and vertical hydraulic conductivities (K_r and K_z, respectively) and storativity (S) of the aquifer and the vertical hydraulic conductivity of the aquitard (K') are required the test will be designed (and the drawdown data interpreted) using a fairly complex conceptual model (Model 7) (Table 2.1). If fewer properties are required, it may be possible to select a simpler conceptual model, thereby simplifying the design procedure and the cost and complexity of the aquifer test. For example, if the aquifer in the previous example is assumed to be isotropic, the test could be designed using a simpler conceptual model (Model 4). Similarly if the vertical hydraulic conductivity of the aquitard is not required, the test could be designed using an even simpler conceptual model (Model 3).

The selection of a simpler conceptual model usually places conditions on the number and location of observation wells, and on the critical times for data collection. For example, consider the choice between Models 3 and 4. The transmissivity (T) and storativity (S) of a leaky, confined aquifer corresponding to the assumptions of Model 4 can be determined using the simpler Model 3 if the observation well(s) are located so that $r < 0.05\sqrt{Tm'/K'}$, where T is the transmissivity of the aquifer and m' and K' are the thickness and hydraulic conductivity of the aquitard, respectively. If site constraints make this impossible, the more complex conceptual model (Model 4) must be used. Note that many of these conditions involve the values of aquifer system properties and these must therefore be estimated before the aquifer test can be designed. Allowance should be made during the design process for uncertainties in these estimates.

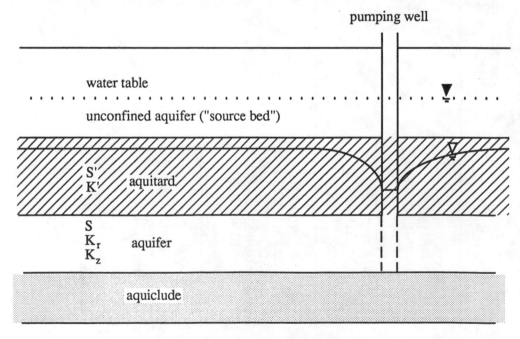

Figure 2.1   A list of required aquifer system properties is needed to select an appropriate conceptual model.

2.3   STEP 3.   ESTIMATE AQUIFER SYSTEM PROPERTIES

As demonstrated in the previous section, estimates of aquifer system properties are essential to proper aquifer test design. Several methods have been developed for estimating these properties from engineering and geologic material descriptions and from correlations between these properties and several commonly measured soil properties. Aquifer system properties can also be determined using laboratory measurements on core specimens.

2.3.1 Estimating Properties from Descriptions of Aquifer Materials

Typical values of hydraulic conductivity, specific yield, and specific storage for a variety of geologic materials are in Figure 2.2, Table 2.3, and Table 2.4, respectively. Unfortunately, estimates of hydraulic conductivity obtained by this method can vary over several orders of magnitude making test design difficult. Estimates for specific yield and specific storage are usually more reliable because these properties are much less variable than hydraulic conductivity.

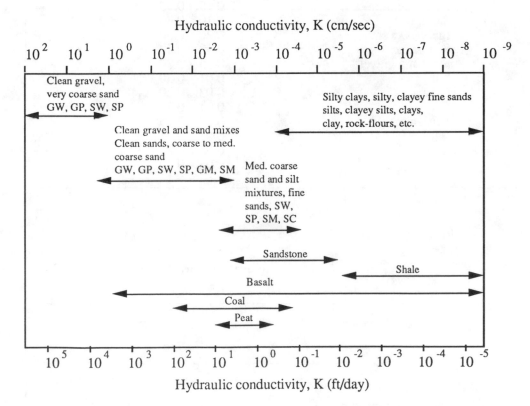

Figure 2.2   Typical values of soil and rock hydraulic conductivity (adapted from Bowles, 1984 and Walton, 1988). Symbols GW, GP, etc. are for the Unified Classification System (see e.g., Bowles, 1984).

Table 2.3   Representative values of specific yield for several aquifer materials (from Walton, 1988).

Material	Specific Yield, S_y (dimensionless)
Peat	0.30 - 0.50
Sand, dune	0.30 - 0.40
Sand, coarse	0.20 - 0.35
Sand, gravelly	0.20 - 0.35
Gravel, fine	0.20 - 0.35
Gravel, coarse	0.10 - 0.25
Gravel, medium	0.15 - 0.25
Loess	0.15 - 0.35
Sand, medium	0.15 - 0.30
Sand, fine	0.10 - 0.30
Igneous, weathered	0.20 - 0.30
Sandstone	0.10 - 0.40
Sand and gravel	0.15 - 0.30
Silt	0.01 - 0.30
Clay, sandy	0.03 - 0.20
Clay	0.01 - 0.20
Volcanic, tuff	0.02 - 0.35
Siltstone	0.01 - 0.35
Limestone	0.01 - 0.25
Till	0.05 - 0.20

Table 2.4   Representative values of specific storage for various materials (from Walton, 1988).

Material	Specific Storage, S_s (ft^{-1})
Clay	10^{-4}
Sand and Gravel	10^{-5}
Rock, fissured	10^{-6}

2.3.2   Estimating Aquifer System Properties from Soil Index Properties

Hydraulic conductivity and specific storage can be computed from estimates of *coefficients of compression* and *consolidation* obtained using correlations between these properties and more easily measured *soil index properties*. These coefficients can also be measured directly on core specimens (see Section 2.3.4). From Terzaghi's one-dimensional theory of consolidation, hydraulic conductivity, K, can be computed from

$$K = c_v m_v \gamma_w = c_v \frac{a_v}{(1 + e_o)} \gamma_w \tag{2.1}$$

where

c_v = coefficient of consolidation, $L^2 T^{-1}$
m_v = coefficient of volume compressibility,
= reciprocal of the constrained modulus of elasticity
a_v = coefficient of compressibility, $T^2 L M^{-1}$
e_o = initial void ratio
γ_w = unit weight of water, $ML^{-2}T^{-2}$

Specific storage, S_s, can be computed from

$$S_s = \gamma_w \left(\frac{\theta}{E_w} + \frac{1}{E_s} \right) \tag{2.2}$$

where

γ_w = unit weight of water, $ML^{-2}T^{-2}$
θ = effective porosity
E_w = bulk modulus of elasticity of water, $ML^{-2}T^{-2}$
E_s = constrained modulus of elasticity of the soil or rock skeleton in the range
of applied effective stress, $ML^{-1}T^{-2}$

It is easy to show that, for small changes in effective stress, the constrained modulus of elasticity can be written

$$E_s = \frac{(1 + e_o) \sigma'}{0.434 C_c} \tag{2.3}$$

where

e_o = initial void ratio
σ' = effective stress in the field
C_c = compression index

For small changes in effective stress the coefficient of compressibility can be computed using

$$a_\upsilon = \frac{e_2 - e_1}{\sigma'_2 - \sigma'_1}$$
(2.4)

where

e_1 = initial void ratio

σ'_1 = initial effective stress, $ML^{-1}T^{-2}$

e_2 = void ratio after consolidation under effective stress σ'_2

σ'_2 = final effective stress, $ML^{-1}T^{-2}$

Similarly the compression index, C_c, can be computed using

$$C_c = \frac{e_2 - e_1}{\log \dfrac{\sigma'_2}{\sigma'_1}}$$
(2.5)

Combining Equations 2.3, 2.4, and 2.5 gives

$$a_\upsilon \approx \frac{0.434C_c}{\sigma'}$$
(2.6)

where σ' represents the effective stress in the field. Equation 2.6 is valid if the change in effective stress during pumping is small.

Several empirical equations have been developed to predict the coefficient of consolidation, c_υ, and the compression index, C_c, from more easily measured soil index properties (Table 2.5). The compression index for clays and silts commonly ranges from 0.1 to 1.0. Recall that C_c applies to clays under effective stresses greater than their preconsolidation stress. Values for C_r, the compression index in the previously consolidated range, are often assumed to be 5 to 10 percent of C_c (Holtz and Kovacs, 1981). Typical values of C_r range from 0.015 to 0.035 (Leonards, 1976). Figure 2.3 gives approximate correlations of the coefficient of consolidation, c_υ, and the liquid limit. Typical values of c_υ range from 4×10^{-3} to 9×10^{-5} cm/sec (Holtz and Kovacs, 1981). Values of γ_w and E_w for a range of temperatures are in Table 2.6.

Table 2.5   Selected Empirical Equations for determining the Compression Index, C_c, from Soil Index Properties*.

Equation	Applicability
$C_c = 0.007 (LL^{(1)} - 7)$	Remolded clays
$C_c = 0.01 w_n^{(2)}$	Chicago clays
$C_c = 1.15 (e_o^{(3)} - 0.35)$	All clays
$C_c = 0.30 (e_o - 0.27)$	Inorganic, cohesive soil; silt with some clay; silty clay; clay
$C_c = 1.15 \times 10 \, w_n$	Organic soils
$C_c = 0.75 (e_o - 0.5)$	Soils of low plasticity
$C_c = 0.009 (LL - 10)$**	Undisturbed clays with low to medium sensitivity

* Selected from Azzouz, Krizek, and Corotis (1976)   ** Terzaghi and Peck (1967)
[1] LL = liquid limit [2] w_n = natural water content [3] e_o = initial void ratio

Table 2.6   Unit weight and bulk modulus of elasticity of water at various temperatures (adapted from Streeter and Wylie, 1979).

Temp (°C)	Unit Weight (N/m³)	Bulk Modulus (x 10^7 N/m²)	Temp (°F)	Unit Weight (lb/ft³)	Bulk Modulus (x 10^5 lb/ft²)
0	9805	204	32	62.42	422
5	9806	206	40	62.43	423
10	9803	211	50	62.41	439
15	9798	214	60	62.37	448
20	9789	220	70	62.30	461
25	9779	222	80	62.22	464
30	9767	223	90	62.11	465
35	9752	224	100	62.00	471
40	9737	227	110	61.86	477
45	9720	229	120	61.71	480
50	9697	230	130	61.55	481
55	9679	231	140	61.38	475
60	9658	228	150	61.20	472
65	9635	226	160	61.00	469
70	9600	225	170	60.80	464
75	9589	223	180	60.58	451
80	9557	221	190	60.36	451
85	9529	217	200	60.12	444
90	9499	216	212	59.83	432
95	9469	211			
100	9438	207			

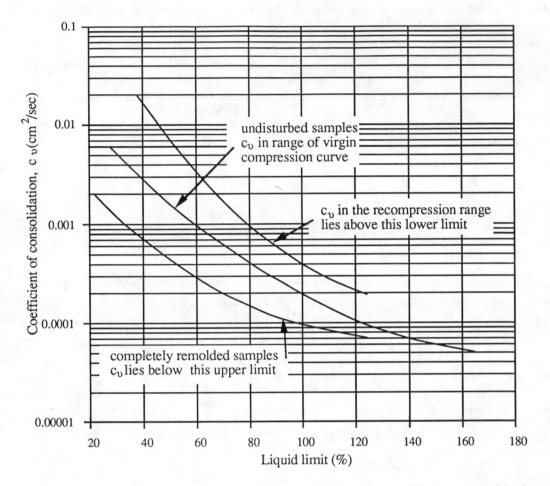

Figure 2.3   **Approximate correlations of the coefficient of consolidation, c_v, with the liquid limit (from Holtz and Kovacs, 1981).**

The void ratio, e, and unit weight, γ, can be measured on core specimens or can be estimated from the material description (Table 2.7). The unit weight can be computed by

$$\gamma = \gamma_d(1 + w) \tag{2.7}$$

where

      w   = gravimetric water content, dimensionless
      γ_d = dry unit weight, $MT^{-2}L^{-2}$

Table 2.7 Typical values of porosity, void ratio, and unit weight of soils (from Peck, Hanson, and Thornburn, 1974).

Material	Porosity n	Void Ratio e	Water Content w (%)	Dry Unit Weight γ_d (g/cm³)(lb/ft³)		Saturated Unit Weight γ_{sat} (g/cm³)(lb/ft³)	
Loose, uniform sand	0.46	0.85	32	1.43	90	1.89	118
Dense, uniform sand	0.34	0.51	19	1.75	109	2.09	130
Loose, nonuniform sand	0.40	0.67	25	1.59	99	1.99	124
Dense, nonuniform sand	0.30	0.43	16	1.86	116	2.16	135
Loess	0.50	0.99	21	1.36	85	1.86	116
Very mixed-grained glacial till	0.20	0.25	9	2.12	132	2.32	145
Soft glacial clay	0.55	1.2	45	1.22	76	1.77	110
Stiff glacial clay	0.37	0.6	22	1.70	106	2.07	129
Soft slightly organic clay	0.66	1.9	70	0.93	58	1.58	98
Soft very organic clay	0.75	3.0	110	0.68	43	1.43	89
Soft montmorillonitic clay	0.84	5.2	194	0.43	27	1.27	80

2.3.3 Estimating Aquifer System Properties using Effective Grain Size

Hydraulic conductivity can also be estimated using correlations between hydraulic conductivity and *effective grain size*, D_{10}, obtained by a sieve analysis of a sample of aquifer or aquitard material. Ten percent of the material has a grain size smaller than D_{10}. Based on tests of clean sand filters, Hazen developed an empirical equation for predicting the hydraulic conductivity of sands with D_{10} sizes between 0.1 and 3.0 mm (Bowles, 1984)

$$K = C(D_{10})^2 \qquad (2.8)$$

where K is in units of cm/sec, D_{10} is in units of cm, and C is a coefficient which varies from 40 to 150 depending on the size and gradation of the sand:

C	Sand Size and Gradation
40 - 80	Very fine, well graded with appreciable lines
80 - 120	Medium coarse, poorly graded; or clean but well graded
120 - 150	Very coarse, very poorly graded, gravelly, clean

Specific yield, S_y, can be computed from

$$S_y = G_a e \qquad (2.9)$$

where the *air space ratio*, G_a, is defined as

$$G_a = \frac{\text{volume of air space}}{\text{volume of total void space}} = 1 - \frac{\text{percent saturation}}{100} \qquad (2.10)$$

and e is the void ratio.

A correlation between the air space ratio and effective grain size was developed from field tests conducted in the central United States (Figure 2.4) (Terzaghi & Peck, 1967). The rate and amount of water which drains from a soil under gravity is dependent not only upon pore size, but pore shape, mineralogy, groundwater characteristics and climate. Thus, actual values of S_y may vary significantly from those computed by use of Figure 2.4 and Equation 2.10.

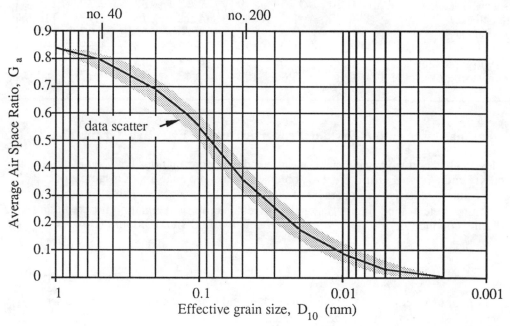

Figure 2.4   Relation between effective grain size and the air space ratio, G_a (from Terzaghi and Peck, 1967).

2.3.4 Determination of Aquifer System Properties from Laboratory Tests on Core Specimens

It is possible to determine aquifer system properties from the results of laboratory tests on core specimens collected from a borehole. However, the results of these tests must be interpreted with caution for several reasons:

1. It is difficult to collect undisturbed core specimens during drilling, especially in cohesionless soils and fractured rock. Disturbance is further increased during the transfer of the specimen from the core barrel or sampling tube to the testing apparatus.

2. Laboratory tests are sometimes conducted at hydraulic gradients that are much higher than occur under field conditions. This can result in turbulent flow within the sample or sample compaction during the test.

3. In most cases, large-scale or "effective" values of aquifer properties are needed for design. Core specimens are necessarily small and may not be representative of geologic materials and structures controlling groundwater flow at the field scale.

Because pumping and slug tests "sample" a much larger volume of aquifer material they provide more representative values of large-scale aquifer properties in heterogeneous systems. Nevertheless, when laboratory data are available they can provide useful preliminary estimates of aquifer system properties for use in aquifer test design.

Hydraulic conductivity can be measured directly on soil samples using standard methods such as the *constant head test* (ASTM D2434 or AASHTO T215) or *falling head test* (Bowles, 1982). The hydraulic conductivity and specific storage of fine-grained materials may also be determined indirectly from the *standard consolidation test* (ASTM D2435 or AASHTO T216) or the *constant rate of strain consolidation test* (ASTM D4186-82). Similarly, the hydraulic conductivity of peat samples can be measured by a falling head test (ASTM D4511) and the hydraulic conductivity of rock cores obtained from results of the *flowing air test* (ASTM D4525). Figure 2.5 shows ranges of hydraulic conductivity typically appropriate to the various laboratory tests and field tests.

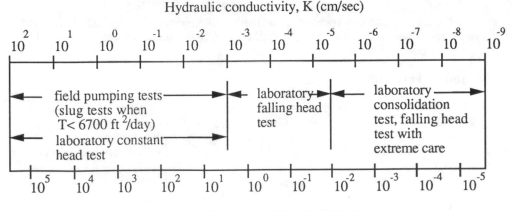

Figure 2.5   Range of applicability of field and laboratory tests for measuring hydraulic conductivity.

The constant head test is preferred for soils, such as sands and gravels with hydraulic conductivities in the range of 10^{-3} to 100 cm/sec (about 3 to 3×10^5 ft/day). The falling head test is preferred for fine-grained soils, usually with hydraulic conductivities in the range of 5×10^{-6} to 10^{-3} cm/sec (about 0.01 to 3 ft/day). This test may last up to several days for fine-grained soils, increasing the errors due to evaporation of water from the core specimen, temperature changes, leakage of water from the test apparatus, etc. In very fine-grained materials it is preferable to compute hydraulic conductivity and specific storage from consolidation test results.

2.4 STEP 4.   SPECIFY TEST CONDITIONS

Using the estimates of the required aquifer system properties and a set of potentially useful conceptual models from Tables 2.1 and 2.2, the test conditions can be specified. These include the type of test (pumping test or slug test), the diameter of the pumping, injection, or withdrawal well, the pumping rate or injection or withdrawal volume, the number and location of the observation wells, the depth and screen length of all wells, and the test duration. It should be emphasized that aquifer test design is an iterative process and earlier steps may need to be repeated as the result of information gained by later work.

After selecting a potential model from Table 2.1 or 2.2, a more detailed model description should be consulted to ensure that test conditions match the assumptions of the conceptual model. In Part II, methods are presented for interpreting pumping test and slug test data using each of the conceptual models in Tables 2.1 and 2.2. Each chapter in Part II describes the methods that are available for a particular conceptual model. For example, Chapter 9 describes the methods that can be used to interpret pumping test data for Model 4 while Chapter 24 describes the methods that can be used to interpret slug test data for Model 19 . The first part of each chapter in Part II contains a list of the assumptions about the geometry, properties, and hydraulic behavior of the aquifer system on which the conceptual model is based. The validity of these assumptions should be critically reviewed using the information from the preliminary site investigation. If a particular assumption seems unreasonable an alternate conceptual model should be selected.

Part II presents methods of analysis for pumping tests (Chapters 6 to 21) and slug tests (Chapters 22 to 24). Slug tests are usually preferable in aquifer materials with relatively small transmissivities less than 6700 ft^2/day (Figure 2.5). If the transmissivity is larger than this value, the piezometric surface or water table recovers too quickly following the injection or withdrawal of water from the well casing, making accurate measurements of buildup or drawdown difficult.

2.4.1 Pumping Test Design

If a pumping test will be used, the designer must specify the pumping rate, the pumping well diameter, depth, and screened interval, the locations and depths of observation wells, and the duration of the test. This section describes the design of constant discharge pumping tests; the design and interpretation of step discharge tests is discussed in Chapter 5.

Selection of the pumping rate and well diameter depend upon the choice of conceptual model and on the estimated values of aquifer properties. The selected pumping rate should be large enough to insure that drawdown can be measured accurately in the pumping well and the observation well(s) (if used). However, the selected pumping rate should not result in excessive drawdown. The water table in unconfined aquifers should not be lowered by more than 25 percent since this is the largest drawdown which can be corrected and analyzed with an analytical solution to the groundwater flow equations (Ferris et al., 1962). In confined aquifers it is desirable to select a pumping rate that does not result in dewatering of the aquifer, although methods are available for interpreting drawdown data when a confined aquifer is partially converted to an unconfined aquifer during a pumping test (see Chapter 17). The pumping rate may be selected in two ways:

1. By using an empirical equation to predict the specific capacity of the pumping well.

2. By using the analytical solutions in Part II to predict drawdown for the pumping and observation well(s) for a range of assumed pumping rates.

Empirical Equations for Specific Capacity

Driscoll (1986) presents empirical equations that may be used to predict the specific capacity, Q/s, of the pumping well. The equations were derived using Cooper and Jacob's (1946) approximation to the Theis solution (Model 3)

$$\frac{Q}{s} = \frac{T}{0.183 \, \log\left(\frac{2.25 \, Tt}{r_w^2 S}\right)} \tag{2.11}$$

where Q is the pumping rate, s is the drawdown in the pumping well, T is the aquifer transmissivity, t is the time since pumping began, r_w is the radius of the pumping well, and S is the aquifer storativity. Driscoll selected the following "typical" values for aquifer properties and test conditions:

transmissivity, T $\quad\quad\quad$ = $30000 \, \frac{\text{gal}}{\text{day} \cdot \text{ft}}$

confined storativity, $S_s m$ = 10^{-3}

unconfined storativity, S_y = 7.5×10^{-2}

radius, r $\quad\quad\quad\quad\quad\quad$ = 0.5 ft

test duration, t $\quad\quad\quad\quad$ = 1 day

Substituting these values into equation 2.11 the following equations can be derived

$$\frac{Q}{s} = \frac{T}{2000} \quad\quad \text{for confined aquifers} \tag{2.12}$$

and

$$\frac{Q}{s} = \frac{T}{1500} \quad\quad \text{for unconfined aquifers} \tag{2.13}$$

where the units of Q, T, and s are gal/min, gal/(day · ft), and feet, respectively. Similar equations can be derived if the test conditions or estimated values of aquifer properties are much different than the values Driscoll used. However, variations in the value of T, t, r_w, and S have relatively small effect on computed values of specific capacity since they are included in a logarithmic term in the equation.

For confined aquifers with partially penetrating wells, J. Kozeny developed the following equation, applicable with any consistent units (Driscoll, 1986, p. 250):

$$\frac{Q}{s_p} = \frac{Q(l-d)}{sm}\left(1 + 7\sqrt{\frac{r}{2(l-d)}} \, \cos\left(\frac{\pi(l-d)}{2m}\right)\right) \tag{2.14}$$

where

s_p $\quad$ = drawdown in a partially penetrating well, L

l - d = length of well screen, L

m $\quad$ = aquifer thickness, L

The equation is based on the assumptions of a small well diameter and no well losses. It may also be inaccurate for small values of aquifer thickness and large ratios of (l - d) to m.

Example 2.1

A confined aquifer has an estimated transmissivity of 20000 gal/(day · ft). What pumping rate is required to produce a drawdown of 15 feet in the pumping well after 1 day of pumping? From Equation 2.12,

$$Q = \frac{sT}{2000} = \frac{15\,(20000)}{2000} = 150\,\frac{gal}{min}$$

Example 2.2

Suppose that a partially penetrating well with a 10-inch (0.83 ft) diameter well is installed in the aquifer of Example 2.1. The aquifer is 50 feet thick and the well is screened over 20 feet of its length. From Equation 2.14,

$$\frac{Q}{s_p} = \frac{\left(150\,\frac{gal}{min}\right)(20\ ft)}{(15\ ft)(50\ ft)}\left(1 + 7\,\sqrt{\frac{0.416\ ft}{2(20\ ft)}}\,\cos\left(\frac{\pi(20\ ft)}{2(50\ ft)}\right)\,\right)$$

$$\frac{Q}{s_p} = 6.8\,\frac{gal}{min \cdot ft}$$

If the desired drawdown is still 15 feet, the required pumping rate is

$$Q = 15\ ft\left(6.8\,\frac{gal}{min \cdot ft}\right) = 102\,\frac{gal}{min}$$

Drawdown Prediction using Analytical Solutions

Drawdown in the pumping and observation wells can also be computed using the analytical solutions presented for each conceptual model in Part II. Appendix B contains computer programs specifically for this purpose but calculations can also be performed using tabulated values of well functions in the chapters in Part II.

Example 2.3

A pumping test is being designed using a conceptual model for a leaky, confined aquifer (Model 4). The selected pumping rate and estimated aquifer properties are listed below (see Chapter 9):

Q = pumping rate = 50 gal/min (9626 gal/day)

T = aquifer transmissivity = 20000 ft^2/day

S = aquifer storativity = 0.0002

m' = aquitard thickness = 50 ft

K' = aquitard vertical hydraulic conductivity = 0.01 ft/day

Predict the drawdown at an observation well located at a radial distance r = 100 ft from the pumping well. What would the predicted drawdown be if the test was designed using Model 3 ?

The appropriate well functions are listed in Tables 8.1 and 9.1, computer programs for predicting drawdown are in Appendix B (Programs DRAW3 and DRAW4). The results are summarized in Figure 2.6 and show that differences in predicted drawdown for the two models are not significant until after one or two days.

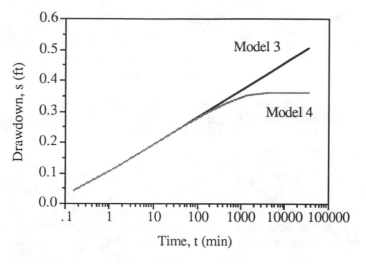

Figure 2.6   Predicted drawdown at observation well.

Pumping Well Casing Diameter Selection

Once a pumping rate has been specified, the diameter of the well casing can be selected. A preliminary selection of casing diameter can be made using Table 2.8; the final selection should be made in consultation with the pump manufacturer.

Table 2.8   Recommended pumping well casing diameters for various pumping rates (after Driscoll, 1986, p. 415).

Pumping Rate		Well Casing Diameter	
$\left(\dfrac{\text{gal}}{\text{min}}\right)$	$\left(\dfrac{\text{m}^3}{\text{day}}\right)$	(in)	(mm)
< 100	< 545	6	152
75 - 175	409 - 954	8	203
150 - 350	818 - 1910	10	254
300 - 700	1640 - 3820	12	305
500 - 1000	2730 - 5450	14	365
800 - 1800	4360 - 9810	16	406
1200 - 3000	6540 - 16400	20	508

Observation Well Location and Screen Placement

Many of the methods for interpreting pumping test data described in Part II require drawdown data from one or more observation wells. The selection of appropriate locations, depths, and screen lengths for these wells is an important part of successful pumping test design. The following is a list of criteria that can be used to guide observation well placement.

1. Predicted drawdown in each observation well should be large enough to be measured accurately. Drawdown can be predicted using the analytical solutions in Part II, the estimated aquifer properties, the selected pumping rate, and the distance from the pumping well to the proposed observation well(s). The calculations can be performed using the tabulated well functions in Part II or the computer programs in Appendix B.

2. If possible, observation wells should be located at logarithmically spaced distances from the pumping well (i.e., 10, 100, and 1000 ft) to permit distance-drawdown plots to be used with the match-point method (see Chapter 4).

3. The only way to determine both horizontal and vertical hydraulic conductivities in a confined aquifer is to: a) install a partially penetrating pumping well, b) have partially penetrating observation wells, and c) place the observation wells at a radial distance $r < 1.5 \; m \sqrt{K_r/K_z}$ from the pumping well, where m is the saturated thickness of the aquifer and K_r and K_z are the horizontal and vertical hydraulic conductivities of the aquifer. Beyond this distance r, flow is essentially horizontal, even in an anisotropic aquifer.

   Both horizontal and vertical hydraulic conductivities may be determined from either fully or partially penetrating wells in an unconfined aquifer as long as accurate data can be collected through a complete time range (i.e., at the beginning, middle, and end of the test, the aquifer response must conform to the conceptual model and sufficient data must be collected in each of the ranges to clearly define a curve shape).

4. If the horizontal (plan view) hydraulic conductivity of the aquifer is assumed to be isotropic, the distribution of the observation wells about the pumping well is arbitrary but an even distribution is desirable so that drawdown measurements are representative of as large a volume of aquifer as possible. If the horizontal hydraulic conductivity of the aquifer is assumed to be anisotropic, drawdown data will be analyzed using the conceptual model in Chapter 16 (Model 11) which requires that no two observation wells be radially aligned with the pumping well.

5. If the test site contains known or suspected lateral discontinuities in the aquifer, observation wells can be placed either to minimize the effect of the boundary on drawdown data or to more precisely identify the location of the discontinuity. The method of *image wells* (see Chapter 4) can be used to predict the effect that aquifer discontinuities, recharge zones, or other pumping wells will have on drawdown at a proposed observation well location. The radial distance beyond which the effects of an aquifer boundary or a pumping well on measured drawdown in an observation well are negligible can be estimated for the Theis conceptual model (Model 3) using an equation developed by Walton (1988). The following equation assumes that drawdown is negligible when u (= $r^2S/(4Tt)$) is greater than 5 (i.e., W(u) < 0.001)

$$r = \sqrt{\frac{20Tt}{S}} \qquad\qquad (2.15)$$

where

> r = distance between observation well and image well beyond which boundary or
> other pumping well impacts are negligible, L
> T = aquifer transmissivity, L^2T^{-1}
> t = time after pumping started, T
> S = aquifer storativity (either confined or unconfined), dimensionless

Example 2.4

Given the soil profile in Figure 2.7, design a pumping test to determine the
properties of the clean sandy gravel layer and the fine sand layer. Soil properties have
already been estimated from hydrogeologic information. There are no other wells in the
area. However, a rock ridge cuts through the soil layers approximately 500 feet from the
proposed site of the pumping well.

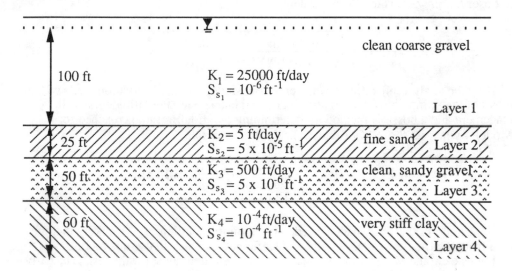

**Figure 2.7 Soil profile for Example 2.4 with estimated hydraulic conductivities, K,
and specific storage coefficients, S_s.**

The steps outlined in this chapter are discussed below:

1. The properties of interest are K_2, S_2, K_3 and S_3. Later in the analysis, the analyst can
   determine if the horizontal and vertical hydraulic conductivities of both layers are
   required.

2. The only physical constraint at the moment is the flow boundary approximately 500
   feet away. Later, to reduce costs, the analyst might add the constraint of a partially
   penetrating well. For now, assume full penetration.

3. The very stiff clay of layer 4 is much less permeable than layer 3, so it can be
   considered impermeable. Layer 1 may be considered as a source bed, while layer 2 is
   an aquitard and layer 3 is an aquifer. Check these assumptions later.

4. The soil properties have been estimated for us.

5. The conceptual model appears to match the description of Model 5. Refer to Chapter 10 to ensure that the example system conforms to the assumptions in Model 5. First, there is negligible source bed drawdown when $100T_{aquifer} \leq T_{source\ bed}$ (Neuman and Witherspoon, 1969b). Second, it is correct to assume that flow is vertical through the confining bed and horizontal through the aquifer if $K/K' > 100$ m/m' (Hantush, 1967). Both of these conditions are met.

Next, check the design tables for the possible use of a simpler model. Table 2.1 shows that storage in the confining bed can be neglected when

$$r < 0.04 m \sqrt{\frac{KS_s}{K'S_s'}}$$

Thus, Model 4 could be used to find all the desired properties except S_2 if the distance to observation wells is less than r, where

$$r = (0.04)\ (50\ ft) \sqrt{\frac{(500\ ft/day)\ (5 \times 10^{-6}\ ft^{-1})}{(5\ ft/day)\ (5 \times 10^{-5}\ ft^{-1})}} = 6.3\ ft$$

Now the analyst must decide whether to use Model 4 with observation wells within 6 feet of the pumping well, or to use Model 5 and have a greater latitude for well placement and a possible method for determining S_2. Equilibrium is reached for Model 4 when $t \geq 8$ m'S/K (Table 2.1). For this example,

$$t \geq \frac{8\ (25\ ft)\ (5 \times 10^{-6}\ ft^{-1})\ (50\ ft)}{\left(5\ \dfrac{ft}{day}\right)\left(\dfrac{1\ day}{1440\ min}\right)} = 14\ min$$

Table 2.1 shows that early data is needed for a unique solution. It is unlikely that the data from the first few minutes will be of sufficient accuracy to use this model.

Let us examine Model 5 to determine the required time range for the test. The early-time solution is applicable when

$$t \leq \frac{m'S'}{10K'} = \frac{(25\ ft)\ (5 \times 10^{-5}\ ft^{-1})\ (25\ ft)}{10\left(5\ \dfrac{ft}{day}\right)\left(\dfrac{1\ day}{1440\ min}\right)} = 1\ min$$

Obviously, the early-time solution will not be applicable to any data unless our estimates of soil properties are wrong by orders of magnitude. Since the late-time solution is indeterminant for *Case B*, the equilibrium drawdown from at least three observation wells must be used to determine soil properties. Thus, it is likely that S_2 (S' in Model 5) will have to be estimated from laboratory tests or hydrogeologic observations.

Check the time for equilibrium drawdown by Model 5. From Chapter 10 the time required to achieve equilibrium is (Equation 10.34)

$$t \approx \frac{8(S + \frac{S'}{3} + S'')}{\frac{K'}{m'} + \frac{K''}{m''}} = \frac{8\left(2.5 \times 10^{-4} + \frac{1.25 \times 10^{-3}}{3} + 0\right)}{\left(\frac{5\ \frac{ft}{day}}{25\ ft}\right) + 0} = 38\ min$$

Therefore, the test should be run for at least 40 minutes or until equilibrium is observed.

Finally, check to see if the boundary will affect the placement of observation wells. The minimum distance between an observation well and image well required to insure that the effects of the boundary will be significant is given by Equation 2.15. Though the equation is meant to be used when conditions conform to those of Model 3, it can be used in this case as a rough approximation. Applying Equation 2.15 with a test duration of 1 hour gives

$$r = \sqrt{\frac{20\ Tt}{S}}$$

$$= \sqrt{\frac{(20)\ (500\ \frac{ft}{day})\ (50\ ft)\ (1\ hr)\ \left(\frac{day}{24\ hr}\right)}{2.5 \times 10^{-4}}}$$

$$\approx 9129\ ft$$

Obviously, the designer would not place observation wells at this distance from the pumping well, so image well theory will be used in the analysis of results.

6. A pumping rate can be chosen by applying Cooper and Jacob's (1946) empirical equation to the desired drawdown at the pumping well. Try a drawdown of 10 feet at the pumping well. From Equation 2.12,

$$Q = \frac{Ts}{2000} = \frac{\left(25000\ \frac{ft^2}{day}\right)(10\ ft)\left(\frac{7.48\ gal}{1\ ft^3}\right)}{2000} = 935\ \frac{gal}{min}$$

From Table 2.8, an acceptable casing diameter for this flow rate is 16 inches. This approximate casing diameter should be used to estimate well storage effects. Table 2.1 shows that the effects of well storage in a leaky aquifer (Model 9) will have dissipated when

$$t > \frac{2.5 \times 10^3\ r_c^2}{T} = \frac{2.5 \times 10^3\ (8/12\ ft)^2}{25000\ \frac{ft^2}{day}\left(\frac{day}{1440\ min}\right)} = 64\ min$$

where r_c is the radius of the well casing.

This time is slightly larger than the time estimated to reach equilibrium for Model 5. However, the test will still be run to equilibrium with a minimum of three observation wells.

Finally, the pumping rate should be used to compute the estimated drawdowns at each observation well. This will ensure that there is measurable drawdown at all wells and identify critical times for measurements.

Example 2.5

Assume that the pumping well of Example 2.3 can only partially penetrate the aquifer. Table 2.1 shows that Model 5 can still be used as long as either:

1. All observation wells fully penetrate the aquifer

2. All observation wells are placed at a distance, r, at least $1.5 \text{ m} \sqrt{\dfrac{K_r}{K_z}}$ from the pumping well.

Typically, the horizontal hydraulic conductivity is at least ten times greater than the vertical hydraulic conductivity. If this typical value is used, then

$$r \geq 1.5 \ (50 \text{ ft}) \ \sqrt{10} \ = 237 \text{ ft}$$

Thus, Model 5 can be used if all observation wells are more than 237 feet from the pumping well. However, at this distance, equilibrium drawdown is less than one foot when pumping at 935 gal/min. Drawdown at two other wells spaced logarithmically, for example, at 350 and 500 feet from the pumping well, will have maximum drawdowns less than 6 inches. These are such small drawdowns and the spacing must be so large that it is preferable to use Model 7 and account for partial penetration.

2.4.2 Slug Test Design

If a slug test will be used, the designer must specify the injection or withdrawal volume, and the well diameter, depth, and screened interval. Selection of the injection or withdrawal volume and well diameter depends upon the choice of conceptual model and on the estimated values of aquifer properties. The selected volume should be large enough to insure that buildup or drawdown can be measured accurately, but small enough so that buildup or drawdown does not result in significant changes in aquifer saturated thickness. The injection or withdrawal volume is most easily selected by using the analytical solutions in Part II to predict buildup or drawdown for a range of assumed aquifer properties.

Example 2.6

An unconfined silty clay aquifer is bounded below by a stiff clay layer of great thickness. The saturated thickness of the silty clay is 20 feet and the estimated hydraulic conductivity is 4 ft/day. The stiff clay has an estimated hydraulic conductivity of 10^{-2} ft/day. Design a slug test to determine the hydraulic conductivity of the aquifer.

The estimated aquifer transmissivity is (4 ft/day)(20 ft) = 80 ft²/day, and referring to Figure 2.5, a slug test is appropriate. The estimated hydraulic conductivity of the stiff clay is more than two orders of magnitude smaller than the silty clay and this layer will be considered to be an aquiclude for purposes of test design.

Referring to Table 2.2, the appropriate conceptual model for this situation appears to be Model 18. To begin, specify a well with a 2-inch diameter ($r_i = 0.08$ ft) casing installed in a 4-inch diameter augered hole ($r_w = 0.17$ ft) and back filled with a coarse sand (porosity = n = 0.40). The well is screened over the entire saturated thickness of the aquifer ($l = l - d = 20$ ft). Design a withdrawal test based on an initial water table drop of 4 ft. The analytical solution is in Equation 23.6. The effective radius of the well casing, r_c, is

$$r_c = \sqrt{r_i^2 + n (r_w^2 - r_i^2)}$$
$$= \sqrt{(0.08 \text{ ft})^2 + 0.4 ((0.17 \text{ ft})^2 - (0.08 \text{ ft})^2)}$$
$$= 0.12 \text{ ft}$$

For a fully penetrating well, read "C" from Figure 23.1, with $(l-d)/r_c = 20 \text{ ft}/0.12 \text{ ft} = 167$ and obtain $C = 5.8$. Using Equation 23.10

$$\ln\left(\frac{R}{r_w}\right) = \left(\frac{1.1}{\ln\left(\dfrac{1 \text{ ft}}{0.17 \text{ ft}}\right)} + \frac{5.8}{\left(\dfrac{20 \text{ ft}}{0.17 \text{ ft}}\right)}\right)^{-1} = 16.6$$

To know how long it will take to recover all but one foot of head drop (so that $H = 1$ ft), rearrange Equation 23.6 as

$$t = \frac{(r_c)^2 \ln (R/r_w) \ln (H_o/H_w)}{2K (l - d)}$$

$$t = \frac{(0.12 \text{ ft})^2 (16.6) \ln\left(\dfrac{18 \text{ ft}}{1 \text{ ft}}\right)}{2 \left(4 \dfrac{\text{ft}}{\text{day}}\right)(20 \text{ ft}) \left(\dfrac{1 \text{ day}}{1440 \text{ min}}\right)} = 6.2 \text{ min}$$

If a pressure transducer or air flow system with continuous readout is available for measuring the depth to water, a sufficient number of accurate measurements can be taken in a 2-inch diameter well with full penetration. However, if a less accurate measuring system is used, the designer may wish to install a casing of larger diameter to increase the test duration.

Chapter 3

CORRECTIONS TO DRAWDOWN DATA

Before pumping test data are analyzed, it may be necessary to correct measured water levels for a variety of factors including trends in water levels, changes in barometric pressure, tidal or river level fluctuations, application or removal of heavy surface loads, and decreases in saturated thickness. These corrections can also be applied to slug test data, although the duration of a slug test is usually short enough that the effects of these factors on drawdown or buildup data can be neglected.

3.1 TRENDS IN WATER LEVELS

The water table or piezometric surface of an aquifer may fluctuate with seasonal changes in the water budget or with daily or weekly changes in the pumping rate at nearby wells. Water levels in the pumping and observation wells should be measured at regular intervals for several days before the test begins. Long term trends can be identified by plotting the pretest water levels vs. time. After pumping begins, the estimated increase (or decrease) in head should be subtracted from (or added to) the observed drawdown.

Example 3.1

Often the water table drops gradually during rainless periods. Figure 3.1 illustrates how drawdown data should be corrected to account for a trend of decreasing head. The apparent trend is projected over the time period of the pumping test. Corrected drawdown ($s_{corrected}$) is computed as the difference between the measured head and the projected head.

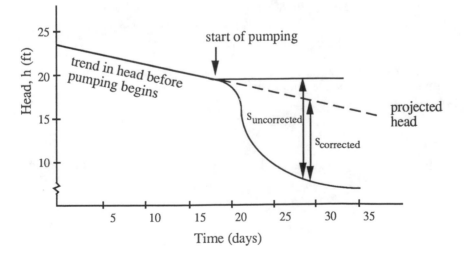

Figure 3.1   Correction of drawdown data for a decreasing trend in head.

3.2   ATMOSPHERIC PRESSURE CHANGES

Changes in atmospheric pressure can cause water levels to fluctuate during a pumping test, making data interpretation difficult. The amount of rise in water levels resulting from a decrease in atmospheric pressure and the amount of fall in water levels resulting from an increase in atmospheric pressure can be determined if the barometric efficiency of the aquifer is known. *Barometric efficiency*, BE, is defined as the rate of change in head with changing atmospheric pressure

$$BE = -\frac{dh}{(dp_a/\gamma_w)} \times 100\%$$ (3.1)

where

dh = change in head, L
dp_a/γ_w = change in atmospheric pressure expressed as a height of water, L
p_a = atmospheric pressure, FL^{-2}
γ_w = unit weight of water, FL^{-3}

BE is always positive. The negative sign in Equation 3.1 occurs because an increase in barometric pressure (dp_a/γ_w is negative) results in a decrease in water level (dh is positive).

If the aquifer and groundwater can be considered to be perfectly elastic, the barometric efficiency can be computed from

$$BE = \frac{1}{1 + \dfrac{E_w}{E_s\theta}} \quad \text{(perfectly elastic)}$$ (3.2)

where

E_w = modulus of elasticity of the water, $M^{-1}T^2L$
E_s = modulus of elasticity of the soil skeleton, $M^{-1}T^2L$
θ = effective porosity of the aquifer, dimensionless

In practice, however, barometric efficiency is computed from measurements of water level and atmospheric pressure made prior to the start of the pumping test

$$BE = \frac{\Delta h}{\Delta(p_a/\gamma_w)} \times 100\%$$ (3.3)

where Δh is the change in head that resulted from a change in atmospheric pressure $\Delta(p_a/\gamma_w)$.

Example 3.2

Several measurements of barometric pressure for a range of water levels are needed to compute the barometric efficiency of an aquifer. Figure 3.2 shows typical data, with measured head plotted as a function of atmospheric pressure in the same units. A straight line was fitted through the data; the barometric efficiency computed from the slope of the line is 60 percent. Barometric efficiency is often greater than 50 percent (Walton, 1988).

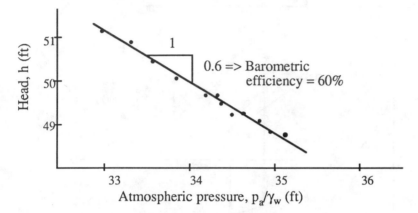

Figure 3.2 **Water levels are plotted against atmospheric pressure to determine barometric efficiency.**

Once the barometric efficiency is determined, drawdown data can be corrected for changes in atmospheric pressure that occur during the test. Figure 3.3 shows water level and atmospheric pressure measurements made during a pumping test. Atmospheric pressure increased during the test thereby lowering the water level, so measured drawdown was decreased by an amount $\Delta h = \dfrac{BE}{100\%} \Delta(p_a/\gamma_w)$.

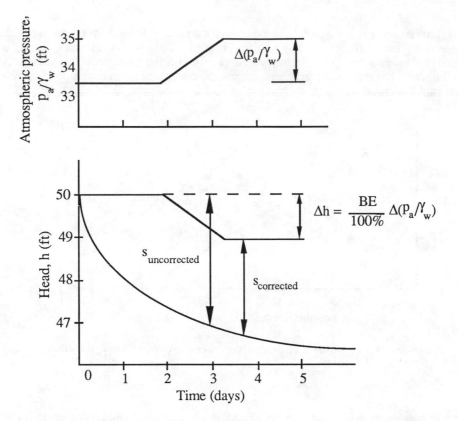

Figure 3.3   Example correction of pumping test data for change in atmospheric pressure.

3.3   TIDAL, RIVER LEVEL, AND SURFACE LOADING CHANGES

As with barometric efficiency, the *tidal, river stage*, or *load change efficiency* of an aquifer must be computed from measurements made prior to the beginning of the test. These efficiencies are defined as

$$TE = -\frac{dh}{d(tide)} \times 100\% \qquad (3.4a)$$

$$RE = -\frac{dh}{d(stage)} \times 100\% \qquad (3.4b)$$

$$LE = -\frac{dh}{[d(load)/\gamma_w]} \times 100\% \qquad (3.4c)$$

where

TE   = tidal efficiency
RE   = river stage efficiency

a well, L

L^{-2}

f Equation 3.3.

.... cia suc aquifer tidal efficiency is related to aquifer compressibility by (Ferris et al., 1962)

$$TE = \frac{\left(\dfrac{E_w}{E_s \theta}\right)}{\left(1 + \dfrac{E_w}{E_s \theta}\right)} \qquad (3.5)$$

and tidal efficiency is related to barometric efficiency by

$$BE + TE = 1 \qquad (3.6)$$

Example 3.3

A pumping test will be conducted in a confined aquifer in a coastal valley. The aquifer and overlying strata extend under a nearby bay. For several days prior to the test, tidal levels and depth to water in a test well are recorded, with the results plotted in Figure 3.4. From the slope of a straight line fitted through the points, the computed tidal efficiency is 0.3 or 30%.

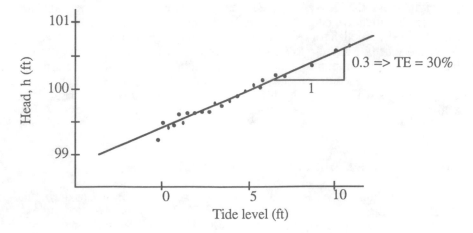

Figure 3.4 Typical data for determining tidal efficiency.

Figure 3.5 shows how the tidal efficiency computed in Figure 3.4 is used with tidal measurements to correct drawdown in a pumping test. Note that in the example, the tidal loading and therefore the drawdown correction change with time.

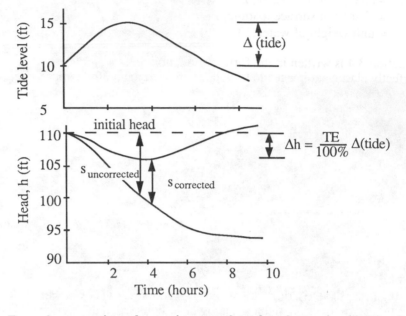

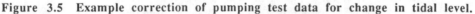

Figure 3.5   Example correction of pumping test data for change in tidal level.

Example 3.4

You are performing a pumping test in a heavy industrial area when, to your horror, a 24-car train carrying aluminum scrap metal chugs onto a nearby side track. The train stays parked for eight hours, then leaves. The train engineer tells you that she repeats this scheduled stop-over with identical cargo every two weeks.

Feeling relieved but foolish for overlooking a vital planning step, you continue the test uninterrupted and schedule a return to the site in two weeks to measure the change in water level that occurs in the pumping well and the observation well(s) as the train arrives at the site. These data can be used to compute the load efficiency for each well. The computed load efficiencies will probably be different for each well and will require a separate correction to the drawdown data collected from each well (the measured drawdowns will be <u>increased</u>).

3.4   DECREASES IN SATURATED THICKNESS

In an unconfined aquifer, the saturated thickness of the aquifer decreases during the pumping test. However, the conceptual models for unconfined aquifers presented in this manual assume that saturated thickness remains constant. This assumption can be accepted, and the methods of analysis presented in Part II can be used to interpret drawdown data from an unconfined aquifer if 1) the saturated thickness does not decrease more than about 25 percent , and 2) the drawdown data is corrected prior to analysis. The correction can be performed using the following equation (Jacob, 1944)

$$s_{corrected} = s - s^2/(2m) \tag{3.7}$$

where

$s_{corrected}$ = corrected drawdown, L
s = observed drawdown, L
m = initial (before pumping begins) saturated aquifer thickness, L.

This correction is based on the *Dupuit-Forchheimer* assumptions for flow toward a well (horizontal groundwater flow and the hydraulic gradient is equal to the slope of the water table). Neuman (1975) showed that the Dupuit-Forchheimer assumptions are not valid in an unconfined aquifer until the latter portion of the test when drawdown data match the Theis type curve (see Chapter 18). Therefore, the correction is not recommended for use with early and intermediate drawdown data. The time range for which this correction may be applied can be computed using a procedure discussed in Chapter 18.

Example 3.5

An unconfined, fine-sand aquifer has a saturated thickness of 75 feet. From the drilling logs, the hydraulic conductivity and specific yield are estimated to be 15 ft/day and 0.1, respectively. An observation well is located 15 feet from the pumping well.

Before data are corrected to account for decreased saturated thickness, the time for delayed yield effects to dissipate should be calculated. From Figure 18.2, the reciprocal of

Boulton's delay index is estimated as 130 min. Next, the quantity $D = \sqrt{\dfrac{T}{\alpha S_y}}$ is computed.

$$D = \sqrt{\frac{\left(15\ \dfrac{ft}{day}\right)(75\ ft)\left(\dfrac{1\ day}{1440\ min}\right)}{\left(\dfrac{1}{130\ min}\right)(0.1)}} = 31.87\ ft$$

Since r = 15 feet, $r/D = \dfrac{15\ ft}{31.87\ ft} = 0.47$

Entering Figure 18.1 with r/D = 0.47 gives

$$\alpha t = 3.25$$

Thus

$$t = 3.25\ (130\ min) = 422.5\ min = 7.0\ hr.$$

Therefore, data collected in the first seven hours should not be corrected.

As a sample calculation, suppose the drawdown after eight hours is 11 feet. The corrected drawdown for this time is

$$s_{corrected} = 11\ ft - \frac{(11\ ft)^2}{2\ (75\ ft)} = 10.2\ ft$$

Chapter 4

ANALYSIS OF PUMPING TEST DATA

This chapter presents a general description of several methods that have been developed to interpret pumping test data; a detailed description of the methods used to interpret pumping test data for each conceptual model in Table 1.1 are in Chapters 8 to 21. Descriptions of the methods used to interpret slug test data for each conceptual model in Table 1.2 are in Chapters 22 to 24.

4.1 GRAPHICAL METHODS OF ANALYSIS

4.1.1 Match-Point Method

The *match-point method*, a graphical procedure for determining aquifer parameters from pumping test data was first developed by Charles Theis (Jacob, 1940). The method was originally developed to compute transmissivity and storativity in a confined aquifer using drawdown data collected during transient flow conditions and was based on a particular conceptual model (Model 3) (Theis, 1935). However, the method can be used to determine aquifer parameters for any other conceptual model if the analytical equation describing the flow of groundwater to the pumping well has a solution that can be written in the form

$$s = \frac{Q}{AT} W(u, x_i) \qquad (4.1)$$

where

Q	= pumping rate, L^3T^{-1}	
A	= a constant	
T	= aquifer transmissivity, L^2T^{-1}	
$W(u, x_i)$	= well function, dimensionless	
u	= $\frac{r^2S}{4Tt}$, dimensionless time parameter	(4.2)
S	= aquifer storativity, dimensionless	
r	= radial distance from the pumping well center to a point of interest, L	
s	= drawdown at a distance r from the pumping well, L	
t	= time since pumping began, T	
x_i	= other dimensionless variables required by the analytical solution	

The variables x_i vary from one conceptual model to the next. For example, in Model 3, the well function is an integral of an exponential function of the single dimensionless variable u while in Model 4 the well function is a function of u and a parameter r/B (see Chapter 9).

For simplicity the development of the match-point method will be presented for Model 3, where the well function is a function of the single parameter u. Similar procedures can be used to develop the match-point method for alternate conceptual models. In Model 3, the well function, W(u) can be interpreted as the "dimensionless drawdown" that would occur at an observation well located a distance r from the pumping well at the

51

"dimensionless time" $1/u$ if the assumptions of the conceptual model are satisfied and if Q, T, and S of the aquifer were known. To determine T and S from measured values of drawdown two plots are prepared: $W(u)$ vs. $1/u$ (the *type curve*) and s vs. t (the *data curve*). If the two curves are plotted on identical (base 10) logarithmic scales Equation 4.1 and 4.2 can be written as

$$\log(s) = \log(Q/(AT) + \log(W(u)) \qquad (4.3)$$

and

$$\log(1/u) = \log(t) + \log(4T/(r^2 S)) \qquad (4.4)$$

When the type curve and data curve are superimposed and moved relative to each other until the curves are aligned (keeping the $W(u)$ axis parallel to the s axis and the $1/u$ axis parallel to the t axis), the values of $W(u)$ and s will be related by constant C1 and the values of $1/u$ and t will be related by constant C2

$$s/W(u) = C1 \qquad (4.5)$$

and

$$t/(1/u) = C2 \qquad (4.6)$$

where

$$C1 = Q/(AT) \qquad (4.7)$$

and

$$C2 = r^2 S/(4T) \qquad (4.8)$$

There is one unique position where the data curve coincides with the type curve. Values of $W(u)$, $1/u$, s, and t recorded from a common point (the *match-point*) when the curves are aligned can be used with Equations 4.1 and 4.2 to compute transmissivity and storativity. Figure 4.1 shows a data curve, type curve, and the alignment of the two curves by the match-point method. The axes have been shifted so that the curves are aligned, and a match-point has been selected. The match-point can be selected at any convenient location, on or off of the data curve, as long as the curves are aligned. Detailed descriptions of the match-point procedures specific to each conceptual model and additional examples are in Part II.

The above relations also apply when the data curve is plotted as s vs. t/r^2. Plotting data in this fashion allows the drawdown from several wells to be viewed on one graph and provides a greater range of values to match with the type curve. Match-point values of $W(u)$, $1/u$, s, and t/r^2 are recorded and substituted into Equations 4.1 and 4.2, exactly as when s vs. t is plotted.

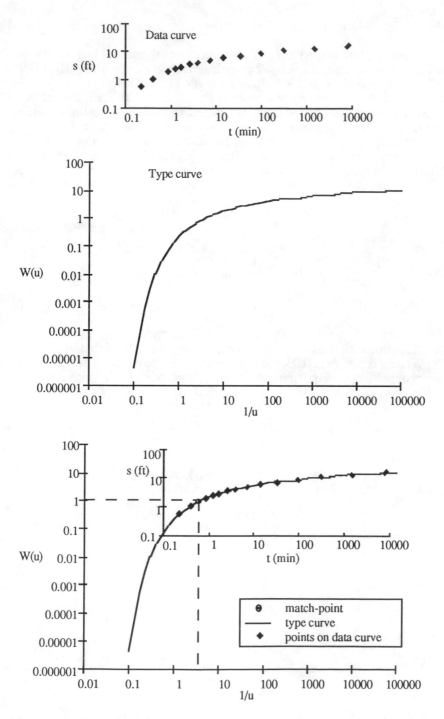

Figure 4.1   Example type curve and data curve alignment for the match-point method.

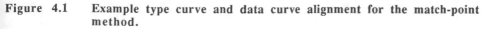

4.1.2   Straight-Line Method

While the match-point method can be applied to any of the conceptual models in Chapters 8 to 21, in some situations a simpler graphical method can be used. The Cooper and Jacob (1946) *straight-line method* is an approximation of the Theis solution. Therefore, flow in the aquifer must conform to the assumptions of the Theis conceptual model (Model 3). However, the method can also be used with other conceptual models under special conditions, for example, with early drawdown data collected before the effects of aquitard leakage or discontinuities in the lateral extent of the aquifer become important or with late drawdown data collected after the effects of well storage or delayed gravity yield have dissipated. The procedure is simple and, as discussed in Section 4.3, is a good way to check drawdown data for the presence of deviations from the assumptions of the selected conceptual model.

The Theis well function in Equation 4.1 and a method of computing it is given below

$$W(u) = \int_u^\infty \left[\frac{e^{-u}}{u} \right] du = -0.577216 - \ln(u) + u - u^2/(2 \cdot 2!) + u^3/(3 \cdot 3!)$$
$$- u^4/(4 \cdot 4!) + \cdots \qquad (4.9)$$

When the r is small or the t is large, the u term in the well function becomes small and the higher order terms of Equation 4.9 become insignificant (see the definition of u given in Equation 4.2). When u is less than 0.03, the error in neglecting all but the first two terms of the polynomial expression in Equation 4.9 is less than one percent. When these terms are neglected, Equation 4.1 can be written (see Appendix A)

$$s = \frac{0.183\,Q}{T} \log\left(\frac{2.25\,Tt}{r^2 S} \right) \qquad (4.10)$$

Thus, plots of s vs. log(t), s vs. log(t/r²), and s vs. log(r) are straight lines.

When time-drawdown data are to be analyzed, Equation 4.10 is written

$$T = \frac{0.183\,Q}{\Delta s} \qquad (4.11)$$

and

$$S = \frac{2.25\,Tt_o}{r^2} \qquad (4.12)$$

or

$$S = 2.25\,T(t/r^2)_o \qquad (4.13)$$

where
Δs = the change in drawdown occurring over one logarithmic cycle
t_o = the zero drawdown intercept of an s vs. log(t) plot
$(t/r^2)_o$ = the zero drawdown intercept of an s vs. log(t/r²) plot.

When distance-drawdown data are to be analyzed, Equation 4.10 is written

$$T = \frac{0.366\,Q}{\Delta s} \qquad\qquad (4.14)$$

and

$$S = \frac{2.25\,Tt}{r_o^2} \qquad\qquad (4.15)$$

where

r_o = the projected distance at which drawdown reaches zero

Figure 4.2 shows an example of time-drawdown data plotted on a semilogarithmic scale. Note that the earliest data, where u is large, plots as a curve above the extension of a line fitted to the data. Data for these values of u are not used to calculate T and S.

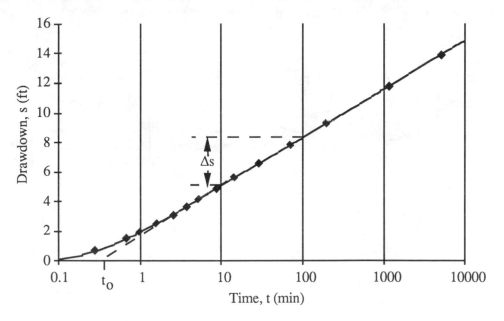

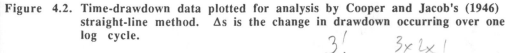

Figure 4.2. Time-drawdown data plotted for analysis by Cooper and Jacob's (1946) straight-line method. Δs is the change in drawdown occurring over one log cycle.

3!   3 x 2 x 1

4.2 THE THEORY OF SUPERPOSITION

All of the conceptual models described in this manual are based on the assumptions that the aquifer extends radially to infinity and that a single pumping well, pumping continuously at a constant rate, is the only cause of groundwater flow in the aquifer system. These assumptions may be relaxed if the pumping test data are analyzed utilizing the *theory of superposition*. Superposition may be used to account for the effects on pumping test drawdown data of other nearby pumping wells, aquifer discontinuities, groundwater recharge from streams or lakes, recovery of water levels after pumping ceases, and intermittent operation of the pumping well. The differential equations which describe groundwater flow are linear (see Part II) in the dependent variable (drawdown). Therefore, a linear combination of individual solutions is also a solution. This means that:

1.   The effects of multiple pumping wells on the predicted drawdown at a point can be computed by summing the predicted drawdowns at the point for each well, and

2.   Drawdown in complex aquifer systems can be predicted by superimposing predicted drawdowns for simpler aquifer systems.

4.2.1 Multiple Well Systems

The effects of multiple pumping wells on predicted drawdown at a point can be obtained by summing the predicted drawdowns for each well. The procedure is illustrated for the case of two pumping wells in Figure 4.3. If the individual effect of pumping well #1 is drawdown s_1 and the individual effect of pumping well #2 is s_2, then the total drawdown, $s_t = s_1 + s_2$. This procedure is valid for any number of pumping wells.

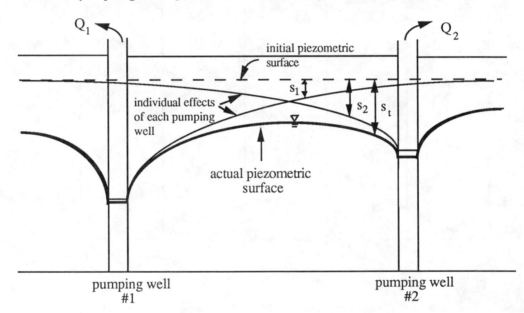

Figure 4.3   The effects of two pumping wells on the piezometric surface. The single-weight lines show the drawdown caused by each well if each were the only well pumping. The double-weight line shows the actual piezometric surface and is obtained by subtracting the sum of the drawdowns for each well from the initial piezometric surface.

4.2.2 Impermeable Boundary

The theory of superposition can also be implemented using the *method of image wells*. For example, the effects of an impermeable aquifer barrier on predicted drawdown can be determined by summing the predicted drawdowns from the (real) pumping well and an image well pumping at the same rate. The image well is located on the opposite side of the barrier at a perpendicular distance identical to that of the real well.

Example 4.1

The locations of a pumping well and an observation well are shown in Figure 4.4. The pumping well is 100 feet west of an impermeable boundary and discharges at 100 gal/min. The transmissivity and storativity of the aquifer are 10000 ft^2/day and 0.001, respectively. Predict the drawdown at the observation well after two days of pumping.

A particular conceptual model (Model 3) is used in this example but the method can be applied with any of the conceptual models described in Part II.

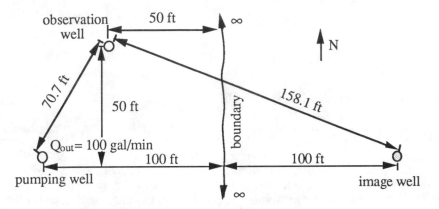

Figure 4.4   Plan view of aquifer in Examples 4.1, 4.2, and 4.3.

To predict the drawdown at the observation well, the predicted drawdown for the pumping well (r = 70.7 ft) is added to the predicted drawdown for an image well placed 100 feet east of the boundary (r = 158.1 ft), also pumping at 100 gal/min

$$s = \frac{Q_p}{4\pi T} W(u)_p + \frac{Q_i}{4\pi T} W(u)_i$$

where the subscript p indicates the pumping well and the subscript i indicates the image well, and

$$u = \frac{r^2 S}{4Tt}$$

Therefore

$$u_p = \frac{(70.7 \text{ ft})^2 \ (0.001)}{4(10000 \ \frac{\text{ft}^2}{\text{day}}) \ (2 \text{ days})} = 6.2 \times 10^{-5}$$

$$u_i = \frac{(158.1 \text{ ft})^2 \ (0.001)}{4(10000 \ \frac{\text{ft}^2}{\text{day}}) \ (2 \text{ days})} = 3.1 \times 10^{-4}$$

$$W(u)_p = 9.111$$
$$W(u)_i = 7.502$$

Since $Q_p = Q_i$, then $s = \dfrac{Q}{4\pi T}\,[W(u)_p + W(u)_i]$

$$s = \frac{(100\ \frac{gal}{min})\ (\frac{ft^3}{7.48\ gal})\ (\frac{1440\ min}{1\ day})}{4\pi\ (10000\ \frac{ft^2}{day})}\ (9.111 + 7.502)$$

$$= 1.40\ ft + 1.15\ ft = 2.55\ ft$$

Thus, the presence of the impermeable barrier increases the predicted drawdown at the observation well by 1.15 feet.

4.2.3   Boundary Between Materials of Different Transmissivities

The effects on predicted drawdown of discontinuities in aquifer properties can also be determined using image well theory. Consider the case of an aquifer separated by a material boundary into two parts, each with a different transmissivity. The effect of this boundary on predicted drawdown due to a pumping well on one side of the boundary can be determined by placing an image well on the opposite side of the boundary. The appropriate pumping rate for the image well is computed from the transmissivities for the two materials (Muskat, 1937)

$$Q_i = Q_p\, D_t \tag{4.16}$$

where

$$D_t = \frac{(T_p - T_d)}{(T_p + T_d)} \tag{4.17}$$

and

Q_i = pumping rate in image well, L^3T^{-1}
Q_p = pumping rate in pumping well, L^3T^{-1}
T_p = transmissivity of the aquifer on the pumping well side of the boundary, L^2T^{-1}
T_d = transmissivity of the aquifer on the image well side of the boundary, L^2T^{-1}

As long as the correct signs are carried through the equations, the effect of the boundary on the predicted drawdown will be correct. If the transmissivity at the image well is larger than at the pumping well computed drawdowns at the image well will be negative, decreasing the predicted drawdown. If the transmissivity at the image well is smaller than at the pumping well, predicted drawdown will be increased.

Example 4.2

Imagine that the aquifer east of the boundary in Figure 4.4 has a transmissivity of 5000 ft²/day. The transmissivity of the aquifer on the pumping well side of the boundary is 10000 ft²/day as in Example 4.1. Predict the drawdown at the observation well after two days of pumping.

From Equations 4.16 and 4.17, the pumping rate of the image well is

$$D_t = \frac{(10000 - 5000)}{(10000 + 5000)} = 0.33$$

$$Q_i = (100 \frac{gal}{min}) (0.33) = 33 \frac{gal}{min}$$

From Example 4.1, $u_p = 6.2 \times 10^{-5}$ and $W(u)_p = 9.111$

$$u_i = \frac{(158.1 \text{ ft})^2 (0.01)}{4 (5000 \frac{ft^2}{day}) (2 \text{ days})} = 6.2 \times 10^{-4}$$

$$W(u_i) = 6.809$$

Therefore,

$$s = \frac{Q_p W(u_p)}{4\pi T_p} + \frac{Q_i W(u_i)}{4\pi T_p}$$

$$= \left(\frac{\left(100 \frac{gal}{min}\right)(9.1)}{4\pi(10000 \frac{ft^2}{day})} + \frac{\left(33 \frac{gal}{min}\right)(6.8)}{4\pi(10000 \frac{ft^2}{day})} \right) \left(\frac{1 \text{ ft}^3}{7.48 \text{ gal}}\right) \left(\frac{1440 \text{ min}}{1 \text{ day}}\right)$$

$$= 1.40 \text{ ft} + 0.34 \text{ ft} = 1.74 \text{ ft}$$

4.2.4   Recharge Boundary

When the aquifer is interrupted by a lake, canal, or some other recharge boundary, drawdown in the aquifer can be predicted using an image well with a negative discharge (i.e., the well is pumping water into the aquifer).

Example 4.3

Given a situation identical to Example 4.1 but with a deep lake east of the boundary, the drawdown in the observation well after two days is summed from the contributions of the two wells computed in Example 4.1 as

$$s = 1.40 \text{ ft} - 1.15 \text{ ft} = 0.25 \text{ ft}$$

4.2.5   Recovery Tests

Aquifer transmissivity and storativity can also be determined from recovery data, drawdown data obtained after pumping stops. The analysis can be performed using image well theory if the pumping rate Q was constant before pumping stopped. The recovery of water levels following the end of pumping is represented by the summation of predicted drawdowns for the pumping well and an image well that injects water into the aquifer at the same rate Q. The term *residual drawdown*, s', refers to the observed drawdown during recovery. Using image well theory, residual drawdown is computed by summing the predicted drawdown due to the pumping well $s_p(t)$ and the build-up due to an injecting image well $s_i(t')$

$$s' = s_p(t) - s_i(t') \tag{4.18}$$

where

    t   = time since pumping began
    t' = time since pumping ended

Equations to predict residual drawdown for any of the conceptual models can be developed by applying image well theory. For example, the residual drawdown for Model 3 can be predicted using

$$s' = Q/(4\pi T)[W(u)-W(u')] \tag{4.19}$$

where

$$u = r^2 S/(4Tt) \tag{4.20}$$

$$u' = r^2 S/(4Tt') \tag{4.21}$$

An example is shown in Figure 4.5.

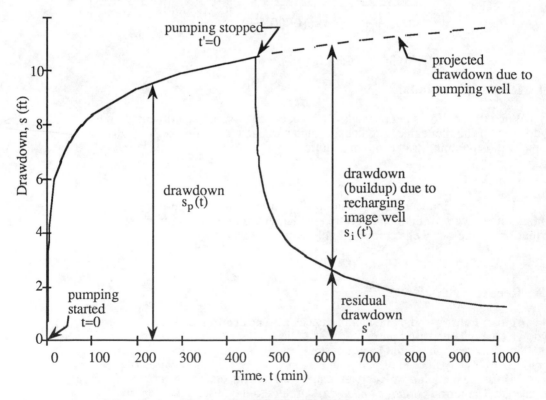

Figure 4.5   Example drawdown and recovery plots for a well pumped at a constant rate.

Recovery test data may be analyzed by the same methods used to analyze drawdown data; the only difference is the use of two times, t and t', and two well functions, W(u) and W(u'). For example, the procedure for the match-point method is as follows:

1. Plot the type curve as the function [W(u)-W(u')] vs. 1/u' on a logarithmic scale.

2. Plot the data curve of s' vs. t' on a scale identical to the type curve.

3. Overlay the plots and adjust the axes relative to each other until a best fit is achieved. Select a match-point within the best fit region and record match-point values [W(u)-W(u')]*, 1/u'*, s'*, and t'*.

4. Substitute [W(u)-W(u')]* and s'* into Equation (4.19) and solve for transmissivity.

5. Substitute the computed value of T and u'* and t'* into Equation (4.21) and solve for storativity.

The straight-line method may also be applied to recovery test results. It may be used with the portion of either the recovery or residual drawdown data for which u or u' is less than 0.03. An analysis of the residual drawdown data is often preferred since computations are independent of extrapolated s(t) values. However, the residual drawdown data cannot be used to find storativity. The steps for analyzing residual drawdown data are:

1. Plot s' vs. log(t/t'). If the assumptions of the flow model are valid, a straight line can be fitted to the data. Determine the change in residual drawdown for one log cycle Δs'.

2. Compute transmissivity using T= 0.183 Q/Δs', where Q is the pumping rate during the test.

The steps for analyzing recovery data are:

1. Plot recovery, (s-s') vs. log(t). If the assumptions of the flow model are valid, a line can be fitted to the data. Determine the change in drawdown for one log cycle.

2. Compute transmissivity using T = 0.183 Q/Δ(s - s').

3. If measurements are taken in at least one observation well, the storage coefficient can be computed. Select t_o' as the time when the straight-line extension of the recovery plot intersects (s - s')=0. Then S= 2.25 T t_o'/r^2, using the transmissivity calculated in step 2.

Recovery tests do not lend themselves to distance-drawdown analyses. Residual drawdown in wells far from the pumping well equalizes soon after pumping stops (Driscoll, 1986).

4.2.6 Intermittent Pumping

The variation of drawdown with time for an intermittently pumped well can be computed by summing the projected residual drawdown with the drawdown produced if the next cycle of pumping had started from equilibrium conditions. Figure 4.6 shows how the residual drawdown from the previous pumping cycle is added to "undisturbed" drawdown from the next cycle to produce the total drawdown.

The method described above can be used to analyze one or two repeating cycles, but the equations quickly become unwieldy for multiple pumping cycles. As an alternative, a procedure outlined by Driscoll (1986) can be used to compute the drawdown at the end of a number of cycles. The procedure is not mathematically precise and cannot be used with a graphical method to solve for aquifer properties from pumping test data. However, it can be used to predict drawdown and produces results within one or two percent of those computed with more exact methods. The method is only applicable when the well is operated on a regular cycle.

The intermittently pumped well is replaced by two image wells. One of the image wells discharges constantly at a rate which would pump the same total volume as pumped by the cycling well. The second image well pumps at a rate equal to the difference in discharge between the real well and the first image well. The real system is simulated by allowing the first image well to pump continuously, while the second well pumps only during the last cycle.

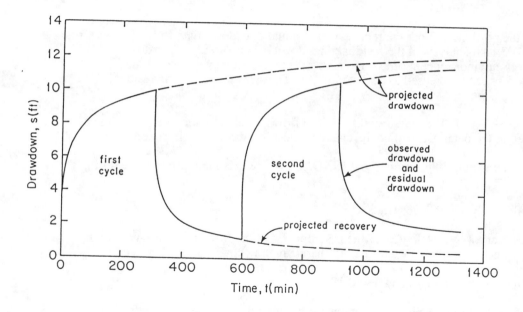

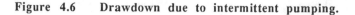

Figure 4.6   Drawdown due to intermittent pumping.

Example 4.4

Suppose that a well pumps 100 gal/min for two days "on" then one day "off" on a regular cycle. To compute the drawdown immediately after the pumping cycle at the end of one month, the real well is replaced by two image wells. The first image well discharges 67 gal/min continuously for 30 days to produce the same volume that the real well would discharge in 30 days. The second image well discharges at 33 gal/min. Drawdown is computed as

$$s_{total} = \text{drawdown from a discharge of} \quad + \quad \text{drawdown from a discharge of}$$
$$\text{67 gal/min for 29 days} \qquad\qquad \text{33 gal/min for 2 days}$$

4.3   VERIFYING CHOICE OF CONCEPTUAL MODEL

It is often difficult to identify an appropriate conceptual model to use to design a pumping test or to interpret drawdown data. The information available from a preliminary site investigation may be inadequate, or several alternative conceptual models may appear to be consistent with the available geological data. The assumptions used to derive an analytical solution for drawdown for a particular conceptual model usually require that specific conditions be satisfied. These conditions are often stated using aquifer properties which are unknown at the start of the analysis. For example, when a large diameter well is used in a pumping test, drawdown data are likely to be affected by well storage. Conceptual models 8, 9, and 15 include the effects of well storage. However, drawdown data collected for times later than $t = 2.5 \times 10^3\, r_c^2/T$, where r_c is the well casing radius and T is transmissivity, are not affected by well storage and these data can be analyzed by a simpler model (Table 2.1). Since T is not known precisely before the analysis begins, how can this time be computed?

Often, the appropriate choice of conceptual model is not apparent until drawdown data for the test are plotted. Cooper and Jacob (1946) showed that drawdown data for an aquifer satisfying the assumptions of the Theis solution (Model 3) will lie on a straight line on a semilogarithmic plot when u is less than about 0.03 (i.e., when t is relatively large compared to r^2S/T) (Figure 4.7a). The slope of the line is smaller in aquifers with a large transmissivity than in aquifers with a small transmissivity. A semilogarithmic plot of drawdown data collected from a pumping test is therefore a useful first step in determining if the assumptions of Model 3 apply to the aquifer. This type of plot may also suggest the presence of characteristics such as aquitard leakage or the presence of aquifer discontinuities (e.g., an impermeable boundary or recharge zone). Predicted drawdown for some conceptual models, such as those that consider well storage, are equivalent to those obtained with Model 3 after prolonged pumping; the transmissivity and storativity of the aquifer may be determined using the analytical solutions for Model 3 if the calculations are made using drawdown data from the later portion of the test after well storage effects have dissipated. Thus, the semilogarithmic plot is also a convenient tool for selecting appropriate ranges of drawdown data for analysis by various methods.

4.3.1 Aquitard Leakage and Storage

The effects of aquitard leakage on drawdown data for a confined aquifer are shown in Figures 4.7b and 4.7c. The following features should be noted:

1. Initially, drawdown in a confined aquifer with aquitard leakage (Figures 4.7b or 4.7c) will be the same as a confined aquifer without aquitard leakage (Figure 4.7a); in the early stages of pumping there has been insufficient time for groundwater to flow ("leak") through the aquitard.

2. As pumping continues, the radius of the cone of depression increases and a larger proportion of the groundwater flowing to the pumping well is derived from aquitard leakage and/or storage and the slope of the drawdown curve decreases; the slope of the curve decreases more if aquitard storage (S') is greater than zero (Figures 4.7b). In practice, however, semilogarithmic plots are of no use in determining whether or not an aquitard has significant storage.

3. After a sufficient period of pumping, additional effects may become apparent. For example:
    a) If the aquitard is bounded above by an impermeable bed (e.g., as in the *Case A* of Model 5), storage in the aquitard will become depleted close to the pumping well

and leakage will decrease (Figure 4.7c). All groundwater flowing to the pumping
well will originate in the aquifer and drawdown data will plot as a line parallel to
drawdown predicted for Model 3.

b) If the aquitard is bounded above by an unconfined aquifer with a large
transmissivity (i.e., the "source bed" of *Cases B* or *C*, Model 5), groundwater
flowing to the pumping well will originate in the source bed and the drawdown
curve becomes a horizontal line (Figure 4.7c).

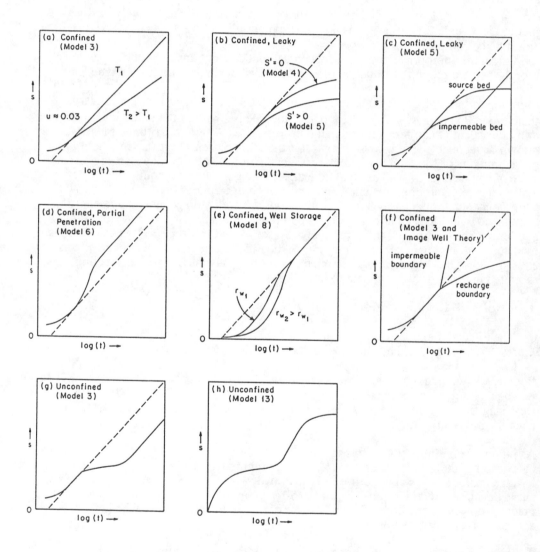

Figure 4.7   The effects of various aquifer characteristics on drawdown.

4.3.2   Partial Penetration

In a confined aquifer, groundwater flow to a fully penetrating pumping well is horizontal (Figure 4.8a). If the pumping well only partially penetrates the aquifer, groundwater flow converges toward the well screen, lengthening the flow path and increasing head losses and drawdown near the well (Figure 4.8b). In many deposits, the vertical flow component causes additional head loss because the vertical hydraulic conductivity is smaller than the horizontal hydraulic conductivity. The effects of partial penetration are only apparent in drawdown data collected within an approximate radial distance r of the pumping well

$$r < 1.5 \text{ m } \sqrt{K_r/K_z}$$

where

m = aquifer thickness, L
K_r = horizontal hydraulic conductivity, LT^{-1}
K_z = vertical hydraulic conductivity, LT^{-1}

beyond this distance groundwater flow is essentially horizontal (Figure 4.8b).

The effect of partial penetration on a semilogarithmic plot of drawdown data can be complex. When pumping begins, flow is primarily horizontal, as though the aquifer base is at the bottom of the well screen, and early drawdown data will be similar to that obtained for a fully penetrating pumping well (Figure 4.7d). As pumping continues, groundwater from the portion of the aquifer below the well screen contributes to the well discharge. The slope of the drawdown curve increases because additional head loss is required to drive groundwater to the well screen. After the flow lines in Figure 4.8b are fully established, the effects of partial penetration on head loss have reached a maximum and the drawdown curve is parallel to the curve that would be obtained for a fully penetrating pumping well (Figure 4.7d).

In practice, the three portions of the drawdown curve described above are not always visible. As the ratio of the radial distance, r, to aquifer saturated thickness, m, increases, or as the ratio of the screened length $(l - d)$ (see Chapter 11) to m decreases the duration of the first two portions decreases. Thus, the effects of partial penetration may easily be mistaken for leakage, well storage, a recharge boundary, or a fully penetrating well in a confined aquifer.

4.3.3   Well Storage

When the radius of the pumping well is very large, some or all of the initial well discharge will come from storage within the well casing. Initial drawdown is less than predicted by conceptual models that assume the pumping well has an infinitesimal diameter (Figure 4.7e). As pumping continues, well storage is depleted and the drawdown curve rejoins the curve predicted with a model that ignores well storage. The effects of well storage are only apparent in drawdown data collected within a time t since the start of pumping:

$$t = 2.5 \text{ x } 10^3 r_c^2/T$$

where

r_c = radius of the pumping well casing, L
T = aquifer transmissivity, L^2T^{-1}

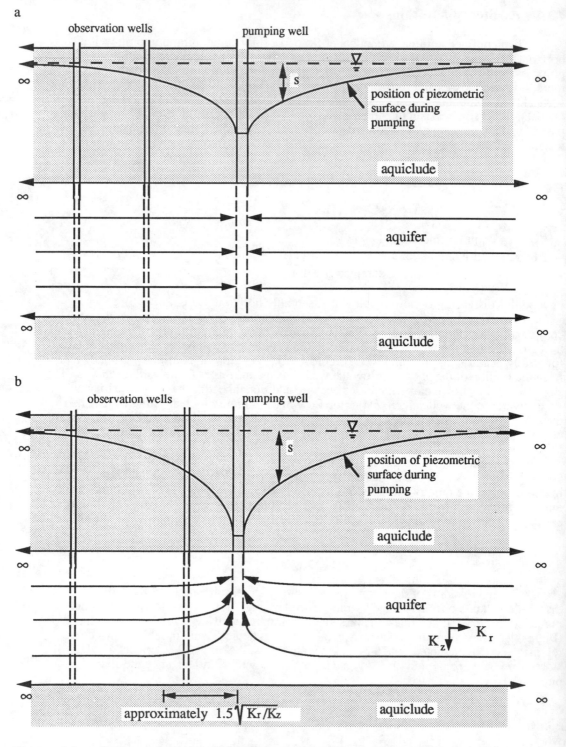

a

observation wells   pumping well

∞   s

position of piezometric
surface during
pumping

aquiclude

∞   ∞

aquifer

∞   ∞

aquiclude

b

observation wells   pumping well

∞   s

position of piezometric
surface during
pumping

aquiclude

∞   ∞

aquifer

K_r
K_z

∞   ∞

approximately $1.5\sqrt{K_r/K_z}$ aquiclude

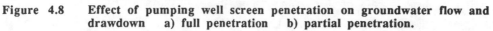

Figure 4.8   Effect of pumping well screen penetration on groundwater flow and
              drawdown   a) full penetration   b) partial penetration.

4.3.4   Aquifer Boundaries

The effect of aquifer boundaries (either impermeable boundaries, zones of higher or lower transmissivity, or recharge zones) are seen on semilogarithmic plots as a change in slope. For example, the presence of a recharge boundary is indicated by a decreasing slope while the presence of an impermeable boundary is indicated by an increasing slope (Figure 4.7f). The following method can be used to compute the distance to the boundary; the case of an impermeable boundary will be used to illustrate the procedure, but it can also be applied to a recharge zone.

From the theory of superposition (Section 4.2), the drawdown at an observation well due to the presence of an aquifer boundary can be represented by the sum of predicted drawdowns for the pumping well and an image well located on the opposite side of the boundary from the pumping well

$$s = s_r + s_i$$

where

s = drawdown at the observation well
s_r = predicted drawdown at the observation well due to the (real) pumping well
s_i = predicted drawdown at the observation well due to the image well

Before the radius of influence reaches the boundary, s_i is zero and $s = s_r$; thus, an extension of the drawdown curve prior to the time when flow is affected by the boundary can be used to compute the distance from the pumping well to the boundary (Figure 4.9). If two times are chosen where s_r equals s_i, then u_r will equal u_i. From the definition of u, it follows that

$$\frac{r_r^2 S}{4Tt_r} = \frac{r_i^2 S}{4Tt_i}$$

where

t_r = total time of pumping which produces drawdown s_r
t_i = total time of pumping which produces drawdown s_i
r_r = distance from the pumping well to the observation well
r_i = distance from the image well to the observation well

which can be simplified to

$$\frac{r_r^2}{t_r} = \frac{r_i^2}{t_i}$$

therefore

$$r_i = r_r \sqrt{\frac{t_i}{t_r}}$$

Values of t_i and t_r can be selected from the semilogarithmic plot by choosing any two locations where $s_r = s_i$ (e.g., t_r^{**} and t_i^{**} in Figure 4.9).

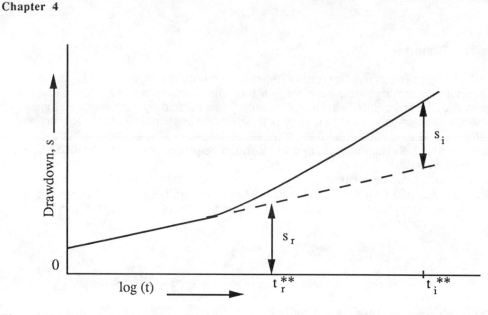

Figure 4.9 **Drawdown in a well near an impermeable boundary. Two times where drawdown from the image well, s_i, equals drawdown from the real well, s_r, are denoted by **.**

4.3.5 Unconfined Flow

In an unconfined aquifer, the groundwater flowing to a pumping well initially originates from the reduction in pore space as the aquifer skeleton compresses elastically and from the expansion of the groundwater in the pores as pressure decreases, similar to a confined aquifer. These processes occur almost instantaneously (the speed at which a pressure wave can be transmitted through an aquifer is very fast). As pumping continues, the majority of well discharge originates from the dewatering of the pore space as the water table falls; this process is relatively slow compared to the elastic compression and is referred to as *delayed yield* (Walton, 1960) or *delayed response* (Neuman, 1974). Consequently, a semilogarithmic plot of drawdown data in an unconfined aquifer has three distinct sections (Figure 4.7g):

1. An initial steep section where well discharge is derived from elastic storage,

2. A section of decreased slope created as water released by the falling water table begins to "catch up" with the initial flow, and

3. A second steep section parallel to the first, where water is, for all practical purposes, derived entirely from the lowering of the water table. The specific yield, S_y, is usually orders of magnitude larger than the elastic storativity, $S_s m$; therefore, there is little error in assuming that all flow at later times is derived from specific yield.

Drawdown data form an "s"-shaped curve on a semilogarithmic plot (Figure 4.7h). In practice, it is often difficult to identify the first section because of its short duration and because it occurs very early in the test when it is difficult to make accurate depth and flow measurements; well storage effects (if present) can also make it difficult to identify this section.

Example 4.5

In practice, the effects of aquifer characteristics appear much less clearly than as illustrated in Figure 4.7. Data scatter and overlapping effects obscure the idealized trends shown. Time and distance can also mask some aquifer characteristics. To illustrate this point, the theoretical drawdowns produced by several aquifer characteristics are plotted in Figures 4.10 through 4.13. Each line represents the effects of a particular characteristic on a homogeneous, isotropic, infinite, confined aquifer with a 100 percent efficient well and the following properties:

$$K_r = K_z = 500 \text{ gal/(day} \cdot \text{ft}^2)$$
$$m = 100 \text{ ft}$$
$$S_s m \text{ (elastic storativity)} = 0.001$$
$$S_y = 0.2.$$

The line labeled "Theis" represents drawdown in the aquifer predicted using Model 3. The line labeled "leakage" represents the drawdown that would occur if the aquifer was overlain by a 100 foot thick aquitard with a vertical hydraulic conductivity of 5 gal/(day $\cdot$ ft^2). The aquitard is overlain by a source bed and there is no significant aquitard storage. The "well storage" line shows drawdown that would occur if the pumping well had the storage capacity of a two-foot diameter well. The effects of reducing the screened length to 50 feet are shown in the "partial penetration" line. The "delayed yield" line shows drawdown if the equilibrium water level was at the top of the aquifer, creating an unconfined aquifer.

The semilogarithmic and logarithmic plots of drawdown at the pumping well are in Figures 4.10 and 4.11 respectively. Figures 4.12 and 4.13 show the computed drawdowns in an observation well 250 feet from the pumping well. The logarithmic plots have been included to show the relative shapes caused by each characteristic and to illustrate the difficulties in identifying the characteristics from these plots alone.

There is a striking difference between the shapes of the drawdown plots at the pumping well and those 250 feet distant. At the pumping well, the effects of leakage, partial penetration, and delayed yield appear very similar. A recharge or more permeable material boundary could have produced a similar shape. These same characteristics produce markedly different shapes at the observation well. At 250 feet, the effects of well storage and partial penetration are not apparent. It takes so long for measurable drawdown to occur at the observation well that well storage has only a small effect. Well storage is further obscured since its effects are only observed during a time when Cooper and Jacob's approximation is invalid, causing the curve in the "Theis" line.

The problems discussed above can be minimized by applying the semilogarithmic plot to data from all observation wells. It is also apparent that one must be careful in interpreting these plots. Once a system model has been defined, the data should be matched to a type curve to verify the model, even if the straight-line plot is sufficient for determining the desired aquifer characteristics.

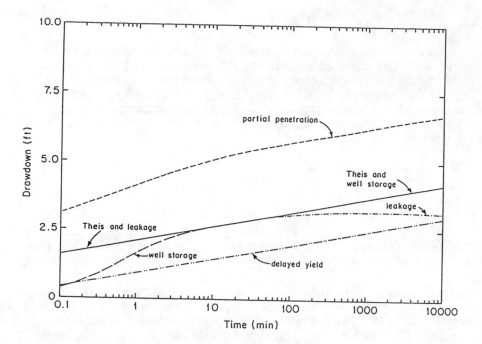

Figure 4.10   Effects of various aquifer characteristics on the semilogarithmic plot
of drawdown vs. time at the pumping well.

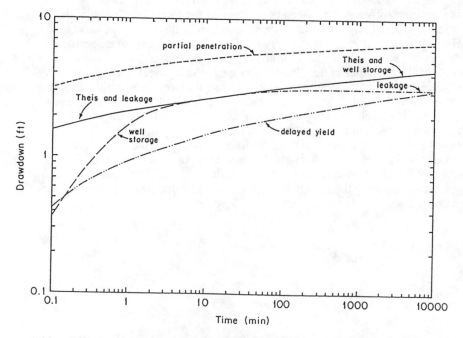

Figure 4.11   Effects of various aquifer characteristics on the logarithmic plot
of drawdown vs. time at the pumping well.

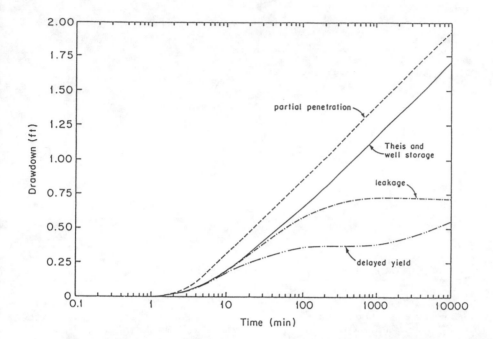

Figure 4.12 Effects of various aquifer characteristics on the semilogarithmic plot of drawdown vs. time at an observation well 250 feet from the pumping well.

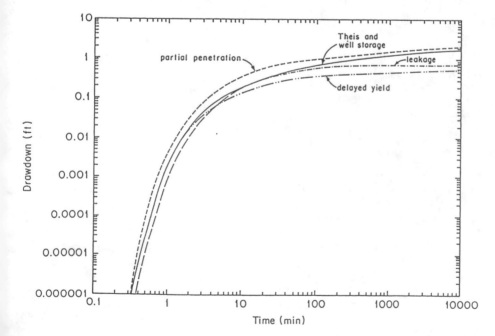

Figure 4.13 Effects of various aquifer characteristics on the logarithmic plot of drawdown vs. time at an observation well 250 feet from the pumping well.

Chapter 5

ANALYSIS OF STEP DISCHARGE PUMPING TESTS

5.1 BACKGROUND

The *step discharge test* can be used to determine the head loss that occurs as water flows through the well screen and pump inlet. The method is based on Cooper and Jacob's (1946) approximate solution for predicting drawdown in a confined aquifer (Model 3) derived in Appendix A

$$s = \frac{0.183\ Q}{T} \log \frac{(2.25\ Tt)}{r^2 S} \tag{5.1}$$

where

s	= drawdown of piezometric surface, L
Q	= pumping rate, $L^3 T^{-1}$
T	= aquifer transmissivity, $L^2 T^{-1}$
t	= time since pumping began, T
r	= radial distance from pumping well to a point on the cone of depression, L
S	= aquifer storativity (dimensionless)

Writing Equation 5.1 for drawdown inside the well casing (i.e., $r = r_w$ where r_w is the well radius) and collecting terms gives

$$s_w = BQ \tag{5.2}$$

where

s_w = drawdown inside the well casing

$B = (0.183/T)\log\dfrac{(2.25Tt)}{(r_w^2 S)}$

= an "aquifer coefficient", a measure of the head loss due to laminar (Darcy) flow in the aquifer

However, if groundwater flow across the well screen is turbulent due to large hydraulic gradients, Darcy's Law does not apply and drawdown cannot be computed using Equation 5.1. For this case Jacob (1946) proposed the following expression for s_w

$$s_w = BQ + CQ^2 \tag{5.3}$$

where

C = a "well coefficient", a measure of the head loss due to turbulent flow in the well screen and pump inlet

73

Others (Rorabaugh, 1953; Bruin and Hudson, 1955) examined the problem and determined that a more exact representation for the turbulent loss term is CQ^n, where n is some number which must be determined for each well. However, Bruin and Hudson (1955) concluded that Equation 5.3 is sufficient for most practical situations. The ratio of laminar to total head losses, L_p (%), is defined as

$$L_p = \frac{BQ}{(BQ + CQ^2)} \cdot 100\% \qquad (5.4)$$

In a step discharge pumping test, pumping rates are increased in a series of "steps" (see below). Because Cooper and Jacob's (1946) equation is only valid when $u = r^2S/(4Tt)$ is small ($u < 0.03$) the pumping rate for each step is held constant long enough to insure that drawdown data for that step plot as a straight line on a semilogarithmic graph. As the pumping rate and duration of pumping increase, the radius of the cone of depression increases in size and may intersect aquifer boundaries, which may be erroneously interpreted as a reduction in specific capacity of the pumping well due to turbulent head loss.

5.2   DETERMINATION OF HEAD LOSS COEFFICIENTS

Bierschenk (1964) developed a graphical method for determining the head loss coefficients B and C. It is based on a plot of specific capacity vs. pumping rate; each point on the plot is obtained from one step. A line is fitted to the data and the head loss coefficients are obtained using Equation 5.3, rearranged in the form

$$s_w/Q = CQ + B \qquad (5.5)$$

The following is a step-by-step description of the procedure:

1. Plot drawdown vs. log(time) as shown in Figure 5.1. If desired, Cooper and Jacob's straight-line method may be used to interpret the drawdown data from each step to determine aquifer transmissivity and storativity (see Chapter 4).

2. For each discharge rate, record the equilibrium drawdown at the pumping well, s_w.

3. Plot s_w/Q vs. Q on an arithmetic scale as shown in Figure 5.2. Fit a straight line through the data and extend the fitted line to a zero pumping rate. The slope of the line is C and the intercept is B.

Example 5.1

A step discharge test is conducted on a confined aquifer which is 100 feet thick. Assume that the appropriate conceptual model is Model 3. The initial discharge is 100 gal/min. Discharge is increased at 200 minute intervals to 200, 300, 425, and 550 gal/min. The pumping well diameter is 12 inches. Determine the aquifer transmissivity and the head loss coefficients B and C.

Drawdown data for the pumping well are plotted in Figure 5.1. From the straight portion of the discharge curve for $Q = 100$ gal/min, the change in drawdown over one logarithmic cycle is 0.53 feet (determined from a larger scale plot which includes measurements taken in the first 10 minutes). Using Cooper and Jacob's equation,

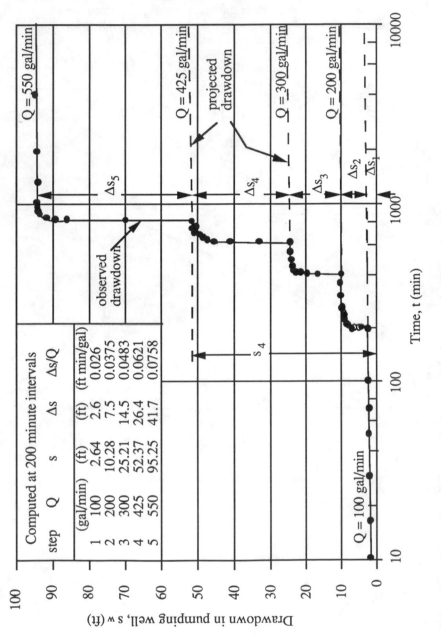

Figure 5.1   Drawdown data for step-discharge pumping test in Example 5.1.

$$T = \frac{2.25 \ (100 \ \frac{gal}{min}) \ (\frac{1440 \ min}{1 \ day})}{4\pi \ (0.53 \ feet)} = 49000 \ gal/day \cdot ft$$

The incremental increase in drawdown, Δs, caused by each new pumping rate is determined by subtracting the projected drawdown at the last pumping rate from the total drawdown at the current pumping rate. Tabular values of Δs appear in the upper left corner of Figure 5.1. Next, the values of $\Delta s/Q$ are plotted against Q on an arithmetic scale (Figure 5.2). A straight line is fitted to the data. The slope of the line is $C \left(0.00011 \ \frac{ft \cdot min^2}{gal^2} \right)$

and the zero-discharge intercept is $B \left(0.0153 \ \frac{ft \cdot min}{gal} \right)$

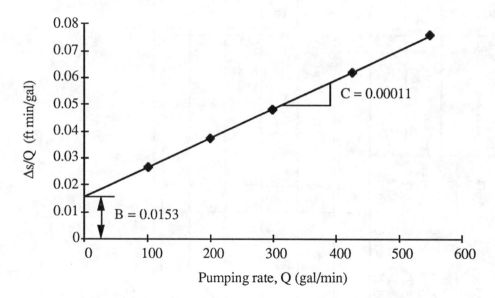

Figure 5.2 $\Delta s/Q$ vs. Q to compute head loss coefficients by Bierschenk's (1964) graphical method.

5.3   DETERMINATION OF WELL EFFICIENCY

Well efficiency, ε, is defined as the ratio of actual to theoretical drawdown at the pumping well (Figure 5.3) for a given pumping rate

$$\varepsilon = \frac{S_{actual}}{S_{theoretical}} \tag{5.6}$$

Two commonly used methods to determine the well efficiency are:

1.   Determine transmissivity and storativity from the drawdown in one or more observation wells. Observation wells are much less likely to experience head losses as water flows through their screens and flow in their vicinities is probably laminar. The theoretical drawdown at the pumping well is predicted using the pumping rate, the pumping well radius, and the computed T and S values; the actual drawdown is measured in the pumping well.

2.   Prepare a distance-drawdown plot using data from at least three observation wells. The theoretical drawdown is determined by fitting a straight line to the data and projecting the fitted line to a distance equal to the pumping well radius; the actual drawdown is measured in the pumping well.

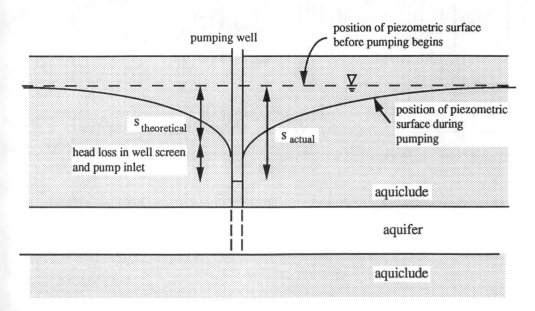

Figure 5.3   Definition of theoretical and actual drawdown.

Example 5.2

A pumping test is conducted using a six-inch diameter pumping well and several observation wells. Measured drawdowns in each well after 12 hours of pumping are in Figure 5.4. Determine the well efficiency.

The extension of a straight line fitted to the drawdown data for the observation wells shows that the theoretical drawdown at the pumping well should be 26 feet. Since the observed drawdown is 45.5 feet, the well efficiency, ε, is (26/45.5) x 100% = 57%.

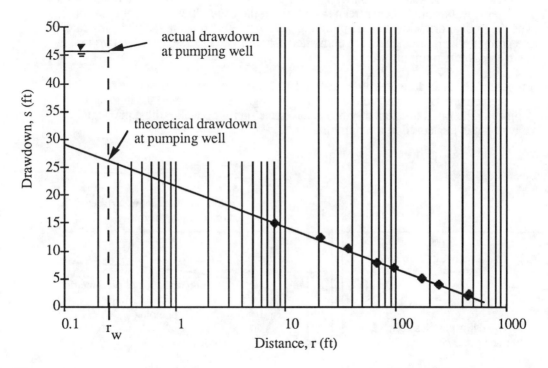

Figure 5.4   Distance-drawdown data used to compute well efficiency in Example 5.2.

Chapter 6

MODEL 1: EQUILIBRIUM, CONFINED

6.1 CONCEPTUAL MODEL

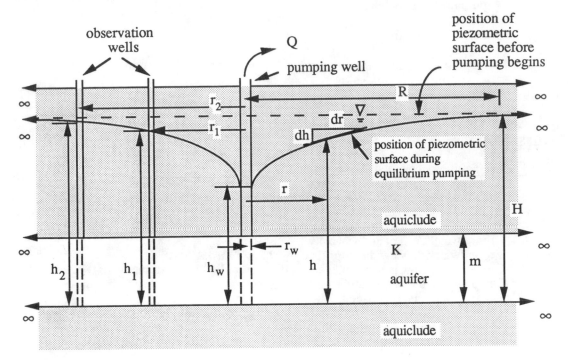

Definition of Terms

dh   = incremental head, L

dr   = incremental radius, L

h   = head at a distance r from the pumping well, L

h_w = head in the pumping well, L

h_1 = head in the first observation well, L

h_2 = head in the second observation well, L

H   = head before pumping begins, L

K   = aquifer hydraulic conductivity, LT^{-1}

m   = aquifer thickness, L

Q   = constant pumping rate, L^3T^{-1}

r   = radial distance from the pumping well to a point on the cone of depression (all distances are measured from the center of wells), L

r_w = pumping well inside radius, L

r_1 = radial distance from the pumping well to the first observation well, L

r_2 = radial distance from the pumping well to the second observation well, L

R   = radius of influence, L

Assumptions

1. The aquifer is bounded above and below by aquicludes.
2. All layers are horizontal and extend infinitely in the radial direction.
3. The initial piezometric surface (before pumping begins) is horizontal and extends infinitely in the radial direction.
4. The aquifer is homogeneous and isotropic.
5. Groundwater density and viscosity are constant.
6. Groundwater flow can be described by Darcy's Law.
7. Groundwater flow is horizontal and is directed radially toward the well.
8. The pumping and observation wells are screened over the entire aquifer thickness.
9. The pumping rate is constant.
10. The aquifer has been pumped long enough that equilibrium has been reached (drawdown is not changing with time).
11. Head losses through the well screen and pump intake are negligible.
12. The pumping well has an infinitesimal diameter.

6.2   MATHEMATICAL MODEL

Governing Equation

Referring to the definition sketch for the conceptual model, groundwater flows radially to the pumping well. The flow through any cylindrical section centered at the pumping well is equal to the pumping rate, Q

$$Q = (2\pi\ rm)\ K\left(\frac{dh}{dr}\right) = 2\pi rT\left(\frac{dh}{dr}\right) \tag{6.1}$$

where r is the radial distance from the pumping well, m is the aquifer thickness, and T = Km is the transmissivity of the aquifer.

Boundary Conditions

Case A
In *Case A* it is assumed that no observation wells are present, so the piezometric head is known only at the pumping well and at the edge of the cone of depression.

- At the pumping well, the piezometric head is h_w

$$h(r = r_w) = h_w \tag{6.2}$$

- At the edge of the cone of depression, the piezometric head is H

$$h(r = R) = H \tag{6.3}$$

Case B
If there are two wells (e.g., the pumping well and an observation well or two observation wells) where the piezometric head can be measured, then

- At the first well, the piezometric head is h_1

$$h(r = r_1) = h_1 \tag{6.4}$$

- At the second well, the piezometric head is h_2

$$h(r = r_2) = h_2 \tag{6.5}$$

Note that <u>any</u> two wells can be used. If the pumping well is used, set r_1 or r_2 equal to r_w and h_1 or h_2 equal to h_w in Equations 6.4 or 6.5.

6.3   ANALYTICAL SOLUTION

Case A
Separating variables, rearranging Equation 6.1, and applying boundary conditions yields

$$\int_{r_w}^{R} \frac{Q}{r} \, dr = 2\pi \, T \int_{h_w}^{H} dh \tag{6.6}$$

which can be solved to obtain

$$T = \frac{Q \ln (R/r_w)}{2\pi(H - h_w)} \tag{6.7}$$

Case B
The above procedure applied with the *Case B* boundary conditions gives

$$T = \frac{Q \ln (r_2/r_1)}{2\pi(h_2 - h_1)} \tag{6.8}$$

6.4   METHODS OF ANALYSIS

Case A

If no observation wells are available, the *Case A* solution (Equation 6.7) must be used to determine transmissivity. Therefore, some method is needed to find the radius of influence, R. Several empirical equations have been proposed for estimating R. One which is widely used is

$$R = C' (H - h_w) \sqrt{K} \tag{6.9}$$

where K is 10 cm/sec, H, h, and R are in feet, and C' is a dimensionless constant with a recommended value of 3 (Mansur and Kaufman, 1962). Equation 6.9 is plotted in Figure 6.1 for C' = 3 and (H - h$_w$) = 10 ft. To compute R for any other drawdown, the results from Figure 6.1 should be multiplied by the ratio of actual drawdown to a drawdown of 10 feet. If no observation wells are available, however, the value of K can probably be estimated as accurately using the material description and one of the methods in Section 2.3. For example, the U.S. Army Corps of Engineers (1956) developed curves relating hydraulic conductivity and effective grain size, D_{10} (Figure 6.1). If the transmissivity is estimated to be less than about 6700 ft^2/day, a slug test may be preferred.

Example 6.1

A pumping test is conducted in a confined sand aquifer with Q = 500 gal/min. The saturated aquifer thickness is 24 feet. The radius of the pumping well is six inches. There are no observation wells, but the sand has an effective grain size, D_{10}, of 0.2 mm. At equilibrium, the drawdown in the pumping well is 20 feet. Compute the hydraulic conductivity.

Since there are no observation wells, the *Case A* solution must be used, requiring an estimate of the radius of influence. Entering Figure 6.1 with D_{10} = 0.2 mm, R for 10 feet of drawdown is 950 feet. For a drawdown of 20 feet, this result is multiplied by two, giving a radius of influence of 1900 feet. Substituting into Equation 6.7,

$$T = \frac{Q \ln (R/r_w)}{2\pi(H - h_w)}$$

$$= \frac{\left(500 \; \frac{gal}{min}\right) \left(1440 \; \frac{min}{day}\right) \ln \left(\frac{1900 \; ft}{0.5 \; ft}\right)}{(2\pi) \, (20 \; ft)}$$

$$= 44088 \; \frac{gal}{day \cdot ft}$$

$$= 5894 \; \frac{ft^2}{day}$$

Case B

If drawdown data are available for at least two wells, the *Case B* solution (Equation 6.8) should be used. One simply substitutes values for Q, r, and h for any pair of wells into Equation 6.8. For more than two wells, T should be computed for each pair and the results averaged.

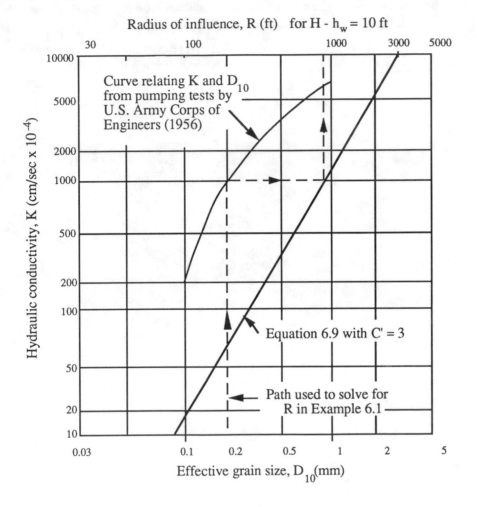

Figure 6.1   Hydraulic conductivity or effective grain size vs. radius of influence for a confined aquifer.

Example 6.2

A pumping test is conducted with $Q = 200$ gal/min. The saturated thickness of the aquifer is 20 feet. The radius of the pumping well is 4 inches. At equilibrium, the drawdown in the pumping well is 8 feet, and the drawdown in an observation well 100 feet from the pumping well is 2 feet. If the initial piezometric head is 50 feet, determine the aquifer transmissivity.

Because there are measurements of head available for two wells, use the *Case B* solution. For the pumping well $r_1 = r_w = 0.333$ ft and $h(r_1) = 50 - 8 = 42$ ft. For the observation well, $r_2 = 100$ ft and $h(r_2) = 50 - 2 = 48$ ft. Substituting into Equation 6.8 yields

$$T = \frac{Q \ln (r_2/r_1)}{2\pi(h_2 - h_1)}$$

$$= \frac{\left(200 \ \frac{gal}{min}\right)\left(1440 \ \frac{min}{day}\right)\ln\left(\frac{100 \ ft}{0.333 \ ft}\right)}{(2\pi) \ (48 \ ft - 42 \ ft)}$$

$$= 43580 \ \frac{gal}{day \cdot ft}$$

$$= 5826 \ \frac{ft^2}{day}$$

Chapter 7

MODEL 2: EQUILIBRIUM, UNCONFINED

7.1 CONCEPTUAL MODEL

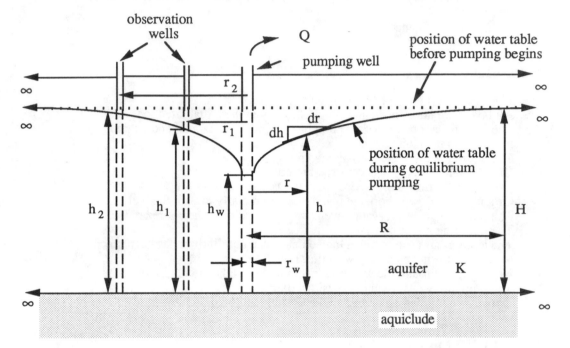

Definition of Terms

dh   = incremental head, L

dr   = incremental radius, L

h   = head at a distance r from the pumping well, L

h_w = head in the pumping well, L

h_1 = head in the first observation well, L

h_2 = head in the second observation well, L

H   = head before pumping begins, L

K   = aquifer hydraulic conductivity, LT^{-1}

Q   = constant pumping rate, L^3T^{-1}

r   = radial distance from the pumping well to a point on the cone of depression (all
             distances are measured from the center of wells), L

r_w = pumping well inside radius, L

r_1 = radial distance from the pumping well to the first observation well, L

r_2 = radial distance from the pumping well to the second observation well, L

R   = radius of influence, L

Assumptions

1. The aquifer is bounded below by an aquiclude.
2. All layers are horizontal and extend infinitely in the radial direction.
3. The initial water table (before pumping begins) is horizontal and extends
    infinitely in the radial direction.
4. The aquifer is homogeneous and isotropic.
5. Groundwater density and viscosity are constant.
6. Groundwater flow can be described by Darcy's Law.
7. Groundwater flow is horizontal and directed radially toward the well.
8. The pumping and observation wells are screened over the entire aquifer thickness.
9. The pumping rate is constant.
10. The aquifer has been pumped long enough that equilibrium has been reached
    (drawdown is not changing with time).
11. Head losses through the well screen and pump intake are negligible.
12. The pumping well has an infinitesimal diameter.
13. Drawdown is small compared to the aquifer saturated thickness.

7.2   MATHEMATICAL MODEL

Governing Equation

Referring to the definition sketch for the conceptual model, groundwater flows
radially to the pumping well. The flow through any cylindrical section centered at the
pumping well is equal to the pumping rate, Q

$$Q = (2\pi\ rh)\ K\left(\frac{dh}{dr}\right) \qquad (7.1)$$

where r is the radial distance from the pumping well, m is the aquifer thickness, and K is
the hydraulic conductivity of the aquifer.

Boundary Conditions

Case A
In *Case A* it is assumed that no observation wells are present, so the piezometric head is known only at the pumping well and at the edge of the cone of depression.

- At the pumping well, the water table height is h_w

$$h(r = r_w) = h_w \tag{7.2}$$

- At the edge of the cone of depression, the water table height is H

$$h(r = R) = H \tag{7.3}$$

Case B
If there are two wells (e.g., the pumping well and an observation well or two observation wells) where the piezometric head can be measured, then

- At the first observation well, the water table height is h_1

$$h(r = r_1) = h_1 \tag{7.4}$$

- At the second observation well, the water table height is h_2

$$h(r = r_2) = h_2 \tag{7.5}$$

Note that any two wells can be used. If the pumping well is used, r_1 or r_2 is set equal to r_w and h_1 or h_2 equal to h_w in Equation 7.4 or 7.5.

7.3   ANALYTICAL SOLUTION

Case A
Separating variables, rearranging Equation 7.1, and applying boundary conditions yields

$$\int_{r_w}^{R} \frac{Q}{r} \, dr = 2\pi \, K \int_{h_w}^{H} h \, dh \tag{7.6}$$

which can be solved to obtain

$$K = \frac{Q \ln (R/r_w)}{\pi(H^2 - h_w^2)} \tag{7.7}$$

Case B
The above procedure applied with the *Case B* boundary conditions gives

$$K = \frac{Q \ln (r_2/r_1)}{\pi(h_2^2 - h_1^2)} \tag{7.8}$$

7.4   METHODS OF ANALYSIS

Case A

If no observation wells are available, the *Case A* solution must be used to determine hydraulic conductivity. Therefore, some method is needed to find the radius of influence, R. From the equations of flow, it can be seen that K is not especially sensitive to the value of R since R will normally be large compared with the radius of the well, r_w.

Several empirical equations have been proposed for estimating R. One which is widely used is

$$R = C' (H - h_w) \sqrt{K} \tag{7.9}$$

where C' is a dimensionless constant that ranges from 1.5 to 2.0 (Mansur and Kaufman, 1962). Hydraulic conductivity, K, is in 10^{-4} cm/sec, and R, H, and h_w are in feet (see Figure 6.1). If no observation wells are available, however, the value of K can probably be estimated as accurately using the material description and one of the methods in Section 2.3.

Example 7.1

A medium sand aquifer is being pumped at $5.0 \ m^3$/sec. The initial saturated thickness is 20 m and the well is pumped until equilibrium is reached, causing 15 m of drawdown at the pumping well. The well diameter is 16 cm. The sand has an effective grain size of 1.0 mm. There are no observation wells. Determine the aquifer hydraulic conductivity.

Since there are no observation wells, the *Case A* solution must be used, requiring an estimate of the radius of influence. With the effective grain size and Hazen's equation (see Chapter 2), the hydraulic conductivity can be estimated to determine R. From Hazen's equation,

$$K \approx C (10_{10})^2$$
$$= 1.0 \ (1.0)^2 = 1.0 \ cm/sec$$

Using Equation 7.9 with C' = 1.5 yields

$$R = C' (H - h_w) \sqrt{K}$$
$$= 1.5 \ (15 \ m) \left(\frac{3.28 \ ft}{1m}\right) \sqrt{1.0} = 74 \ ft = 23 \ m$$

From Equation 7.7,

$$K = \frac{Q \ ln \ (R/r_w)}{\pi (H^2 - h_w^2)}$$

$$= \frac{\left(5.0 \ \frac{m^3}{s}\right) \left(\frac{100 \ cm}{1 \ m}\right) ln \left(\frac{23 \ m}{0.08 \ m}\right)}{(\pi) \ ((20 \ m)^2 - (5 \ m)^2)} = 2.4 \ cm/sec$$

Note that the estimate for K used to determine R has a relatively small effect on the hydraulic conductivity computed by Equation 7.7. If the original estimate of K was 10 cm/sec, Equation 7.9 would have given R = 73 m and the hydraulic conductivity computed by Equation 7.7 would have been 2.9 cm/sec.

Example 7.2

A silty sand aquifer with a saturated thickness of 30 feet is pumped at 10 gal/min. At equilibrium, the following heads are measured in two observation wells:

	equilibrium head (ft)	distance from pumping well (ft)
observation well # 1	10	15
observation well # 2	20	75

From Equation 7.8,

$$K = \frac{Q \ln (r_2/r_1)}{\pi(h_2^2 - h_1^2)}$$

$$= \frac{\left(25 \ \frac{gal}{min}\right)\left(1440 \ \frac{min}{day}\right) \ln \left(\frac{75 \ ft}{15 \ ft}\right)}{(\pi) \ ((20 \ ft)^2 - (10 \ ft)^2)}$$

$$= 61.5 \ \frac{gal}{day \cdot ft^2}$$

$$= 8.2 \ ft/day$$

Chapter 8

MODEL 3: TRANSIENT, CONFINED

8.1   CONCEPTUAL MODEL

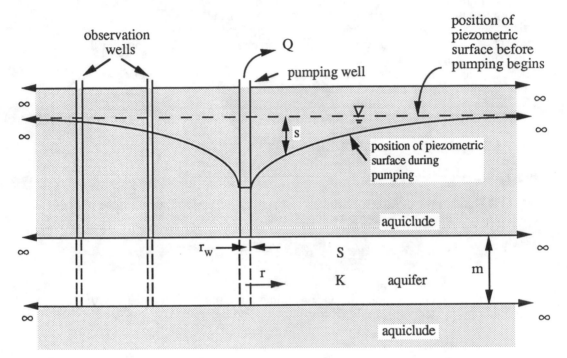

Definition of Terms

K   = aquifer hydraulic conductivity, LT^{-1}

m   = aquifer thickness, L

Q   = constant pumping rate, L^3T^{-1}

r   = radial distance from the pumping well to a point on the cone of depression (all distances are measured from the center of wells), L

s   = drawdown of piezometric surface during pumping, L

S   = aquifer storativity, dimensionless

Assumptions

1. The aquifer is bounded above and below by aquicludes.
2. All layers are horizontal and extend infinitely in the radial direction.
3. The initial piezometric surface (before pumping begins) is horizontal and extends infinitely in the radial direction.
4. The aquifer is homogeneous and isotropic.
5. Groundwater density and viscosity are constant.
6. Groundwater flow can be described by Darcy's Law.
7. Groundwater flow is horizontal and is directed radially toward the well.
8. The pumping and observation wells are screened over the entire aquifer thickness.
9. The pumping rate is constant.
10. Head losses through the well screen and pump intake are negligible.
11. The pumping well has an infinitesimal diameter.
12. The aquifer is compressible and completely elastic.

8.2   MATHEMATICAL MODEL

Governing Equation

The governing equation is derived by combining Darcy's Law with the principle of conservation of mass in a radial coordinate system

$$\frac{\partial^2 s}{\partial r^2} + \left(\frac{1}{r}\right)\frac{\partial s}{\partial r} = \left(\frac{S}{T}\right)\frac{\partial s}{\partial t} \tag{8.1}$$

where s is drawdown, r is radial distance from the pumping well, S and $T = Km$ are aquifer storativity and transmissivity, respectively, and t is time.

Initial Conditions

• Before pumping begins drawdown is zero everywhere

$$s(r, t = 0) = 0 \text{ for } r \geq 0 \tag{8.2}$$

Boundary Conditions

• At an infinite distance from the pumping well, drawdown is zero

$$s(r = \infty, t) = 0 \text{ for } t \geq 0 \tag{8.3}$$

• Groundwater flow to the well is constant and uniform over the aquifer thickness (which is a result of the assumption of horizontal groundwater flow)

$$\lim_{r \to 0} r\,\frac{\partial s(r, t)}{\partial r} = -\frac{Q}{2\pi T} \tag{8.4}$$

8.3   ANALYTICAL SOLUTION

8.3.1 General Solution

The solution to Equation 8.1 was obtained by Theis (1935). Drawdown at any time t and radial distance r is given by

$$s = \frac{Q}{4\pi T} W(u) \qquad\qquad (8.5)$$

where

$$W(u) = \int_{u}^{\infty} \left(\frac{e^{-u}}{u} \right) du \qquad\qquad (8.6)$$

and

$$u = \frac{r^2 S}{4Tt} \qquad\qquad (8.7)$$

W(u) is commonly referred to as the *Theis well function*. The function may be expressed as the series

$$\int_{u}^{\infty} \left(\frac{e^{-u}}{u} \right) du = -0.577216 - \ln(u) + u - \frac{u^2}{(2)\,2!} + \frac{u^3}{(3)\,3!} - \frac{u^4}{(4)\,4!} + \ldots \qquad (8.8)$$

Selected values of W(u) are in Table 8.1. Values of W(u) can also be computed using the computer program TYPE3; drawdown can be computed using the computer program DRAW3 (Appendix B).

Table 8.1   Values of the Theis well function, W(u) (after Ferris et al., 1962).

u	W(u)	u	W(u)
1.0×10^{-8}	17.8435	4.0×10^{-4}	7.2472
1.5×10^{-8}	17.4380	5.0×10^{-4}	7.0242
2.0×10^{-8}	17.1503	6.0×10^{-4}	6.8420
2.5×10^{-8}	16.9272	7.0×10^{-4}	6.6879
3.0×10^{-8}	16.7449	8.0×10^{-4}	6.5545
3.5×10^{-8}	16.5907	9.0×10^{-4}	6.4368
4.0×10^{-8}	16.4572	1.0×10^{-3}	6.3315
5.0×10^{-8}	16.2340	1.5×10^{-3}	5.9266
6.0×10^{-8}	16.0517	2.0×10^{-3}	5.6394
7.0×10^{-8}	15.8976	2.5×10^{-3}	5.4167
8.0×10^{-8}	15.7640	3.0×10^{-3}	5.2349
9.0×10^{-8}	15.6462	3.5×10^{-3}	5.0813
1.0×10^{-7}	15.5409	4.0×10^{-3}	4.9482

Table 8.1   (Continued).

u	W(u)	u	W(u)
1.5×10^{-7}	15.1354	5.0×10^{-3}	4.7261
2.0×10^{-7}	14.8477	6.0×10^{-3}	4.5448
2.5×10^{-7}	14.6246	7.0×10^{-3}	4.3916
3.0×10^{-7}	14.4423	8.0×10^{-3}	4.2591
3.5×10^{-7}	14.2881	9.0×10^{-3}	4.1423
4.0×10^{-7}	14.1546	1.0×10^{-2}	4.0379
5.0×10^{-7}	13.9314	1.5×10^{-2}	3.6374
6.0×10^{-7}	13.7491	2.0×10^{-2}	3.3547
7.0×10^{-7}	13.5950	2.5×10^{-2}	3.1365
8.0×10^{-7}	13.4614	3.0×10^{-2}	2.9591
9.0×10^{-7}	13.3437	3.5×10^{-2}	2.8099
1.0×10^{-6}	13.2383	4.0×10^{-2}	2.6813
1.5×10^{-6}	12.8328	5.0×10^{-2}	2.4679
2.0×10^{-6}	12.5451	6.0×10^{-2}	2.2953
2.5×10^{-6}	12.3220	7.0×10^{-2}	2.1508
3.0×10^{-6}	12.1397	8.0×10^{-2}	2.0269
3.5×10^{-6}	11.9855	9.0×10^{-2}	1.9187
4.0×10^{-6}	11.8520	1.0×10^{-1}	1.8229
5.0×10^{-6}	11.6280	1.5×10^{-1}	1.4645
6.0×10^{-6}	11.4465	2.0×10^{-1}	1.2227
7.0×10^{-6}	11.2924	2.5×10^{-1}	1.0443
8.0×10^{-6}	11.1589	3.0×10^{-1}	0.9057
9.0×10^{-6}	11.0411	3.5×10^{-1}	0.7942
1.0×10^{-5}	10.9357	4.0×10^{-1}	0.7024
1.5×10^{-5}	10.5303	5.0×10^{-1}	0.5598
2.0×10^{-5}	10.2426	6.0×10^{-1}	0.4554
2.5×10^{-5}	10.0194	7.0×10^{-1}	0.3738
3.0×10^{-5}	9.8371	8.0×10^{-1}	0.3106
3.5×10^{-5}	9.6830	9.0×10^{-1}	0.2602
4.0×10^{-5}	9.5495	1.0×10^{-0}	0.2194
5.0×10^{-5}	9.3263	1.5×10^{-0}	0.1000
6.0×10^{-5}	9.1440	2.0×10^{-0}	0.0489
7.0×10^{-5}	8.9899	2.5×10^{-0}	0.0249
8.0×10^{-5}	8.8563	3.0×10^{-0}	0.0131
9.0×10^{-5}	8.7386	3.5×10^{-0}	0.0070
1.0×10^{-4}	8.6332	4.0×10^{-0}	0.0038
1.5×10^{-4}	8.2278	5.0×10^{-0}	0.0011
2.0×10^{-4}	7.9402	6.0×10^{-0}	0.0004
2.5×10^{-4}	7.7172	7.0×10^{-0}	0.0001
3.0×10^{-4}	7.5348	8.0×10^{-0}	
3.5×10^{-4}	7.3807	9.0×10^{-0}	

8.3.2 Special Case Solution

Small Radius or Large Time

When the radius is small or the time is large the value of u is small. For u less than 0.03, all but the first two terms on the right side of Equation 8.8 may be neglected, resulting in an error in the calculation of W(u) of less than one percent (Cooper and Jacob, 1946). It is then convenient to use an alternate, approximate expression for Equation 8.5 derived in Appendix A

$$s = \frac{0.183Q}{T} \log \left(\frac{2.25\ Tt}{r^2 S} \right)$$ (8.9)

8.4   METHODS OF ANALYSIS

8.4.1 General Solution

Match-Point Method

The theoretical basis for this method of analysis is described in Chapter 4. The specific steps are as follows:

i   Prepare a plot of W(u) vs. 1/u on logarithmic paper. This plot is called the type curve (Figure 8.1). The type curve can be plotted using the values of W(u) in Table 8.1 or values computed by the computer program TYPE3 (Appendix B).

ii   Plot s vs. t (or t/r^2 for more than one observation well) using the same logarithmic scales used to prepare the type curve. This plot is called the data curve.

iii   Overlay the data curve on the type curve. Shift the plots relative to each other keeping the axes parallel until a best fit is found using as much of the two curves as possible.

iv   From the best fit portion of the curves, select a match-point and record the match-point values W(u)*, 1/u*, s*, and t* or t/r^2 *.

v   Substitute W(u)* and s* into Equation 8.5 and solve for T.

vi   Substitute u*, t* (or t/r^2 *), and the computed value of T into Equation 8.7 and solve for S.

Example 8.1

A pumping test was conducted on a sand and gravel aquifer with conditions approximating those of Model 3, except that there is a boundary between 700 and 1000 feet north of the production well. Observation wells 1, 2, and 3 are 30, 200, and 1000 feet west of the production well, respectively. All wells are fully penetrating. The production well is pumped at 150 gal/min. Table 8.2 lists drawdown data for the three observation wells. Determine the aquifer transmissivity and storativity.

Table 8.2   Drawdown data for Example 8.1.

Time, t (min)	Drawdown, s (ft)		
	r = 30 ft	r = 200 ft	r = 1000 ft
1	1.53	0.07	0.00
2	1.87	0.21	0.00
3	2.07	0.33	0.00
4	2.22	0.43	0.00
5	2.33	0.51	0.00
6	2.42	0.58	0.00
7	2.50	0.65	0.00
8	2.56	0.70	0.00
9	2.62	0.74	0.00
10	2.67	0.80	0.01
20	3.02	1.08	0.05
30	3.22	1.26	0.11
40	3.37	1.40	0.16
50	3.48	1.51	0.22
60	3.57	1.62	0.28
70	3.64	1.73	0.33
80	3.71	1.81	0.37
90	3.77	1.90	0.43
100	3.82	1.97	0.48
200	4.17	2.45	0.91
300	4.37	2.77	1.22
400	4.52	3.06	1.43
500	4.63	3.24	1.65
600	4.72	3.50	1.72
700	4.80	3.60	1.95
800	4.86	3.82	2.06
900	4.92	3.96	2.17
1000	4.97	4.02	2.22
1440	5.16	4.45	2.58

Figure 8.1 shows the data curve from the observation well at r = 200 ft placed over the type curve. The axes have been shifted relative to each other to achieve a best fit. Later data were not matched since they could include effects from the barrier. Match-point values chosen from the figure are

$$s^* = 1.0 \text{ ft}$$
$$t^* = 12 \text{ min}$$
$$W(u)^* = 2$$
$$1/u^* = 10$$

Substituting these values into Equation 8.5 gives

$$T = \frac{Q\ W(u)^*}{4\pi s^*} = \frac{\left(150\ \frac{gal}{min}\right)(2)\left(\frac{1440\ min}{day}\right)}{4\pi\ (1\ ft)} = 34380\ \frac{gal}{day \cdot ft}$$

and substituting into Equation 8.7 gives

$$S = \frac{u^*4Tt^*}{r^2} = \frac{\left(\frac{1}{10}\right)(4)\left(34380\ \frac{gal}{day \cdot ft}\right)(12\ min)\left(\frac{1\ ft^3}{7.48\ gal}\right)\left(\frac{1\ day}{1440\ min}\right)}{(200\ ft)^2}$$

$$= 0.00038$$

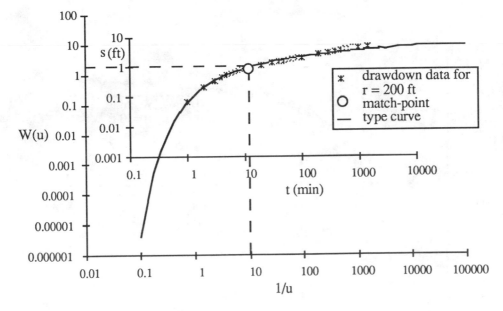

Figure 8.1   Type curve for the Theis well function, W(u), overlain by drawdown vs. time data at r = 200 ft for Example 8.1.

Example 8.2

When data from more than one observation well are available, the above procedure may be repeated for each well and the results combined. Another approach is to plot the data curve as s vs. t/r^2 using data from all observation wells. Figure 8.2 shows this plot for the drawdown data of Example 8.1. The match-point values are

$$s^* = 0.51$$
$$t^* = 0.00014 \text{ min/ft}^2$$
$$W(u)^* = 1$$
$$1/u^* = 6$$

Substituting into Equations 8.5 and 8.6 gives

$$T = \frac{\left(150 \frac{\text{gal}}{\text{min}}\right)(1.0)\left(\frac{1440 \text{ min}}{\text{day}}\right)}{4\pi \ (0.51 \text{ ft})} = 33700 \frac{\text{gal}}{\text{day} \cdot \text{ft}}$$

$$S = \left(\frac{1}{6}\right)(4)\left(33700 \frac{\text{gal}}{\text{day} \cdot \text{ft}}\right)\left(0.00014 \frac{\text{min}}{\text{ft}^2}\right)\left(\frac{1 \text{ ft}^3}{7.48 \text{ gal}}\right)\left(\frac{1 \text{ day}}{1440 \text{ min}}\right)$$

$$= 0.00029$$

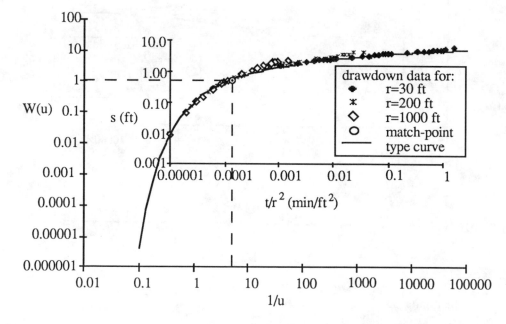

Figure 8.2   Type curve for the Theis well function, W(u), overlain by distance-drawdown data for Example 8.2.

8.4.2 Special Case Solutions

Straight-Line Time-Drawdown Method

Cooper and Jacob (1946) have shown that drawdown, s, varies with $\log(t/r^2)$ when t is relatively large. Thus, the aquifer properties can be determined from the slope of the s vs. $\log(t)$ plot. The procedure is as follows:

i Plot s vs. log(t) on a scale appropriate for the full range of data.

ii Fit a straight-line to the data. Ignore points where t is very small. Recall that when u = 0.03, the error caused by neglecting the higher order terms in the Theis well function is less than one percent. As u increases (and therefore t decreases) the error becomes larger.

iii For the fitted line, compute the change in drawdown, Δs, for one log cycle.

iv Compute T using (see Appendix A)

$$T = \frac{0.183\ Q}{\Delta s} \tag{8.10}$$

v Extend the fitted line to obtain the value of time at zero drawdown, t_o. Compute S using

$$S = \frac{2.25\ T\ t_o}{r^2} \tag{8.11}$$

The effects of aquitard leakage, well storage, partial penetration, and delayed yield may affect drawdown by a fixed amount after a sufficient time. Under these conditions, drawdown can be written in general form as

$$s = \frac{Q}{4\pi T}\ [W(u) + f] \tag{8.12}$$

where

$f =$ a function independent of time and varying according to aquifer and well properties and the model description

Thus, after a certain time, the transmissivity of these more complex aquifers can be computed by Equation 8.10. Storativity can be computed as

$$S = \frac{2.25\ T t_o e^f}{r^2} \tag{8.13}$$

Other conditions, such as encountering a boundary or lowering the water level below the top of a confined aquifer, affect late drawdown while early data conform to the Theis model. In these cases, data gathered before the cone of depression intersects the boundary or before the aquifer is converted to undefined conditions can be analyzed by Equations 8.10 and 8.11.

Example 8.3

Figure 8.3 illustrates the straight-line method with time-drawdown data from Example 8.1. Drawdown data are plotted against the log of time for each observation well. A straight line was fitted to the early data (i.e., before boundary effects were apparent). The change in drawdown for one log cycle, Δs, was obtained for each well and an average value computed

$$\Delta s_{ave} = (\Delta s_{r = 30\ ft} + \Delta s_{r = 200\ ft} + \Delta s_{r = 1000\ ft})/3$$

$$= (1.3\ ft + 1.15\ ft + 0.9\ ft)/3 = 1.12$$

From Equation 8.10,

$$T = \frac{0.183\ Q}{\Delta s} = \frac{0.183 \left(150\ \frac{gal}{min}\right) \left(1440\ \frac{min}{day}\right)}{1.12\ ft} = 35293\ \frac{gal}{day \cdot ft}$$

Storativity is computed using this transmissivity and the zero drawdown intercept, t_o. Using observation well 1 intercept, $t_{o,1}$, the storativity was determined from Equation 8.11

$$S = \frac{2.25\ Tt_o}{r^2} = \frac{2.25 \left(35293\ \frac{gal}{day \cdot ft}\right)(0.05\ min)\left(\frac{1\ ft^3}{7.48\ gal}\right)\left(\frac{1\ day}{1440\ min}\right)}{(30\ ft)^2}$$

$$= 0.00041$$

This process can be repeated for the other observation wells and the computed values of S and T averaged by some method. A convenient alternative is to plot s vs. $\log(t/r^2)$. Equations 8.10 and 8.11 still apply but the zero drawdown intercept $(t/r^2)_o$ is substituted for t_o.

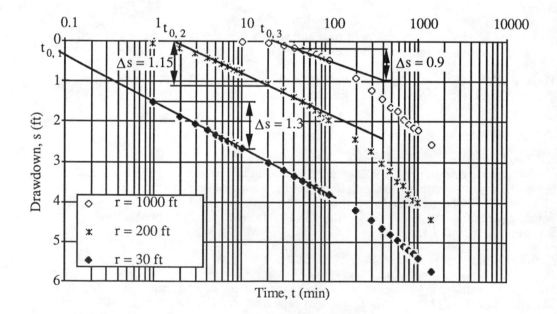

Figure 8.3   Illustration of straight-line method with time-drawdown data from Example 8.3.

Straight-Line Distance-Drawdown Method

This method is similar to the straight-line time-drawdown method except drawdown is plotted against the log of distance for a selected time t for three or more observation wells. The procedure is:

i Plot s vs. log(r) using the data from all observation wells for a time where the ratio r/t is relatively large (i.e., u < 0.03) .

ii Fit a straight line to the data and compute the change in drawdown over one log cycle, Δs.

iii Compute T using (see Appendix A)

$$T = \frac{0.366\,Q}{\Delta s} \tag{8.14}$$

iv For the case of distance-drawdown data, Equation 8.9 can be written (see Appendix A)

$$S = \frac{2.25\,Tt}{r_o^2} \tag{8.15}$$

Extend the fitted line to obtain the zero drawdown intercept, r_o, then use Equation 8.15 to solve for S.

As with the straight-line time-drawdown method, the above procedure can be used to compute T for other model systems as long as the effects of well storage, aquitard leakage, etc. have dissipated and u < 0.03. As with the straight-line time-drawdown method, Equation 8.15 cannot be used to solve for S unless a constant is added to account for the initial deviation from Theis conditions.

Example 8.4

Figure 8.4 illustrates the straight-line method with distance-drawdown data from Example 8.1 at t = 40 minutes. Although there are limited data for this example, the change in drawdown over one logarithmic cycle, Δs, and the zero drawdown intercept, r_o, are:

$$\Delta s = 2.2 \text{ ft}$$
$$r_o = 1100 \text{ ft}$$

Substituting these values into Equations 8.14 and 8.15 gives,

$$T = \frac{0.366\,Q}{\Delta s} = \frac{0.366\left(150\,\frac{\text{gal}}{\text{min}}\right)\left(1440\,\frac{\text{min}}{\text{day}}\right)}{2.2 \text{ ft}} = 35934\,\frac{\text{gal}}{\text{day}\cdot\text{ft}}$$

$$S = \frac{2.25 \, Tt}{r_o^2} = \frac{2.25 \left(35934 \, \frac{\text{gal}}{\text{day} \cdot \text{ft}}\right)(0.05 \text{ min})\left(\frac{1 \text{ ft}^3}{7.48 \text{ gal}}\right)\left(\frac{1 \text{ day}}{1440 \text{ min}}\right)}{(30 \text{ ft})^2}$$

$$= 0.00025$$

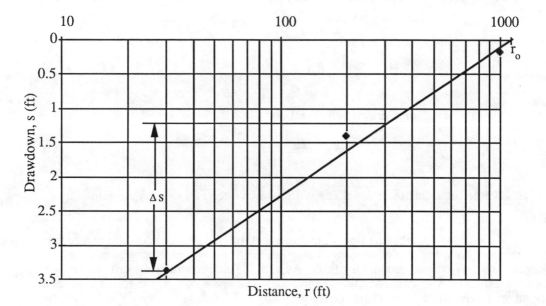

Figure 8.4   Straight-line method with distance-drawdown data at t = 40 minutes from Example 8.4.

Chapter 9

MODEL 4: TRANSIENT, CONFINED, LEAKY

9.1   CONCEPTUAL MODEL

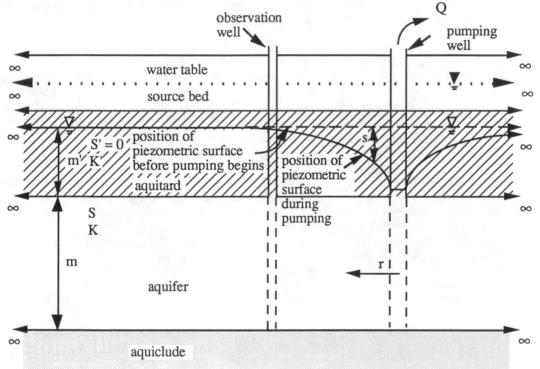

Definition of Terms

K   = aquifer hydraulic conductivity, LT^{-1}

K'   = aquitard vertical hydraulic conductivity, LT^{-1}

m   = aquifer thickness, L

m'   = aquitard thickness, L

Q   = constant pumping rate, L^3T^{-1}

r   = radial distance from the pumping well to a point on the cone of depression
          (all distances are measured from the center of wells), L

s   = drawdown of piezometric surface during pumping, L

S   = aquifer storativity, dimensionless

S'   = aquitard storativity, dimensionless

Assumptions

1. The aquifer is bounded above by an aquitard and an unconfined aquifer (the "source bed") and bounded below by an aquiclude.
2. All layers are horizontal and extend infinitely in the radial direction.
3. The initial piezometric surface (before pumping begins) is horizontal and extends infinitely in the radial direction. The water table in the source bed is horizontal, extends infinitely in the radial direction, and remains constant during pumping (zero drawdown). Drawdown of the water table in the source bed can be neglected when either $t < \dfrac{S'(m')^2}{10mK'}$ or when the transmissivity of the source bed is greater than 100 times the aquifer transmissivity (Neuman and Witherspoon, 1969b).
4. The aquifer and aquitard are homogeneous and isotropic.
5. Groundwater density and viscosity are constant.
6. Groundwater flow can be described by Darcy's Law.
7. Groundwater flow in the aquitard is vertical. Groundwater flow in the aquifer is horizontal and directed radially toward the well. This assumption is valid when m/B < 0.1 (Hantush, 1967), where B is defined in Equation 9.2.
8. The pumping and observation wells are screened over the entire aquifer thickness.
9. The pumping rate is constant.
10. Head losses through the well screen and pump intake are negligible.
11. The pumping well has an infinitesimal diameter.
12. The aquifer is compressible and completely elastic. The aquitard is incompressible (i.e., no water is released from aquitard storage during pumping). This assumption is valid when t > 0.036 m'S'/K' (Hantush, 1960) or when $r < 0.04m \sqrt{\dfrac{KS_s}{K'S_s'}}$ (Neuman and Witherspoon, 1969b).

9.2   MATHEMATICAL MODEL

Governing Equation

The governing equation is derived by combining Darcy's Law with the principle of conservation of mass in a radial coordinate system

$$\frac{\partial^2 s}{\partial r^2} + \left(\frac{1}{r}\right)\frac{\partial s}{\partial r} - \frac{s}{B^2} = \left(\frac{S}{T}\right)\frac{\partial s}{\partial t} \tag{9.1}$$

where s is drawdown, r is radial distance from the pumping well, S and T = Km are aquifer storativity and transmissivity, respectively, t is time, and

$$B^2 = \frac{Tm'}{K'} \tag{9.2}$$

Referring to the definition sketch for the conceptual model and recalling Darcy's Law, it is easy to see that sK'/m' represents the rate of vertical groundwater flow ("leakage") through the aquitard (the gradient is the change in head, s divided by the length of the flow path which is equal to the aquitard thickness, m'). If the aquitard is considered impermeable, $K' = 0$, $s/B^2 = 0$ and Equation 9.1 reduces to Equation 8.1 (Model 3).

Initial Conditions

- Before pumping begins drawdown is zero everywhere

$$s(r, t = 0) = 0 \tag{9.3}$$

Boundary Conditions

- At an infinite distance from the pumping well, drawdown is zero

$$s(r = \infty, t) = 0 \tag{9.4}$$

- Groundwater flow to the pumping well is constant and uniform over the aquifer thickness (which is a result of the assumption of horizontal groundwater flow in the aquifer)

$$\lim_{r \to 0} \frac{r \, \partial s \, (r, \, t)}{\partial r} = -\frac{Q}{2\pi T} \tag{9.5}$$

9.3   ANALYTICAL SOLUTION

9.3.1 General Solution

The solution developed by Hantush and Jacob (1955) can be used to predict drawdown when

$$t > 30 \, r^2 \left(\frac{S}{T}\right)\left[1 - \left(\frac{10r}{B}\right)^2\right] \tag{9.6}$$

and

$$\frac{r}{B} < 0.1 \tag{9.7}$$

The solution is

$$s = \frac{Q}{4\pi T} W(u, r/B) \tag{9.8}$$

where

$$W(u, r/B) = \int_u^\infty \frac{1}{y} \exp\left[-y - \frac{r^2}{4B^2 y}\right] dy \tag{9.9}$$

is the *Hantush-Jacob well function* and

$$u = \frac{r^2 S}{4Tt} \tag{9.10}$$

Selected values of W(u, r/B) are in Table 9.1. Values of W(u, r/B) can also be computed using the computer program TYPE4; drawdown can be computed using the computer program DRAW4 (Appendix B).

A series expansion for W(u, r/B) is (Hunt, 1983)

$$W(u, r/B) = \sum_{n=0}^{\infty} \frac{E_{n+1}(u)}{n!} \left[-\frac{r^2}{4uB^2} \right]^n$$

which may be used when

$$\frac{r^2}{4uB^2} \geq 0 \tag{9.11}$$

$E_n(u)$ is the exponential integral

$$E_n(u) = \int_1^{\infty} \frac{e^{-ux}}{x^n} \, dx \tag{9.12}$$

It is only necessary to calculate $E_1(u)$ since other values can be computed from the recursive formula

$$E_{n+1}(u) = \frac{1}{n} [e^{-u} - uE_n(u)] \quad \text{with } n = 1, 2, 3, \ldots \tag{9.13}$$

and $E_1(u) = W(u)$, the Theis well function of Model 3.

Table 9.1   Values of the well function W(u, r/B) for a leaky aquifer with no
aquitard storage (after Hantush, 1964, pp. 322-324).

$$W(u, r/B)$$

u	1.0×10^{-3}	5.0×10^{-3}	1.0×10^{-2}	2.5×10^{-2}	5.0×10^{-2}
1.0×10^{-6}	13.0031	10.8283	9.4425	7.6111	6.2285
2.0×10^{-6}	12.4240	10.8174	9.4425	7.6111	6.2285
3.0×10^{-6}	12.0581	10.7849	9.4425	7.6111	6.2285
4.0×10^{-6}	11.7905	10.7374	9.4422	7.6111	6.2285
5.0×10^{-6}	11.5795	10.6822	9.4413	7.6111	6.2285
6.0×10^{-6}	11.4053	10.6240	9.4394	7.6111	6.2285
7.0×10^{-6}	11.2570	10.5652	9.4361	7.6111	6.2285
8.0×10^{-6}	11.1279	10.5072	9.4313	7.6111	6.2285
9.0×10^{-6}	11.0135	10.4508	9.4251	7.6111	6.2285
1.0×10^{-5}	10.9109	10.3963	9.4176	7.6111	6.2285
2.0×10^{-5}	10.2301	9.9530	9.2961	7.6111	6.2285
3.0×10^{-5}	9.8288	9.6392	9.1499	7.6101	6.2285
4.0×10^{-5}	9.5432	9.3992	9.0102	7.6069	6.2285
5.0×10^{-5}	9.3213	9.2052	8.8827	7.6000	6.2285
6.0×10^{-5}	9.1398	9.0426	8.7673	7.5894	6.2285
7.0×10^{-5}	8.9863	8.9027	8.6625	7.5754	6.2285
8.0×10^{-5}	8.8532	8.7798	8.5669	7.5589	6.2284
9.0×10^{-5}	8.7358	8.6703	8.4792	7.5402	6.2283
1.0×10^{-4}	8.6308	8.5717	8.3983	7.5199	6.2282
2.0×10^{-4}	7.9390	7.9092	7.8192	7.2898	6.2173
3.0×10^{-4}	7.5340	7.5141	7.4534	7.0759	6.1848
4.0×10^{-4}	7.2466	7.2317	7.1859	6.8929	6.1373
5.0×10^{-4}	7.0237	7.0118	6.9750	6.7357	6.0821
6.0×10^{-4}	6.8416	6.8316	6.8009	6.5988	6.0239
7.0×10^{-4}	6.6876	6.6790	6.6527	6.4777	5.9652
8.0×10^{-4}	6.5542	6.5467	6.5237	6.3695	5.9073
9.0×10^{-4}	6.4365	6.4299	6.4094	6.2716	5.8509
1.0×10^{-3}	6.3313	6.3253	6.3069	6.1823	5.7965
2.0×10^{-3}	5.6393	5.6363	5.6271	5.5638	5.3538
3.0×10^{-3}	5.2348	5.2329	5.2267	5.1845	5.0408
4.0×10^{-3}	4.9482	4.9467	4.9421	4.9105	4.8016
5.0×10^{-3}	4.7260	4.7249	4.7212	4.6960	4.6084
6.0×10^{-3}	4.5448	4.5438	4.5407	4.5197	4.4467
7.0×10^{-3}	4.3916	4.3908	4.3882	4.3702	4.3077
8.0×10^{-3}	4.2590	4.2583	4.2561	4.2404	4.1857
9.0×10^{-3}	4.1423	4.1416	4.1396	4.1258	4.0772
1.0×10^{-2}	4.0379	4.0373	4.0356	4.0231	3.9795

Table 9.1   (Continued).

W(u, r/B)

u	r/B				
	1.0×10^{-3}	5.0×10^{-3}	1.0×10^{-2}	2.5×10^{-2}	5.0×10^{-2}
2.0×10^{-2}	3.3547	3.3544	3.3536	3.3476	3.3264
3.0×10^{-2}	2.9591	2.9589	2.9584	2.9545	2.9409
4.0×10^{-2}	2.6812	2.6811	2.6807	2.6779	2.6680
5.0×10^{-2}	2.4679	2.4678	2.4675	2.4653	2.4576
6.0×10^{-2}	2.2953	2.2952	2.2950	2.2932	2.2870
7.0×10^{-2}	2.1508	2.1508	2.1506	2.1491	2.1439
8.0×10^{-2}	2.0269	2.0269	2.0267	2.0255	2.0210
9.0×10^{-2}	1.9187	1.9187	1.9185	1.9174	1.9136
1.0×10^{-1}	1.8229	1.8229	1.8227	1.8218	1.8184
2.0×10^{-1}	1.2226	1.2226	1.2226	1.2222	1.2209
3.0×10^{-1}	0.9057	0.9057	0.9056	0.9054	0.9047
4.0×10^{-1}	0.7024	0.7024	0.7024	0.7022	0.7018
5.0×10^{-1}	0.5598	0.5598	0.5598	0.5597	0.5594
6.0×10^{-1}	0.4544	0.4544	0.4544	0.4543	0.4541
7.0×10^{-1}	0.3738	0.3738	0.3738	0.3737	0.3735
8.0×10^{-1}	0.3106	0.3106	0.3106	0.3106	0.3104
9.0×10^{-1}	0.2602	0.2602	0.2602	0.2602	0.2601
1.0×10^{0}	0.2194	0.2194	0.2194	0.2194	0.2193
2.0×10^{0}	0.0489	0.0489	0.0489	0.0489	0.0489
3.0×10^{0}	0.0130	0.0130	0.0130	0.0130	0.0130
4.0×10^{0}	0.0038	0.0038	0.0038	0.0038	0.0038
5.0×10^{0}	0.0011	0.0011	0.0011	0.0011	0.0011
6.0×10^{0}	0.0004	0.0004	0.0004	0.0004	0.0004
7.0×10^{0}	0.0001	0.0001	0.0001	0.0001	0.0001

W(u, r/B)

u	r/B				
	7.5×10^{-2}	1.5×10^{-1}	3.0×10^{-1}	5.0×10^{-1}	7.0×10^{-1}
1.0×10^{-4}	5.4228	4.0601	2.7449	1.8488	1.3210
2.0×10^{-4}	5.4227	4.0601	2.7449	1.8488	1.3210
3.0×10^{-4}	5.4212	4.0601	2.7449	1.8488	1.3210
4.0×10^{-4}	5.4160	4.0601	2.7449	1.8488	1.3210
5.0×10^{-4}	5.4062	4.0601	2.7449	1.8488	1.3210
6.0×10^{-4}	5.3921	4.0601	2.7449	1.8488	1.3210

Table 9.1   (Continued).

$$W(u, r/B)$$

u	r/B				
	7.5×10^{-2}	1.5×10^{-1}	3.0×10^{-1}	5.0×10^{-1}	7.0×10^{-1}
7.0×10^{-4}	5.3745	4.0600	2.7449	1.8488	1.3210
8.0×10^{-4}	5.3542	4.0599	2.7449	1.8488	1.3210
9.0×10^{-4}	5.3317	4.0598	2.7449	1.8488	1.3210
1.0×10^{-3}	5.3078	4.0595	2.7449	1.8488	1.3210
2.0×10^{-3}	5.0517	4.0435	2.7449	1.8488	1.3210
3.0×10^{-3}	4.8243	4.0092	2.7448	1.8488	1.3210
4.0×10^{-3}	4.6335	3.9551	2.7444	1.8488	1.3210
5.0×10^{-3}	4.4713	3.8821	2.7428	1.8488	1.3210
6.0×10^{-3}	4.3311	3.8384	2.7398	1.8488	1.3210
7.0×10^{-3}	4.2078	3.7529	2.7350	1.8488	1.3210
8.0×10^{-3}	4.0980	3.6903	2.7284	1.8488	1.3210
9.0×10^{-3}	3.9991	3.6302	2.7202	1.8487	1.3210
1.0×10^{-2}	3.9091	3.5725	2.7104	1.8486	1.3210
2.0×10^{-2}	3.2917	3.1158	2.5688	1.8379	1.3207
3.0×10^{-2}	2.9183	2.8017	2.4110	1.8062	1.3177
4.0×10^{-2}	2.6515	2.5655	2.2661	1.7603	1.3094
5.0×10^{-2}	2.4448	2.3776	2.1371	1.7075	1.2955
6.0×10^{-2}	2.2766	2.2218	2.0227	1.6524	1.2770
7.0×10^{-2}	2.1352	2.0894	1.9206	1.5973	1.2551
8.0×10^{-2}	2.0136	1.9745	1.8290	1.5436	1.2310
9.0×10^{-2}	1.9072	1.8732	1.7460	1.4918	1.2054
1.0×10^{-1}	1.8128	1.7829	1.6704	1.4422	1.1791
2.0×10^{-1}	1.2186	1.2066	1.1602	1.0592	0.9284
3.0×10^{-1}	0.9035	0.8969	0.8713	0.8142	0.7369
4.0×10^{-1}	0.7010	0.6969	0.6809	0.6446	0.5943
5.0×10^{-1}	0.5588	0.5561	0.5453	0.5206	0.4860
6.0×10^{-1}	0.4537	0.4518	0.4441	0.4266	0.4018
7.0×10^{-1}	0.3733	0.3719	0.3663	0.3534	0.3351
8.0×10^{-1}	0.3102	0.3092	0.3050	0.2953	0.2815
9.0×10^{-1}	0.2599	0.2591	0.2559	0.2485	0.2378
1.0×10^{0}	0.2191	0.2186	0.2161	0.2103	0.2020
2.0×10^{0}	0.0489	0.0488	0.0485	0.0477	0.0467
3.0×10^{0}	0.0130	0.0130	0.0130	0.0128	0.0126
4.0×10^{0}	0.0038	0.0038	0.0038	0.0037	0.0037
5.0×10^{0}	0.0011	0.0011	0.0011	0.0011	0.0011
6.0×10^{0}	0.0004	0.0004	0.0004	0.0004	0.0004
7.0×10^{0}	0.0001	0.0001	0.0001	0.0001	0.0001

Table 9.1   (Continued).

$$W(u, r/B)$$

u	r/B				
	8.5×10^{-1}	1.0×10^{-0}	1.5×10^{-0}	2.0×10^{-0}	2.5×10^{-0}
1.0×10^{-2}	1.0485	0.8420	0.4276	0.2278	0.1247
2.0×10^{-2}	1.0484	0.8420	0.4276	0.2278	0.1247
3.0×10^{-2}	1.0481	0.8420	0.4276	0.2278	0.1247
4.0×10^{-2}	1.0465	0.8418	0.4276	0.2278	0.1247
5.0×10^{-2}	1.0426	0.8409	0.4276	0.2278	0.1247
6.0×10^{-2}	1.0362	0.8391	0.4276	0.2278	0.1247
7.0×10^{-2}	1.0272	0.8360	0.4276	0.2278	0.1247
8.0×10^{-2}	1.0161	0.8316	0.4275	0.2278	0.1247
9.0×10^{-2}	1.0032	0.8259	0.4274	0.2278	0.1247
1.0×10^{-1}	0.9890	0.8190	0.4271	0.2278	0.1247
2.0×10^{-1}	0.8216	0.7148	0.4135	0.2268	0.1247
3.0×10^{-1}	0.6706	0.6010	0.3812	0.2211	0.1240
4.0×10^{-1}	0.5501	0.5024	0.3411	0.2096	0.1217
5.0×10^{-1}	0.4550	0.4210	0.3007	0.1944	0.1174
6.0×10^{-1}	0.3793	0.3543	0.2630	0.1774	0.1112
7.0×10^{-1}	0.3183	0.2996	0.2292	0.1602	0.1040
8.0×10^{-1}	0.2687	0.2543	0.1994	0.1436	0.0961
9.0×10^{-1}	0.2280	0.2168	0.1734	0.1281	0.0881
1.0×10^{-0}	0.1943	0.1855	0.1509	0.1139	0.0803
2.0×10^{0}	0.0456	0.0444	0.0394	0.0335	0.0271
3.0×10^{0}	0.0124	0.0122	0.0112	0.0100	0.0086
4.0×10^{0}	0.0036	0.0036	0.0034	0.0031	0.0027
5.0×10^{0}	0.0011	0.0011	0.0010	0.0010	0.0009
6.0×10^{0}	0.0004	0.0004	0.0003	0.0003	0.0003
7.0×10^{0}	0.0001	0.0001	0.0001	0.0001	0.0001

9.3.2 Special Case Solutions

Late-Time Solution

After pumping has continued for a long time, most of the groundwater flowing to the pumping well will originate in the source bed (i.e., aquitard leakage will be approximately equal to the pumping rate, Q). The following solution can be used if

$t > \dfrac{8m'S}{K'}$ (Hantush and Jacob, 1955)

$$s = \frac{Q}{2\pi T} K_o \ (r/B) \tag{9.14}$$

where K_o is the zero-order modified Bessel function of the second kind.

Negligible Aquitard Leakage

When $r/B < 0.01$, aquitard leakage can be neglected and the solution for Model 3 applies (Equation 8.5).

9.4   METHODS OF ANALYSIS

9.4.1 General Solution

Match-Point Method

The theoretical basis for this method of analysis is described in Chapter 4. The specific steps are:

i Prepare a plot of $W(u, r/B)$ vs. $1/u$ on logarithmic paper. This plot is called the type curve (Figure 9.1). The type curve can be plotted using the values of $W(u, r/B)$ in Table 9.1 or values computed by the computer program TYPE4 (Appendix B).

ii Plot s vs. t (or s vs. t/r^2 for more than one observation well) using the same logarithmic scales used to prepare the type curve. This plot is called the data curve.

iii Overlay the data curve on the type curves. Shift the plots relative to each other, keeping the axes parallel, until a position of best fit is found between the data curve and one of the type curves. Try to use as much early drawdown data as possible. Unless a significant number of points fall within the period where leakage effects are insignificant, (i.e., where $t < 0.25t_i$ with t_i as the inflection point in the curve, usually estimated by inspection) a unique solution is impossible (Hantush, 1964).

iv From the best fit portion of the curves, select a match-point and record match-point values $W(u, r/B)^*$, u^*, r/B^*, s^*, and t^* (or t/r^{2*}).

v Substitute $W(u, r/B)^*$ and s^* into Equation 9.8 and solve for T.

vi Substitute the computed value of T and u^* and t^* (or t/r^{2*}) into Equation 9.10 and solve for S.

vii Substitute r/B^*, the computed T, and known values of r and m' into Equation 9.2 and solve for K'. Note that this computation can be applied only to s vs. t plots unless s vs. t/r^2 data are plotted so that individual values of t and r can be recalled.

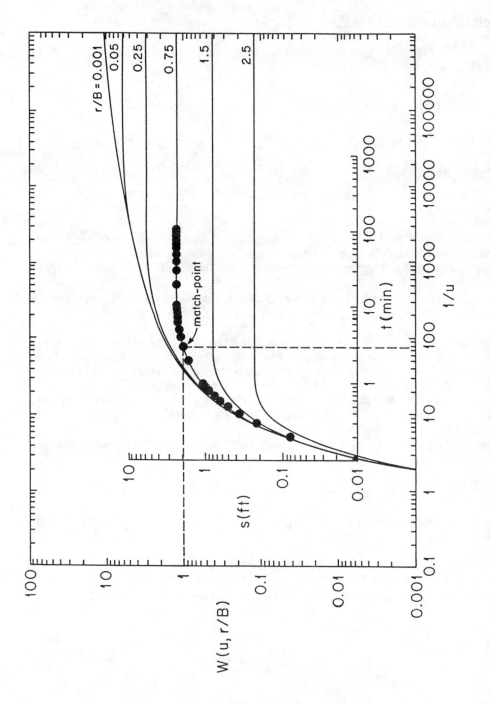

Figure 9.1   Type curves for W(u, r/B) overlain by the data curve from Example 9.1.

Example 9.1

The data in Table 9.2 were collected from a pumping test in a leaky aquifer shown conceptually in Figure 9.2. The well was pumped at a rate of 70 gal/min. Determine the transmissivity and storativity of the coarse sand aquifer and the hydraulic conductivity of the silty sand aquitard.

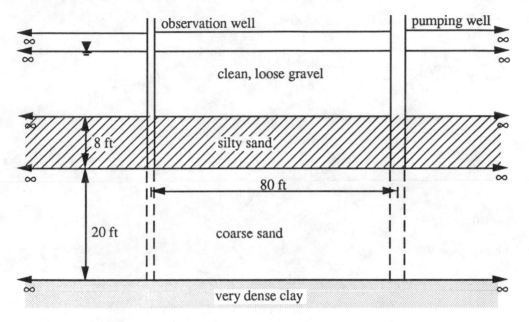

Figure 9.2   Aquifer for Example 9.1.

Table 9.2   Time-drawdown data for Example 9.1.

Time, t (min)	Drawdown, s (ft)	Time, t (min)	Drawdown, s (ft)
0.1	0.01	6	2.22
0.2	0.08	7	2.25
0.3	0.22	8	2.27
0.4	0.37	9	2.28
0.5	0.51	10	2.29
0.6	0.65	20	2.30
0.7	0.77	30	2.30
0.8	0.89	40	2.30
0.9	0.99	50	2.30
1	1.08	60	2.30
2	1.67	70	2.30
3	1.95	80	2.30
4	2.10	90	2.30
5	2.18	100	2.30

Figure 9.1 shows the data curve and the selected match-point; the match-point values are:

$$
\begin{aligned}
s^* &= 1.95 \text{ ft} \\
t^* &= 3.0 \text{ min} \\
W(u, r/B)^* &= 1.0 \\
1/u^* &= 75 \\
r/B^* &= 0.75
\end{aligned}
$$

Substituting these values into Equation 9.8 gives

$$
T = \frac{QW(u,r/B)^*}{4\pi s^*} = \frac{\left(70 \ \frac{\text{gal}}{\text{min}}\right)(1.0)\left(1440 \ \frac{\text{min}}{\text{day}}\right)}{4\pi(1.95 \text{ ft})} = 4114 \ \frac{\text{gal}}{\text{day} \cdot \text{ft}}
$$

from Equation 9.10

$$
S = \frac{4u^* \ Tt^*}{r^2} = \frac{4\left(\frac{1}{75}\right)\left(4114 \ \frac{\text{gal}}{\text{day} \cdot \text{ft}}\right)(3.0 \text{ min})\left(\frac{1 \text{ ft}^3}{7.48 \text{ gal}}\right)\left(\frac{1 \text{ day}}{1440 \text{ min}}\right)}{(80 \text{ ft})^2}
$$

$$
= 9.5 \times 10^{-6}
$$

and from Equation 9.2

$$
K' = \frac{Tm'}{B^2} = \frac{\left(4114 \ \frac{\text{gal}}{\text{day} \cdot \text{ft}}\right)(8 \text{ ft})}{\left(\frac{80 \text{ ft}}{0.75}\right)^2} = 2.9 \ \frac{\text{gal}}{\text{day} \cdot \text{ft}^2}
$$

Hantush Inflection-Point Method

For a leaky aquifer, a plot of s vs. log (t) will show an inflection point as illustrated in Figure 9.3. At the inflection point the following equations apply (Hantush, 1956)

$$
u_i = \frac{r^2 S}{4Tt_i} = \frac{r}{2B} \tag{9.15}
$$

$$
m_i = \left(\frac{2.3 \ Q}{4\pi T}\right)(e^{-r/B}) \tag{9.16}
$$

$$
s_i = 0.5 s_m = \frac{Q}{4\pi T} K_o \ (r/B) \tag{9.17}
$$

$$
2.3 \ \frac{s_i}{m_i} = \exp \ (r/B) \ K_o \ (r/B) \tag{9.18}
$$

or for r/B < 0.01,

$$\frac{s_i}{m_i} = \log\left(\frac{2B}{r} - 0.251\right) \tag{9.19}$$

where the subscript i indicates values of the variables at the inflection point value, and

s_m = maximum drawdown
m_i = slope of the drawdown curve at the inflection point (usually, this can be approximated by the slope of the straight-line portion of the curve on which the inflection point lies)
K_o = zero order modified Bessel function of the second kind

These equations are the basis for the following method of analysis:

i Plot s vs. log t.

ii Estimate the maximum drawdown, s_m, and compute $s_i = 0.5\ s_m$.

iii Using the value of s_i, locate the inflection point and record the value of t_i.

iv Fit a straight line to the drawdown data through the inflection point.

v Using the fitted line determine m_i by measuring the change in drawdown occurring over one log cycle.

vi Calculate s_i/m_i and use Equation 9.18 or 9.19 to compute B. Tabular values of $\exp(x)K_o(x)$ are in Table 9.3.

vii Substitute inflection point values into Equations 9.16 or 9.17, 9.15, and 9.2 to compute T, S, and K' respectively.

Example 9.2

The data from Example 9.1 are plotted in Figure 9.3. From this figure, the following inflection point values were determined

$$s_m \quad = 2.3\ \text{ft}$$
$$s_i \quad = 1.15\ \text{ft}$$
$$t_i \quad = 1.1\ \text{min}$$
$$m_i \quad = 2.05\ \text{ft}$$

Using Equation 9.18 gives

$$\exp(r/B)\ K_o\ (r/B) = 2.3\left(\frac{s_i}{m_i}\right) = 2.3\left(\frac{1.15\ \text{ft}}{2.05\ \text{ft}}\right) = 1.29$$

From Table 9.3, when $\exp(x)K_o(x) = 1.29$, x = 0.754. Therefore, r/B = 0.754 and B = 80 ft/0.754 = 106.1 ft. Using Equation 9.17 and a value of 0.607 for $K_o(0.754)$ from Table 9.3 the transmissivity is

$$T = \frac{QK_o(r/B)^*}{4\pi s_i} = \frac{\left(70\ \frac{gal}{min}\right)(0.607)\left(1440\ \frac{min}{day}\right)}{4\pi(1.15\ ft)} = 4234\ \frac{gal}{day \cdot ft}$$

From Equation 9.15,

$$S = \frac{2Tt_i}{rB} = \frac{2\left(4234\ \frac{gal}{day \cdot ft}\right)(1.1\ min)\left(\frac{1\ day}{1440\ min}\right)\left(\frac{1\ ft^3}{7.48\ gal}\right)}{80\ ft\ (106.1\ ft)}$$

$$= 0.0001$$

From Equation 9.2,

$$K' = \frac{Tm'}{B^2} = \frac{\left(4234\ \frac{gal}{day \cdot ft}\right)(8\ ft)}{(106.1\ ft)^2} = 3.0\ \frac{gal}{day \cdot ft^2}$$

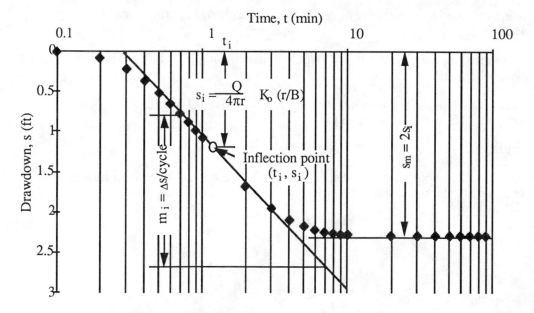

Figure 9.3   The Hantush inflection-point method for unsteady drawdown applied to data from Example 9.1.

9.4.2 Equilibrium Solution

Match-Point Method

This procedure can be used to compute T and K' when r/B > 0.05 and early data are unavailable or model assumptions are not met for early times (Hantush, 1964). Drawdown data for three or more observation wells are required.

The procedure is as follows:

  i Plot K_o(r/B) vs. r/B on logarithmic paper. This is the type curve for equilibrium drawdown data (Figure 9.4). The type curve can be plotted using the values of $K_o(x)$ in Table 9.3.

  ii Plot s vs. r using the same logarithmic scales used to prepare the type curve. This curve is called the data curve.

  iii Overlay the data curve on the type curve. Shift the plots relative to each other, keeping respective axes parallel, until a position of best fit is found between the data curve and the type curve.

  iv From the best fit portions of the curves, select a match-point and record match-point values r/B*, K_o(r/B)*, s*, and r*.

  v Substitute the match-point values into the following equations and solve for T and K'.

$$T = \left(\frac{Q}{2\pi s^*}\right) K_o(r/B)^* \qquad (9.20)$$

and

$$K' = \frac{m' \, T}{B^2} \qquad (9.21)$$

where

$$B = \frac{r^*}{(r/B)^*} \qquad (9.22)$$

Example 9.3

Realistically, it would be impossible to gather data as shown in Table 9.2. In order to pump 70 gallons per minute, a six-inch diameter well is probably needed. For the given hydraulic conductivity, casing storage effects would influence all of the early drawdown data. Papadopulos and Cooper (1967) note that well storage effects are noticeable until $t > 2.5 \times 10^3 \, r_c^2/T$, where r_c is the well casing radius. For the conditions of Examples 9.1 and 9.2 with a six-inch diameter well,

$$\text{time for well storage effects to dissipate} = \frac{2.5 \times 10^2 \, (3 \text{ in})^2 \left(\dfrac{\text{ft}}{12 \text{ in}}\right)^2}{4200 \, \dfrac{\text{gal}}{\text{day} \cdot \text{ft}} \left(\dfrac{1 \text{ ft}^3}{7.48 \text{ gal}}\right) \left(\dfrac{1 \text{ day}}{1440 \text{ min}}\right)} = 40 \text{ min}$$

Thus, the aquifer properties could be obtained with this model only by examining equilibrium drawdown data. For example, imagine that the following distance-drawdown data were obtained at equilibrium for a pumping test in the aquifer of Examples 9.1 and 9.2. The pumping rate was 70 gal/min. Determine T and S.

Table 9.4   Equilibrium distance-drawdown data for Example 9.3.

Distance from pumping well (ft)	Drawdown at equilibrium (ft)
10	9.41
30	5.41
55	3.39
80	2.30
120	1.33
180	0.63

The data curve plotted from Table 9.4 and the r/B vs. $K_o(r/B)$ type curve are shown in Figure 9.4. From the figure, the following match-point values were selected:

$$
\begin{aligned}
r/B^* &= 1.0 \text{ ft} \\
K_o(r/B)^* &= 0.421 \\
s^* &= 1.6 \text{ ft} \\
r^* &= 106 \text{ ft}
\end{aligned}
$$

Substituting s^* and $K_o(r/B)^*$ into Equation 9.20 gives

$$
T = \left(\frac{\left(70 \, \frac{\text{gal}}{\text{min}} \right) \left(1440 \, \frac{\text{min}}{\text{day}} \right)}{2\pi \, (1.6 \text{ ft})} \right) (0.421) = 4221 \, \frac{\text{gal}}{\text{day} \cdot \text{ft}}
$$

Substituting r/B^*, r^*, and T into Equations 9.21 and 9.22 gives

$$
K' = m' \, T \, \frac{(r/B)^{*2}}{r^{*2}} = \frac{(8 \text{ ft}) \left(4221 \, \frac{\text{gal}}{\text{day} \cdot \text{ft}} \right) (1.0)^2}{(106 \text{ ft})^2} = 3 \, \frac{\text{gal}}{\text{day} \cdot \text{ft}^2}
$$

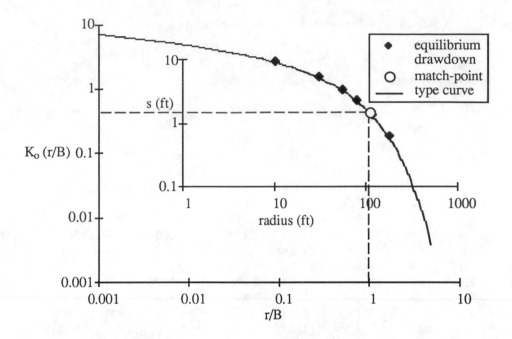

Figure 9.4   Equilibrium type curve for a leaky aquifer and distance-drawdown data curve from Example 9.3.

Straight-Line Method

If $r/B < 0.05$, an approximate solution for drawdown is

$$s = \frac{2.3Q}{2\pi T} \log\left(1.12\,\frac{B}{r}\right) \tag{9.23}$$

A plot of s vs. log(r) will define a straight line with slope b

$$b = \frac{\Delta s}{\Delta \log(r)} = \frac{2.3Q}{2\pi T} \tag{9.24}$$

and zero-drawdown-intercept, r_o

$$r_o = 1.12B \tag{9.25}$$

Thus, T and K' may be calculated using

$$T = \frac{2.3\,Q}{2\pi b} \tag{9.26}$$

$$K' = \frac{1.25\,Tm'}{r_o^2} \tag{9.27}$$

Table 9.3   Values of the functions $K_0(x)$ and $exp(x)K_0(x)$.

(x)	$K_0(x)$	$exp(x)K_0(x)$	(x)	$K_0(x)$	$exp(x)K_0(x)$	(x)	$K_0(x)$	$exp(x)K_0(x)$
0.000	101.449	101.449	0.044	3.242	3.387	0.088	2.553	2.788
0.001	7.024	7.031	0.045	3.219	3.367	0.089	2.542	2.779
0.002	6.331	6.343	0.046	3.197	3.348	0.090	2.531	2.769
0.003	5.925	5.943	0.047	3.176	3.329	0.091	2.520	2.760
0.004	5.637	5.660	0.048	3.155	3.310	0.092	2.509	2.751
0.005	5.414	5.441	0.049	3.134	3.292	0.093	2.499	2.742
0.006	5.232	5.263	0.050	3.114	3.274	0.094	2.488	2.733
0.007	5.078	5.114	0.051	3.095	3.256	0.095	2.478	2.725
0.008	4.944	4.984	0.052	3.075	3.239	0.096	2.467	2.716
0.009	4.827	4.870	0.053	3.056	3.223	0.097	2.457	2.707
0.010	4.721	4.769	0.054	3.038	3.206	0.098	2.447	2.699
0.011	4.626	4.677	0.055	3.019	3.190	0.099	2.437	2.691
0.012	4.539	4.594	0.056	3.001	3.174	0.100	2.427	2.682
0.013	4.459	4.517	0.057	2.984	3.159	0.101	2.417	2.674
0.014	4.385	4.447	0.058	2.967	3.144	0.102	2.408	2.666
0.015	4.316	4.381	0.059	2.950	3.129	0.103	2.398	2.658
0.016	4.251	4.320	0.060	2.933	3.114	0.104	2.388	2.650
0.017	4.191	4.263	0.061	2.916	3.100	0.105	2.379	2.642
0.018	4.134	4.209	0.062	2.900	3.086	0.106	2.370	2.635
0.019	4.080	4.158	0.063	2.884	3.072	0.107	2.360	2.627
0.020	4.028	4.110	0.064	2.869	3.058	0.108	2.351	2.619
0.021	3.980	4.064	0.065	2.853	3.045	0.109	2.342	2.612
0.022	3.933	4.021	0.066	2.838	3.032	0.110	2.333	2.605
0.023	3.889	3.979	0.067	2.823	3.019	0.111	2.324	2.597
0.024	3.846	3.940	0.068	2.809	3.006	0.112	2.316	2.590
0.025	3.806	3.902	0.069	2.794	2.994	0.113	2.307	2.583
0.026	3.766	3.866	0.070	2.780	2.981	0.114	2.298	2.576
0.027	3.729	3.831	0.071	2.766	2.969	0.115	2.290	2.569
0.028	3.692	3.797	0.072	2.752	2.957	0.116	2.281	2.562
0.029	3.657	3.765	0.073	2.738	2.946	0.117	2.273	2.555
0.030	3.624	3.734	0.074	2.725	2.934	0.118	2.264	2.548
0.031	3.591	3.704	0.075	2.711	2.923	0.119	2.256	2.541
0.032	3.559	3.675	0.076	2.698	2.911	0.120	2.248	2.534
0.033	3.528	3.647	0.077	2.685	2.900	0.121	2.240	2.528
0.034	3.499	3.620	0.078	2.673	2.889	0.122	2.232	2.521
0.035	3.470	3.593	0.079	2.660	2.879	0.123	2.224	2.515
0.036	3.442	3.568	0.080	2.647	2.868	0.124	2.216	2.508
0.037	3.414	3.543	0.081	2.635	2.858	0.125	2.208	2.502
0.038	3.388	3.519	0.082	2.623	2.847	0.126	2.200	2.496
0.039	3.362	3.495	0.083	2.611	2.837	0.127	2.192	2.489
0.040	3.337	3.473	0.084	2.599	2.827	0.128	2.185	2.483
0.041	3.312	3.451	0.085	2.588	2.817	0.129	2.177	2.477
0.042	3.288	3.429	0.086	2.576	2.807	0.130	2.170	2.471
0.043	3.264	3.408	0.087	2.565	2.798	0.131	2.162	2.465

Table 9.3   (Continued).

(x)	$K_0(x)$	$\exp(x)K_0(x)$	(x)	$K_0(x)$	$\exp(x)K_0(x)$	(x)	$K_0(x)$	$\exp(x)K_0(x)$
0.132	2.155	2.459	0.176	1.875	2.236	0.220	1.662	2.071
0.133	2.147	2.453	0.177	1.870	2.232	0.221	1.658	2.068
0.134	2.140	2.447	0.178	1.864	2.228	0.222	1.653	2.064
0.135	2.133	2.441	0.179	1.859	2.223	0.223	1.649	2.061
0.136	2.125	2.435	0.180	1.854	2.219	0.224	1.645	2.058
0.137	2.118	2.429	0.181	1.848	2.215	0.225	1.641	2.055
0.138	2.111	2.424	0.182	1.843	2.211	0.226	1.637	2.051
0.139	2.104	2.418	0.183	1.838	2.207	0.227	1.632	2.048
0.140	2.097	2.412	0.184	1.833	2.203	0.228	1.628	2.045
0.141	2.090	2.407	0.185	1.827	2.199	0.229	1.624	2.042
0.142	2.083	2.401	0.186	1.822	2.195	0.230	1.620	2.039
0.143	2.077	2.396	0.187	1.817	2.191	0.231	1.616	2.036
0.144	2.070	2.390	0.188	1.812	2.187	0.232	1.612	2.033
0.145	2.063	2.385	0.189	1.807	2.183	0.233	1.608	2.030
0.146	2.056	2.380	0.190	1.802	2.179	0.234	1.604	2.026
0.147	2.050	2.374	0.191	1.797	2.175	0.235	1.600	2.023
0.148	2.043	2.369	0.192	1.792	2.171	0.236	1.596	2.020
0.149	2.037	2.364	0.193	1.787	2.167	0.237	1.592	2.017
0.150	2.030	2.359	0.194	1.782	2.163	0.238	1.588	2.014
0.151	2.024	2.353	0.195	1.777	2.159	0.239	1.584	2.011
0.152	2.017	2.348	0.196	1.772	2.156	0.240	1.580	2.008
0.153	2.011	2.343	0.197	1.767	2.152	0.241	1.576	2.005
0.154	2.004	2.338	0.198	1.762	2.148	0.242	1.572	2.002
0.155	1.998	2.333	0.199	1.757	2.144	0.243	1.568	1.999
0.156	1.992	2.328	0.200	1.753	2.141	0.244	1.564	1.997
0.157	1.986	2.323	0.201	1.748	2.137	0.245	1.560	1.994
0.158	1.980	2.318	0.202	1.743	2.133	0.246	1.557	1.991
0.159	1.973	2.314	0.203	1.738	2.130	0.247	1.553	1.988
0.160	1.967	2.309	0.204	1.734	2.126	0.248	1.549	1.985
0.161	1.961	2.304	0.205	1.729	2.123	0.249	1.545	1.982
0.162	1.955	2.299	0.206	1.725	2.119	0.250	1.542	1.979
0.163	1.949	2.295	0.207	1.720	2.115	0.251	1.538	1.977
0.164	1.944	2.290	0.208	1.715	2.112	0.252	1.534	1.974
0.165	1.938	2.285	0.209	1.711	2.108	0.253	1.530	1.971
0.166	1.932	2.281	0.210	1.706	2.105	0.254	1.527	1.968
0.167	1.926	2.276	0.211	1.702	2.101	0.255	1.523	1.965
0.168	1.920	2.272	0.212	1.697	2.098	0.256	1.519	1.963
0.169	1.914	2.267	0.213	1.693	2.095	0.257	1.516	1.960
0.170	1.909	2.263	0.214	1.688	2.091	0.258	1.512	1.957
0.171	1.903	2.258	0.215	1.684	2.088	0.259	1.508	1.954
0.172	1.898	2.254	0.216	1.679	2.084	0.260	1.505	1.952
0.173	1.892	2.249	0.217	1.675	2.081	0.261	1.501	1.949
0.174	1.886	2.245	0.218	1.671	2.078	0.262	1.498	1.946
0.175	1.881	2.241	0.219	1.666	2.074	0.263	1.494	1.944

Table 9.3   (Continued).

(x)	$K_0(x)$	$exp(x)K_0(x)$	(x)	$K_0(x)$	$exp(x)K_0(x)$	(x)	$K_0(x)$	$exp(x)K_0(x)$
0.264	1.491	1.941	0.308	1.348	1.835	0.352	1.228	1.746
0.265	1.487	1.938	0.309	1.345	1.833	0.353	1.225	1.744
0.266	1.484	1.936	0.310	1.342	1.830	0.354	1.223	1.742
0.267	1.480	1.933	0.311	1.340	1.828	0.355	1.220	1.740
0.268	1.477	1.930	0.312	1.337	1.826	0.356	1.218	1.738
0.269	1.473	1.928	0.313	1.334	1.824	0.357	1.215	1.736
0.270	1.470	1.925	0.314	1.331	1.822	0.358	1.213	1.734
0.271	1.466	1.923	0.315	1.328	1.820	0.359	1.210	1.733
0.272	1.463	1.920	0.316	1.325	1.817	0.360	1.208	1.731
0.273	1.459	1.918	0.317	1.322	1.815	0.361	1.205	1.729
0.274	1.456	1.915	0.318	1.319	1.813	0.362	1.203	1.727
0.275	1.453	1.913	0.319	1.316	1.811	0.363	1.200	1.725
0.276	1.449	1.910	0.320	1.314	1.809	0.364	1.198	1.724
0.277	1.446	1.907	0.321	1.311	1.807	0.365	1.195	1.722
0.278	1.443	1.905	0.322	1.308	1.805	0.366	1.193	1.720
0.279	1.439	1.902	0.323	1.305	1.803	0.367	1.190	1.718
0.280	1.436	1.900	0.324	1.302	1.801	0.368	1.188	1.716
0.281	1.433	1.898	0.325	1.299	1.799	0.369	1.186	1.715
0.282	1.429	1.895	0.326	1.297	1.796	0.370	1.183	1.713
0.283	1.426	1.893	0.327	1.294	1.794	0.371	1.181	1.711
0.284	1.423	1.890	0.328	1.291	1.792	0.372	1.178	1.709
0.285	1.420	1.888	0.329	1.288	1.790	0.373	1.176	1.708
0.286	1.416	1.885	0.330	1.286	1.788	0.374	1.174	1.706
0.287	1.413	1.883	0.331	1.283	1.786	0.375	1.171	1.704
0.288	1.410	1.881	0.332	1.280	1.784	0.376	1.169	1.702
0.289	1.407	1.878	0.333	1.277	1.782	0.377	1.167	1.701
0.290	1.404	1.876	0.334	1.275	1.780	0.378	1.164	1.699
0.291	1.400	1.873	0.335	1.272	1.778	0.379	1.162	1.697
0.292	1.397	1.871	0.336	1.269	1.776	0.380	1.160	1.696
0.293	1.394	1.869	0.337	1.267	1.774	0.381	1.157	1.694
0.294	1.391	1.866	0.338	1.264	1.772	0.382	1.155	1.692
0.295	1.388	1.864	0.339	1.261	1.770	0.383	1.153	1.691
0.296	1.385	1.862	0.340	1.259	1.768	0.384	1.150	1.689
0.297	1.382	1.859	0.341	1.256	1.767	0.385	1.148	1.687
0.298	1.379	1.857	0.342	1.253	1.765	0.386	1.146	1.686
0.299	1.376	1.855	0.343	1.251	1.763	0.387	1.143	1.684
0.300	1.372	1.853	0.344	1.248	1.761	0.388	1.141	1.682
0.301	1.369	1.850	0.345	1.246	1.759	0.389	1.139	1.681
0.302	1.366	1.848	0.346	1.243	1.757	0.390	1.137	1.679
0.303	1.363	1.846	0.347	1.240	1.755	0.391	1.134	1.677
0.304	1.360	1.844	0.348	1.238	1.753	0.392	1.132	1.676
0.305	1.357	1.841	0.349	1.235	1.751	0.393	1.130	1.674
0.306	1.354	1.839	0.350	1.233	1.749	0.394	1.128	1.672
0.307	1.351	1.837	0.351	1.230	1.747	0.395	1.126	1.671

Table 9.3   (Continued).

(x)	$K_O(x)$	$\exp(x)K_O(x)$	(x)	$K_O(x)$	$\exp(x)K_O(x)$	(x)	$K_O(x)$	$\exp(x)K_O(x)$
0.396	1.123	1.669	0.440	1.032	1.603	0.484	0.951	1.544
0.397	1.121	1.667	0.441	1.030	1.601	0.485	0.950	1.543
0.398	1.119	1.666	0.442	1.028	1.600	0.486	0.948	1.541
0.399	1.117	1.664	0.443	1.026	1.598	0.487	0.946	1.540
0.400	1.115	1.663	0.444	1.024	1.597	0.488	0.945	1.539
0.401	1.112	1.661	0.445	1.022	1.596	0.489	0.943	1.538
0.402	1.110	1.659	0.446	1.021	1.594	0.490	0.941	1.536
0.403	1.108	1.658	0.447	1.019	1.593	0.491	0.940	1.535
0.404	1.106	1.656	0.448	1.017	1.591	0.492	0.938	1.534
0.405	1.104	1.655	0.449	1.015	1.590	0.493	0.936	1.533
0.406	1.102	1.653	0.450	1.013	1.589	0.494	0.934	1.531
0.407	1.099	1.652	0.451	1.011	1.587	0.495	0.933	1.530
0.408	1.097	1.650	0.452	1.009	1.586	0.496	0.931	1.529
0.409	1.095	1.649	0.453	1.007	1.584	0.497	0.929	1.528
0.410	1.093	1.647	0.454	1.005	1.583	0.498	0.928	1.527
0.411	1.091	1.645	0.455	1.004	1.582	0.499	0.926	1.525
0.412	1.089	1.644	0.456	1.002	1.580	0.500	0.924	1.524
0.413	1.087	1.642	0.457	1.000	1.579	0.501	0.923	1.523
0.414	1.085	1.641	0.458	0.998	1.578	0.502	0.921	1.522
0.415	1.082	1.639	0.459	0.996	1.576	0.503	0.919	1.521
0.416	1.080	1.638	0.460	0.994	1.575	0.504	0.918	1.519
0.417	1.078	1.636	0.461	0.992	1.574	0.505	0.916	1.518
0.418	1.076	1.635	0.462	0.991	1.572	0.506	0.915	1.517
0.419	1.074	1.633	0.463	0.989	1.571	0.507	0.913	1.516
0.420	1.072	1.632	0.464	0.987	1.570	0.508	0.911	1.515
0.421	1.070	1.630	0.465	0.985	1.568	0.509	0.910	1.513
0.422	1.068	1.629	0.466	0.983	1.567	0.510	0.908	1.512
0.423	1.066	1.627	0.467	0.981	1.566	0.511	0.906	1.511
0.424	1.064	1.626	0.468	0.980	1.564	0.512	0.905	1.510
0.425	1.062	1.624	0.469	0.978	1.563	0.513	0.903	1.509
0.426	1.060	1.623	0.470	0.976	1.562	0.514	0.902	1.508
0.427	1.058	1.621	0.471	0.974	1.560	0.515	0.900	1.506
0.428	1.056	1.620	0.472	0.973	1.559	0.516	0.898	1.505
0.429	1.054	1.618	0.473	0.971	1.558	0.517	0.897	1.504
0.430	1.052	1.617	0.474	0.969	1.557	0.518	0.895	1.503
0.431	1.050	1.615	0.475	0.967	1.555	0.519	0.894	1.502
0.432	1.048	1.614	0.476	0.965	1.554	0.520	0.892	1.501
0.433	1.046	1.613	0.477	0.964	1.553	0.521	0.891	1.499
0.434	1.044	1.611	0.478	0.962	1.551	0.522	0.889	1.498
0.435	1.042	1.610	0.479	0.960	1.550	0.523	0.887	1.497
0.436	1.040	1.608	0.480	0.958	1.549	0.524	0.886	1.496
0.437	1.038	1.607	0.481	0.957	1.548	0.525	0.884	1.495
0.438	1.036	1.605	0.482	0.955	1.546	0.526	0.883	1.494
0.439	1.034	1.604	0.483	0.953	1.545	0.527	0.881	1.493

Table 9.3 (Continued).

(x)	$K_O(x)$	$\exp(x)K_O(x)$	(x)	$K_O(x)$	$\exp(x)K_O(x)$	(x)	$K_O(x)$	$\exp(x)K_O(x)$
0.528	0.880	1.491	0.572	0.815	1.444	0.616	0.757	1.402
0.529	0.878	1.490	0.573	0.814	1.443	0.617	0.756	1.401
0.530	0.877	1.489	0.574	0.812	1.442	0.618	0.755	1.400
0.531	0.875	1.488	0.575	0.811	1.441	0.619	0.753	1.399
0.532	0.873	1.487	0.576	0.810	1.440	0.620	0.752	1.398
0.533	0.872	1.486	0.577	0.808	1.439	0.621	0.751	1.397
0.534	0.870	1.485	0.578	0.807	1.438	0.622	0.750	1.396
0.535	0.869	1.484	0.579	0.806	1.437	0.623	0.748	1.395
0.536	0.867	1.483	0.580	0.804	1.436	0.624	0.747	1.394
0.537	0.866	1.481	0.581	0.803	1.435	0.625	0.746	1.393
0.538	0.864	1.480	0.582	0.801	1.434	0.626	0.745	1.393
0.539	0.863	1.479	0.583	0.800	1.433	0.627	0.743	1.392
0.540	0.861	1.478	0.584	0.799	1.432	0.628	0.742	1.391
0.541	0.860	1.477	0.585	0.797	1.431	0.629	0.741	1.390
0.542	0.858	1.476	0.586	0.796	1.430	0.630	0.740	1.389
0.543	0.857	1.475	0.587	0.795	1.429	0.631	0.739	1.388
0.544	0.855	1.474	0.588	0.793	1.428	0.632	0.737	1.387
0.545	0.854	1.473	0.589	0.792	1.427	0.633	0.736	1.386
0.546	0.852	1.472	0.590	0.791	1.426	0.634	0.735	1.385
0.547	0.851	1.471	0.591	0.789	1.425	0.635	0.734	1.384
0.548	0.850	1.469	0.592	0.788	1.424	0.636	0.732	1.384
0.549	0.848	1.468	0.593	0.787	1.423	0.637	0.731	1.383
0.550	0.847	1.467	0.594	0.785	1.423	0.638	0.730	1.382
0.551	0.845	1.466	0.595	0.784	1.422	0.639	0.729	1.381
0.552	0.844	1.465	0.596	0.783	1.421	0.640	0.728	1.380
0.553	0.842	1.464	0.597	0.781	1.420	0.641	0.726	1.379
0.554	0.841	1.463	0.598	0.780	1.419	0.642	0.725	1.378
0.555	0.839	1.462	0.599	0.779	1.418	0.643	0.724	1.377
0.556	0.838	1.461	0.600	0.778	1.417	0.644	0.723	1.376
0.557	0.836	1.460	0.601	0.776	1.416	0.645	0.722	1.376
0.558	0.835	1.459	0.602	0.775	1.415	0.646	0.721	1.375
0.559	0.834	1.458	0.603	0.774	1.414	0.647	0.719	1.374
0.560	0.832	1.457	0.604	0.772	1.413	0.648	0.718	1.373
0.561	0.831	1.456	0.605	0.771	1.412	0.649	0.717	1.372
0.562	0.829	1.455	0.606	0.770	1.411	0.650	0.716	1.371
0.563	0.828	1.454	0.607	0.768	1.410	0.651	0.715	1.370
0.564	0.826	1.453	0.608	0.767	1.409	0.652	0.714	1.370
0.565	0.825	1.452	0.609	0.766	1.408	0.653	0.712	1.369
0.566	0.824	1.451	0.610	0.765	1.407	0.654	0.711	1.368
0.567	0.822	1.449	0.611	0.763	1.406	0.655	0.710	1.367
0.568	0.821	1.448	0.612	0.762	1.405	0.656	0.709	1.366
0.569	0.819	1.447	0.613	0.761	1.404	0.657	0.708	1.365
0.570	0.818	1.446	0.614	0.760	1.404	0.658	0.707	1.364
0.571	0.817	1.445	0.615	0.758	1.403	0.659	0.705	1.364

Table 9.3   (Continued).

(x)	$K_0(x)$	$\exp(x)K_0(x)$	(x)	$K_0(x)$	$\exp(x)K_0(x)$	(x)	$K_0(x)$	$\exp(x)K_0(x)$
0.660	0.704	1.363	0.704	0.656	1.327	0.748	0.612	1.294
0.661	0.703	1.362	0.705	0.655	1.326	0.749	0.612	1.293
0.662	0.702	1.361	0.706	0.654	1.325	0.750	0.611	1.293
0.663	0.701	1.360	0.707	0.653	1.325	0.751	0.610	1.292
0.664	0.700	1.359	0.708	0.652	1.324	0.752	0.609	1.291
0.665	0.699	1.359	0.709	0.651	1.323	0.753	0.608	1.290
0.666	0.698	1.358	0.710	0.650	1.322	0.754	0.607	1.290
0.667	0.696	1.357	0.711	0.649	1.322	0.755	0.606	1.289
0.668	0.695	1.356	0.712	0.648	1.321	0.756	0.605	1.288
0.669	0.694	1.355	0.713	0.647	1.320	0.757	0.604	1.288
0.670	0.693	1.354	0.714	0.646	1.319	0.758	0.603	1.287
0.671	0.692	1.354	0.715	0.645	1.319	0.759	0.602	1.286
0.672	0.691	1.353	0.716	0.644	1.318	0.760	0.601	1.285
0.673	0.690	1.352	0.717	0.643	1.317	0.761	0.600	1.285
0.674	0.689	1.351	0.718	0.642	1.316	0.762	0.599	1.284
0.675	0.687	1.350	0.719	0.641	1.315	0.763	0.598	1.283
0.676	0.686	1.349	0.720	0.640	1.315	0.764	0.597	1.283
0.677	0.685	1.349	0.721	0.639	1.314	0.765	0.597	1.282
0.678	0.684	1.348	0.722	0.638	1.313	0.766	0.596	1.281
0.679	0.683	1.347	0.723	0.637	1.312	0.767	0.595	1.281
0.680	0.682	1.346	0.724	0.636	1.312	0.768	0.594	1.280
0.681	0.681	1.345	0.725	0.635	1.311	0.769	0.593	1.279
0.682	0.680	1.344	0.726	0.634	1.310	0.770	0.592	1.278
0.683	0.679	1.344	0.727	0.633	1.309	0.771	0.591	1.278
0.684	0.678	1.343	0.728	0.632	1.309	0.772	0.590	1.277
0.685	0.677	1.342	0.729	0.631	1.308	0.773	0.589	1.276
0.686	0.675	1.341	0.730	0.630	1.307	0.774	0.588	1.276
0.687	0.674	1.340	0.731	0.629	1.306	0.775	0.587	1.275
0.688	0.673	1.340	0.732	0.628	1.306	0.776	0.587	1.274
0.689	0.672	1.339	0.733	0.627	1.305	0.777	0.586	1.274
0.690	0.671	1.338	0.734	0.626	1.304	0.778	0.585	1.273
0.691	0.670	1.337	0.735	0.625	1.304	0.779	0.584	1.272
0.692	0.669	1.336	0.736	0.624	1.303	0.780	0.583	1.272
0.693	0.668	1.336	0.737	0.623	1.302	0.781	0.582	1.271
0.694	0.667	1.335	0.738	0.622	1.301	0.782	0.581	1.270
0.695	0.666	1.334	0.739	0.621	1.301	0.783	0.580	1.270
0.696	0.665	1.333	0.740	0.620	1.300	0.784	0.579	1.269
0.697	0.664	1.332	0.741	0.619	1.299	0.785	0.578	1.268
0.698	0.663	1.332	0.742	0.618	1.298	0.786	0.578	1.268
0.699	0.662	1.331	0.743	0.617	1.298	0.787	0.577	1.267
0.700	0.661	1.330	0.744	0.616	1.297	0.788	0.576	1.266
0.701	0.659	1.329	0.745	0.615	1.296	0.789	0.575	1.266
0.702	0.658	1.329	0.746	0.614	1.295	0.790	0.574	1.265
0.703	0.657	1.328	0.747	0.613	1.295	0.791	0.573	1.264

Table 9.3 (Continued).

(x)	$K_O(x)$	$\exp(x)K_O(x)$	(x)	$K_O(x)$	$\exp(x)K_O(x)$	(x)	$K_O(x)$	$\exp(x)K_O(x)$
0.792	0.572	1.264	0.836	0.535	1.235	0.880	0.501	1.209
0.793	0.571	1.263	0.837	0.535	1.235	0.881	0.501	1.208
0.794	0.571	1.262	0.838	0.534	1.234	0.882	0.500	1.207
0.795	0.570	1.262	0.839	0.533	1.233	0.883	0.499	1.207
0.796	0.569	1.261	0.840	0.532	1.233	0.884	0.498	1.206
0.797	0.568	1.260	0.841	0.531	1.232	0.885	0.498	1.206
0.798	0.567	1.260	0.842	0.531	1.231	0.886	0.497	1.205
0.799	0.566	1.259	0.843	0.530	1.231	0.887	0.496	1.205
0.800	0.565	1.258	0.844	0.529	1.230	0.888	0.495	1.204
0.801	0.564	1.258	0.845	0.528	1.230	0.889	0.495	1.203
0.802	0.564	1.257	0.846	0.527	1.229	0.890	0.494	1.203
0.803	0.563	1.256	0.847	0.527	1.228	0.891	0.493	1.202
0.804	0.562	1.256	0.848	0.526	1.228	0.892	0.493	1.202
0.805	0.561	1.255	0.849	0.525	1.227	0.893	0.492	1.201
0.806	0.560	1.254	0.850	0.524	1.227	0.894	0.491	1.201
0.807	0.559	1.254	0.851	0.523	1.226	0.895	0.490	1.200
0.808	0.559	1.253	0.852	0.523	1.225	0.896	0.490	1.199
0.809	0.558	1.252	0.853	0.522	1.225	0.897	0.489	1.199
0.810	0.557	1.252	0.854	0.521	1.224	0.898	0.488	1.198
0.811	0.556	1.251	0.855	0.520	1.223	0.899	0.487	1.198
0.812	0.555	1.250	0.856	0.520	1.223	0.900	0.487	1.197
0.813	0.554	1.250	0.857	0.519	1.222	0.901	0.486	1.197
0.814	0.553	1.249	0.858	0.518	1.222	0.902	0.485	1.196
0.815	0.553	1.248	0.859	0.517	1.221	0.903	0.485	1.195
0.816	0.552	1.248	0.860	0.516	1.220	0.904	0.484	1.195
0.817	0.551	1.247	0.861	0.516	1.220	0.905	0.483	1.194
0.818	0.550	1.247	0.862	0.515	1.219	0.906	0.482	1.194
0.819	0.549	1.246	0.863	0.514	1.219	0.907	0.482	1.193
0.820	0.548	1.245	0.864	0.513	1.218	0.908	0.481	1.193
0.821	0.548	1.245	0.865	0.513	1.217	0.909	0.480	1.192
0.822	0.547	1.244	0.866	0.512	1.217	0.910	0.480	1.192
0.823	0.546	1.243	0.867	0.511	1.216	0.911	0.479	1.191
0.824	0.545	1.243	0.868	0.510	1.216	0.912	0.478	1.190
0.825	0.544	1.242	0.869	0.510	1.215	0.913	0.478	1.190
0.826	0.543	1.241	0.870	0.509	1.215	0.914	0.477	1.189
0.827	0.543	1.241	0.871	0.508	1.214	0.915	0.476	1.189
0.828	0.542	1.240	0.872	0.507	1.213	0.916	0.475	1.188
0.829	0.541	1.240	0.873	0.507	1.213	0.917	0.475	1.188
0.830	0.540	1.239	0.874	0.506	1.212	0.918	0.474	1.187
0.831	0.539	1.238	0.875	0.505	1.212	0.919	0.473	1.187
0.832	0.539	1.238	0.876	0.504	1.211	0.920	0.473	1.186
0.833	0.538	1.237	0.877	0.504	1.210	0.921	0.472	1.185
0.834	0.537	1.236	0.878	0.503	1.210	0.922	0.471	1.185
0.835	0.536	1.236	0.879	0.502	1.209	0.923	0.471	1.184

Table 9.3   (Continued).

(x)	$K_O(x)$	$\exp(x)K_O(x)$	(x)	$K_O(x)$	$\exp(x)K_O(x)$
0.924	0.470	1.184	0.968	0.441	1.161
0.925	0.469	1.183	0.969	0.440	1.160
0.926	0.469	1.183	0.970	0.440	1.160
0.927	0.468	1.182	0.971	0.439	1.159
0.928	0.467	1.182	0.972	0.438	1.158
0.929	0.466	1.181	0.973	0.438	1.158
0.930	0.466	1.181	0.974	0.437	1.157
0.931	0.465	1.180	0.975	0.436	1.157
0.932	0.464	1.179	0.976	0.436	1.156
0.933	0.464	1.179	0.977	0.435	1.156
0.934	0.463	1.178	0.978	0.435	1.155
0.935	0.462	1.178	0.979	0.434	1.155
0.936	0.462	1.177	0.980	0.433	1.154
0.937	0.461	1.177	0.981	0.433	1.154
0.938	0.460	1.176	0.982	0.432	1.153
0.939	0.460	1.176	0.983	0.431	1.153
0.940	0.459	1.175	0.984	0.431	1.152
0.941	0.458	1.175	0.985	0.430	1.152
0.942	0.458	1.174	0.986	0.430	1.151
0.943	0.457	1.174	0.987	0.429	1.151
0.944	0.456	1.173	0.988	0.428	1.150
0.945	0.456	1.173	0.989	0.428	1.150
0.946	0.455	1.172	0.990	0.427	1.149
0.947	0.454	1.171	0.991	0.426	1.149
0.948	0.454	1.171	0.992	0.426	1.148
0.949	0.453	1.170	0.993	0.425	1.148
0.950	0.452	1.170	0.994	0.425	1.147
0.951	0.452	1.169	0.995	0.424	1.147
0.952	0.451	1.169	0.996	0.423	1.146
0.953	0.450	1.168	0.997	0.423	1.146
0.954	0.450	1.168	0.998	0.422	1.145
0.955	0.449	1.167	0.999	0.422	1.145
0.956	0.449	1.167	1.000	0.421	1.144
0.957	0.448	1.166			
0.958	0.447	1.166			
0.959	0.447	1.165			
0.960	0.446	1.165			
0.961	0.445	1.164			
0.962	0.445	1.164			
0.963	0.444	1.163			
0.964	0.443	1.163			
0.965	0.443	1.162			
0.966	0.442	1.162			
0.967	0.441	1.161			

Chapter 10

MODEL 5: TRANSIENT, CONFINED, LEAKY, AQUITARD STORAGE

10.1 CONCEPTUAL MODEL

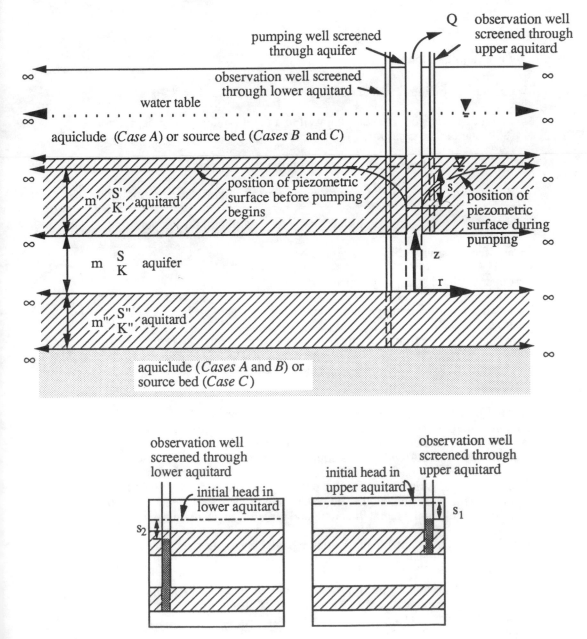

Definition of Terms

K   = aquifer hydraulic conductivity, LT^{-1}

K'   = upper aquitard vertical hydraulic conductivity, LT^{-1}

K"   = lower aquitard vertical hydraulic conductivity, LT^{-1}

m   = aquifer thickness, L

m'   = upper aquitard thickness, L

m"   = lower aquitard thickness, L

Q   = constant pumping rate, L^3T^{-1}

r   = radial distance from the pumping well to a point on the cone of depression (all distances are measured from the center of wells), L

s   = drawdown of piezometric surface in aquifer during pumping, L

s_1 = drawdown of piezometric surface in the upper aquitard during pumping, L

s_2 = drawdown of piezometric surface in the lower aquitard during pumping, L

S   = aquifer storativity, dimensionless

S'   = upper aquitard storativity, dimensionless

S"   = lower aquitard storativity, dimensionless

z   = vertical distance from the aquifer base to a point of interest, L

Assumptions

1. The aquifer is bounded above by an aquitard and either:
   an aquiclude *(Case A)*, or an unconfined aquifer (the "source bed") *(Cases B and C)*. The aquifer is bounded below by an aquitard and either:
   an aquiclude *(Cases A and B)*, or a source bed *(Case C)*.
2. All layers are horizontal and extend infinitely in the radial direction.
3. The initial piezometric surface (before pumping begins) is horizontal and extends infinitely in the radial direction. The water table in the source bed(s) is horizontal, extends infinitely in the radial direction, and remains constant during pumping (zero drawdown).
   Drawdown of the water table in the source bed(s) can be neglected when either
   $t < \dfrac{0.1m'S'}{K'}$, $t < \dfrac{0.1m"S"}{K"}$, or when the transmissivity of the source bed(s) is greater than 100 times the aquifer transmissivity (Neuman and Witherspoon, 1969b).
4. The aquifer and aquitard are homogeneous and isotropic.
5. Groundwater density and viscosity are constant.
6. Groundwater flow can be described by Darcy's Law.

7. Groundwater flow in the aquitard(s) is vertical. Groundwater flow in the aquifer is horizontal and directed radially toward the well. This assumption is valid when m/B < 0.1 (Hantush, 1967), where B is defined in Equation 9.2.
8. The pumping and observation wells are screened over the entire aquifer thickness.
9. The pumping rate is constant.
10. Head losses through the well screen and pump intake are negligible.
11. The pumping well has an infinitesimal diameter.
12. The aquifer and aquitard(s) are compressible and completely elastic.

10.2 MATHEMATICAL MODEL

10.2.1 In the Upper Aquitard

Governing Equation

The governing equation is derived by combining Darcy's Law with the principle of conservation of mass for the case of one-dimensional (vertical) flow across the upper aquitard

$$\frac{\partial^2 s_1}{\partial z^2} = \frac{S'}{K'm'} \frac{\partial s_1}{\partial t} \tag{10.1}$$

where s_1 is the drawdown in the upper aquitard, z is the vertical distance from the aquifer base, S', K', and m' are the storativity, vertical hydraulic conductivity, and thickness of the upper aquitard, and t is time.

Initial Conditions

• Before pumping begins drawdown in the upper aquitard is zero

$$s_1(r, z, t = 0) = 0 \tag{10.2}$$

Boundary Conditions

• The drawdown in the upper aquitard equals the drawdown in the aquifer at the aquifer/aquitard interface

$$s_1(r, z = m, t \geq 0) = s(r, t \geq 0) \tag{10.3}$$

• For *Case A,* where the aquitard is bounded above by an aquiclude, the change in drawdown with depth is zero at the top of the upper aquitard

$$\frac{\partial s_1(r, z = m + m', t \geq 0)}{\partial z} = 0 \tag{10.4}$$

• For *Cases B* and *C,* where the aquitard is bounded above by an unconfined aquifer (the "source bed"), drawdown is zero at the top of the upper aquitard

$$s_1(r, z = m + m', t \geq 0) = 0 \tag{10.5}$$

10.2.2   In the Lower Aquitard

Governing Equation

The governing equation is derived by combining Darcy's Law with the principle of conservation of mass for one-dimensional (vertical) flow across the lower aquitard

$$\frac{\partial^2 s_2}{\partial z^2} = \frac{S''}{K''m''} \frac{\partial s_2}{\partial t} \tag{10.6}$$

where s_2 is the drawdown in the lower aquitard, z is vertical distance from the aquifer base, S'', K'', and m'' are the storativity, vertical hydraulic conductivity, and thickness of the lower aquitard, and t is time.

Initial Conditions

• Before pumping begins drawdown in the lower aquitard is zero

$$s_2(r, z, t = 0) = 0 \tag{10.7}$$

Boundary Conditions

• The drawdown in the lower aquitard equals the drawdown in the aquifer at the aquifer/aquitard interface

$$s_2(r, z = 0, t \geq 0) = s(r, t \geq 0) \tag{10.8}$$

• For *Cases A and B*, the change in head with depth is zero at the base of the lower aquitard

$$\frac{\partial s_2(r, z = -m'', t \geq 0)}{\partial z} = 0 \tag{10.9}$$

• For *Case C*, the drawdown at the base of the lower aquitard is zero

$$s_2(r, z = -m, t \geq 0) = 0 \tag{10.10}$$

10.2.3   In the Aquifer

Governing Equation

The governing equation is derived by combining Darcy's Law with the principle of conservation of mass in a radial coordinate system

$$\frac{\partial^2 s}{\partial r^2} + \left(\frac{1}{r}\right)\frac{\partial s}{\partial r} + \left(\frac{K'}{T}\right)\frac{\partial s_1(r, z = m, t)}{\partial z} - \left(\frac{K''}{T}\right)\frac{\partial s_2(r, z = 0, t)}{\partial z} = \left(\frac{S}{T}\right)\frac{\partial s}{\partial t} \tag{10.11}$$

where s is the drawdown in the aquifer, r is radial distance from the pumping well, K' is the vertical hydraulic conductivity of the upper aquitard, T = Km is the transmissivity of the aquifer, s_1 is the drawdown in the upper aquitard, z is vertical distance from the aquifer base, K'' is the hydraulic conductivity of the lower aquitard, s_2 is the drawdown in the lower aquitard, and S is the storativity of the aquifer. Referring to the definition sketch for the conceptual model and recalling Darcy's Law, it is easy to see that the terms

$$+ K' \frac{\partial s_1(r, z = m, t)}{\partial z}$$

$$- K'' \frac{\partial s_2(r, z = 0, t)}{\partial z}$$

represent the rate of vertical groundwater flow ("leakage") through the two aquitards. Note that for flow <u>from</u> the lower aquitard <u>to</u> the aquifer (groundwater flow is upward), the gradient in the lower aquitard is negative.

Initial Conditions

• Before pumping begins drawdown is zero everywhere

$$s(r, t = 0) = 0 \tag{10.12}$$

Boundary Conditions

• At an infinite distance from the pumping well, drawdown is zero

$$s(r = \infty, t \geq 0) = 0 \tag{10.13}$$

• Groundwater flow to the pumping well is constant and uniform over the aquifer thickness (which is a result of the assumption of horizontal groundwater flow)

$$\lim_{r \to 0} \frac{r \partial s(r, t \geq 0)}{\partial r} = -\frac{Q}{2\pi T} \tag{10.14}$$

10.3 ANALYTICAL SOLUTION

It is not practical to develop a general solution which is applicable for all time following the start of pumping. However, solutions have been developed for both early and late time periods.

10.3.1 Early-Time Solution

This solution applies to all three cases for early times or for relatively thick, impermeable aquitards, defined as when $t < \frac{m'S'}{10K'}$ and $t < \frac{m''S''}{10K''}$ (Hantush, 1960 and 1964). Under these conditions, drawdown is given by

$$s = \frac{Q}{4\pi T} H(u, \beta) \tag{10.15}$$

where

$$u = \frac{r^2 S}{4Tt} \tag{10.16}$$

and

$$\beta = \frac{r}{4}\left[\left(\frac{K'S'}{TSm'}\right)^{1/2} + \left(\frac{K''S''}{TSm''}\right)^{1/2}\right] \tag{10.17}$$

The well function $H(u, \beta)$ for leaky aquifers with aquitard storage is defined as

$$H(u, \beta) = \int_u^\infty \frac{e^{-y}}{y} \, \text{erfc}\left[\frac{\beta\sqrt{u}}{\sqrt{y\,(y-u)}}\right] dy \tag{10.18}$$

where $\text{erfc}(x)$ is the complimentary error function defined by

$$\text{erfc}(x) = 1 - \text{erf}(x) = \frac{2}{\sqrt{\pi}} \int_0^x e^{-y} \, dy$$

which may be approximated by

i for $u > 10^4\beta^2$,

$$H(u, \beta) \approx W(u) - \left(\frac{4\beta}{\sqrt{\pi u}}\right)(0.258 + 0.693e^{-0.5u}) \tag{10.19}$$

ii for $u < \dfrac{10^{-5}}{\beta^2}$ and $u < 10^{-4}\beta^2$

$$H(u, \beta) \approx \frac{1}{2}\ln\left(\frac{0.044}{\beta^2 u}\right) \tag{10.20}$$

Values of the function $H(u, \beta)$ are given in Table 10.1. Values of $H(u, \beta)$ may also be computed with the program TYPE5; drawdown can be computed with the computer program DRAW5 (Appendix B).

Table 10.1 Values of the well function $H(u, \beta)$ for a leaky aquifer with aquitard storage (after Hantush, 1964, p. 313).

	β				
u	1.0×10^{-2}	5.0×10^{-2}	1.0×10^{-1}	2.0×10^{-1}	5.0×10^{-1}
1.0×10^{-6}	9.9259	8.3395	7.6497	6.9590	6.0463
2.0×10^{-6}	9.5677	7.9908	7.3024	6.6126	5.7012
3.0×10^{-6}	9.3561	7.7864	7.0991	6.4100	5.4996
4.0×10^{-6}	9.2047	7.6412	6.9547	6.2663	5.3567
5.0×10^{-6}	9.0866	7.5284	6.8427	6.1548	5.2459
6.0×10^{-6}	8.9894	7.4362	6.7512	6.0637	5.1555
7.0×10^{-6}	8.9069	7.3581	6.6737	5.9867	5.0790
8.0×10^{-6}	8.8350	7.2904	6.6066	5.9200	5.0129
9.0×10^{-6}	8.7714	7.2306	6.5474	5.8611	4.9545
1.0×10^{-5}	8.7142	7.1771	6.4944	5.8085	4.9024
2.0×10^{-5}	8.3315	6.8238	6.1453	5.4623	4.5598
3.0×10^{-5}	8.1013	6.6159	5.9406	5.2597	4.3600
4.0×10^{-5}	7.9346	6.4677	5.7951	5.1160	4.2185
5.0×10^{-5}	7.8031	6.3523	5.6821	5.0045	4.1090
6.0×10^{-5}	7.6941	6.2576	5.5896	4.9134	4.0196
7.0×10^{-5}	7.6007	6.1773	5.5113	4.8364	3.9442
8.0×10^{-5}	7.5190	6.1076	5.4434	4.7697	3.8789
9.0×10^{-5}	7.4461	6.0459	5.3834	4.7108	3.8214
1.0×10^{-4}	7.3803	5.9906	5.3297	4.6581	3.7700
2.0×10^{-4}	6.9321	5.6226	4.9747	4.3115	3.4334
3.0×10^{-4}	6.6563	5.4035	4.7655	4.1086	3.2379
4.0×10^{-4}	6.4541	5.2459	4.6161	3.9645	3.0999
5.0×10^{-4}	6.2934	5.1223	4.4996	3.8527	2.9933
6.0×10^{-4}	6.1596	5.0203	4.4040	3.7612	2.9065
7.0×10^{-4}	6.0447	4.9333	4.3228	3.6838	2.8334
8.0×10^{-4}	5.9439	4.8573	4.2523	3.6167	2.7702
9.0×10^{-4}	5.8539	4.7898	4.1898	3.5575	2.7146
1.0×10^{-3}	5.7727	4.7290	4.1337	3.5045	2.6650
2.0×10^{-3}	5.2203	4.3184	3.7598	3.1549	2.3419
3.0×10^{-3}	4.8837	4.0683	3.5363	2.9494	2.1559
4.0×10^{-3}	4.6396	3.8859	3.3750	2.8030	2.0253
5.0×10^{-3}	4.4474	3.7415	3.2483	2.6891	1.9250
6.0×10^{-3}	4.2888	3.6214	3.1436	2.5957	1.8437
7.0×10^{-3}	4.1536	3.5185	3.0542	2.5165	1.7754
8.0×10^{-3}	4.0357	3.4282	2.9762	2.4478	1.7166
9.0×10^{-3}	3.9313	3.3478	2.9068	2.3870	1.6651

Table 10.1   (Continued).

u	β				
	1.0×10^{-2}	5.0×10^{-2}	1.0×10^{-1}	2.0×10^{-1}	5.0×10^{-1}
1.0×10^{-2}	3.8374	3.2752	2.8443	2.3325	1.6193
2.0×10^{-2}	3.2133	2.7829	2.4227	1.9741	1.3239
3.0×10^{-2}	2.8452	2.4844	2.1680	1.7579	1.1570
4.0×10^{-2}	2.5842	2.2691	1.9841	1.6056	1.0416
5.0×10^{-2}	2.3826	2.1007	1.8401	1.4872	0.9540
6.0×10^{-2}	2.2188	1.9626	1.7217	1.3905	0.8838
7.0×10^{-2}	2.0812	1.8458	1.6213	1.3088	0.8255
8.0×10^{-2}	1.9630	1.7448	1.5343	1.2381	0.7758
9.0×10^{-2}	1.8595	1.6559	1.4577	1.1760	0.7327
1.0×10^{-1}	1.7677	1.5768	1.3893	1.1207	0.6947
2.0×10^{-1}	1.1895	1.0714	0.9497	0.7665	0.4603
3.0×10^{-1}	0.8825	0.7986	0.7103	0.5739	0.3390
4.0×10^{-1}	0.6850	0.6218	0.5543	0.4482	0.2619
5.0×10^{-1}	0.5463	0.4969	0.4436	0.3591	0.2083
6.0×10^{-1}	0.4437	0.4041	0.3613	0.2927	0.1688
7.0×10^{-1}	0.3651	0.3330	0.2980	0.2415	0.1386
8.0×10^{-1}	0.3035	0.2770	0.2481	0.2012	0.1151
9.0×10^{-1}	0.2543	0.2323	0.2082	0.1690	0.0010
1.0×10^{0}	0.2144	0.1961	0.1758	0.1427	0.0008
2.0×10^{0}	0.0005	0.0004	0.0004	0.0003	0.0002
3.0×10^{0}	0.0001	0.0001	0.0001	0.0001	

u	β				
	1.0×10^{0}	2.0×10^{0}	5.0×10^{0}	1.0×10^{1}	2.0×10^{1}
1.0×10^{-6}	5.3575	4.6721	3.7756	3.1110	2.4671
2.0×10^{-6}	5.0141	4.3312	3.4412	2.7857	2.1568
3.0×10^{-6}	4.8136	4.1327	3.2474	2.5984	1.9801
4.0×10^{-6}	4.6716	3.9922	3.1109	2.4671	1.8571
5.0×10^{-6}	4.5617	3.8836	3.0055	2.3661	1.7633
6.0×10^{-6}	4.4719	3.7951	2.9199	2.2844	1.6877
7.0×10^{-6}	4.3962	3.7204	2.8478	2.2158	1.6246
8.0×10^{-6}	4.3306	3.6558	2.7856	2.1568	1.5706
9.0×10^{-6}	4.2728	3.5989	2.7309	2.1050	1.5234
1.0×10^{-5}	4.2212	3.5481	2.6822	2.0590	1.4816
2.0×10^{-5}	3.8827	3.2162	2.3660	1.7632	1.2170

Table 10.1   (Continued).

$$\beta$$

u	1.0×10^0	2.0×10^0	5.0×10^0	1.0×10^1	2.0×10^1
3.0×10^{-5}	3.6858	3.0241	2.1850	1.5965	1.0716
4.0×10^{-5}	3.5468	2.8889	2.0588	1.4815	0.9730
5.0×10^{-5}	3.4394	2.7848	1.9622	1.3943	0.8994
6.0×10^{-5}	3.3519	2.7002	1.8841	1.3244	0.8412
7.0×10^{-5}	3.2781	2.6290	1.8189	1.2664	0.7934
8.0×10^{-5}	3.2143	2.5677	1.7629	1.2169	0.7530
9.0×10^{-5}	3.1583	2.5138	1.7139	1.1739	0.7182
1.0×10^{-4}	3.1082	2.4658	1.6704	1.1359	0.6878
2.0×10^{-4}	2.7819	2.1549	1.3937	0.8992	0.5044
3.0×10^{-4}	2.5937	1.9778	1.2401	0.7721	0.4111
4.0×10^{-4}	2.4617	1.8545	1.1352	0.6875	0.3514
5.0×10^{-4}	2.3601	1.7604	1.0564	0.6252	0.3089
6.0×10^{-4}	2.2778	1.6846	0.9937	0.5765	0.2766
7.0×10^{-4}	2.2087	1.6212	0.9420	0.5370	0.2510
8.0×10^{-4}	2.1492	1.5670	0.8982	0.5040	0.2300
9.0×10^{-4}	2.0971	1.5196	0.8603	0.4758	0.2125
1.0×10^{-3}	2.0506	1.4776	0.8271	0.4513	0.1976
2.0×10^{-3}	1.7516	1.2116	0.6238	0.3084	0.1164
3.0×10^{-3}	1.5825	1.0652	0.5182	0.2394	0.0008
4.0×10^{-3}	1.4656	0.9658	0.4496	0.1970	0.0006
5.0×10^{-3}	1.3767	0.8915	0.4001	0.1677	0.0005
6.0×10^{-3}	1.3054	0.8327	0.3620	0.1460	0.0004
7.0×10^{-3}	1.2460	0.7843	0.3315	0.1292	0.0003
8.0×10^{-3}	1.1953	0.7435	0.3064	0.1158	0.0003
9.0×10^{-3}	1.1512	0.7083	0.2852	0.1047	0.0003
1.0×10^{-2}	1.1122	0.6675	0.2670	0.0010	0.0002
2.0×10^{-2}	0.8677	0.4910	0.1653	0.0005	0.0001
3.0×10^{-2}	0.7353	0.3965	0.1197	0.0004	
4.0×10^{-2}	0.6467	0.3357	0.0009	0.0003	
5.0×10^{-2}	0.5812	0.2923	0.0008	0.0002	
6.0×10^{-2}	0.5298	0.2593	0.0006	0.0001	
7.0×10^{-2}	0.4880	0.2332	0.0005	0.0001	
8.0×10^{-2}	0.4530	0.2119	0.0005	0.0001	
9.0×10^{-2}	0.4230	0.1941	0.0004	0.0001	
1.0×10^{-1}	0.3970	0.1789	0.0004	0.0001	
2.0×10^{-1}	0.2452	0.0010	0.0001		
3.0×10^{-1}	0.1729	0.0006	0.0001		
4.0×10^{-1}	0.1296	0.0004	0.0001		
5.0×10^{-1}	0.1006	0.0003			
6.0×10^{-1}	0.0008	0.0002			
7.0×10^{-1}	0.0006	0.0002			

Table 10.1   (Continued).

$$\beta$$

u	1.0×10^0	2.0×10^0
8.0×10^{-1}	0.0005	0.0002
9.0×10^{-1}	0.0004	0.0001
1.0×10^{-0}	0.0004	0.0001
2.0×10^{-0}	0.0001	0.0001
3.0×10^{-0}	0.0001	

10.3.2   Late-Time Solution

For relatively late times or for thin, relatively permeable aquitards, solutions for *Cases A, B,* and *C* are considered separately.

Case A - aquicludes above upper aquitard and below lower aquitard

When $t > \dfrac{2m'S'}{K'}$, $t > \dfrac{30\delta_1 r^2 S}{T}$, and $t > \dfrac{2m''S''}{K''}$

where

$$\delta_1 = 1 + \frac{S'}{S} \tag{10.21}$$

the solution is

$$s = \frac{Q}{4\pi T} W(\delta_1, u) \tag{10.22}$$

where $W(\delta_1, u)$ is the Theis well function of Model 3 with $\delta_1 u$ replacing u (Hantush, 1964).

Case B - source bed above upper aquitard, aquiclude below lower aquitard

When $t > \dfrac{2m'S'}{K'}$, $t > \dfrac{30\delta_2 r_w^2 S}{T}\left[1 - \left(\dfrac{10 r_w}{B}\right)^2\right]$, and $t > \dfrac{2m''S''}{K''}$

where

$$\delta_2 = 1 + \frac{(S'/3 + S'')}{S} \tag{10.23}$$

and

$$B = \sqrt{\frac{Tm'}{K'}} \tag{10.24}$$

the solution is

$$s = \frac{Q}{4\pi T} W(\delta_2 u, r/B) \tag{10.25}$$

where $W(\delta_2 u, r/B)$ is the Hantush - Jacob leaky well function (defined in Equation 9.9) with $\delta_2 u$ replacing u. Thus

$$W(\delta_2 u, r/B) = \int_{\delta_2 u}^{\infty} \frac{1}{y} \exp\left[-y - \frac{r^2}{4B^2 y} \right] dy \tag{10.26}$$

Case C - aquitards bounded above and below by source beds

When $t > \frac{5m'S'}{K'}$ and $t > \frac{5m''S''}{K''}$, the solution is

$$s = \frac{Q}{4\pi T} W(\delta_3 u, \alpha) \tag{10.27}$$

where

$$\delta_3 = 1 + \frac{(S' + S'')}{3S} \tag{10.28}$$

and

$$\alpha = r \sqrt{\frac{K'}{m' T} + \frac{K''}{m'' T}} \tag{10.29}$$

$W(u\delta_3, \alpha)$ is the Hantush-Jacob leaky well function (defined in Equation 9.9) with $u\delta_3$ replacing u (Hantush, 1960). Thus

$$W(\delta_3 u, r/B) = \int_{\delta_3 u}^{\infty} \frac{1}{y} \exp\left[-y - \frac{\alpha^2}{y} \right] dy \tag{10.30}$$

10.3.3   Intermediate-Time Solution

There is no analytical solution available for intermediate times. However, asymptotic values from the early- and late-time solutions can be used to obtain an approximate solution for intermediate times.

10.3.4   Equilibrium Solution

Case A - aquicludes above upper aquitard and below lower aquitard

After prolonged pumping, drawdown will continue to follow the solution of Equation 10.22, i.e.,

$$s = \frac{Q}{4\pi T} W(\delta_1, u)$$

The cone of depression will continue to expand, with flow originating from the aquifer and aquitards.

Cases B and C - source beds above upper aquitard and below lower aquitard

Storage in the aquitards and aquifer in the vicinity of the pumping well will be depleted and groundwater flow to the well will originate entirely in the source bed(s). The solution is

$$s = \frac{Q}{2\pi T} \left[\frac{K_0(r/B)}{(r_w/B) \, K_1(r_w/B)} \right] \qquad (10.31)$$

where $K_0(x)$ and $K_1(x)$ are the zero-order and first-order modified Bessel functions of the second kind.

In practice, r_w/B is usually less than 0.01 and $xK_1(x) \approx 1$ for $x < 0.1$, so Equation 10.31 can be simplified to

$$s = \frac{Q}{2\pi T} \left[K_0(r/B) \right] \qquad (10.32)$$

which is identical to the late-time solution for Model 4 (Equation 9.14), except that B is defined as

$$B = \frac{1}{\sqrt{\dfrac{K'}{m'T} + \dfrac{K''}{m''T}}} \qquad (10.33)$$

The second term in the denominator is zero for *Case B*. The approximate times required to reach equilibrium are given below:

Case B
$$t \approx \frac{8 \left(S + \dfrac{S'}{3} + S'' \right)}{\left(\dfrac{K'}{m'} + \dfrac{K''}{m''} \right)} \qquad (10.34)$$

Case C
$$t \approx \frac{8 \, (3S + S' + S'')}{3 \left(\dfrac{K'}{m'} + \dfrac{K''}{m''} \right)} \qquad (10.35)$$

10.4 METHODS OF ANALYSIS

If both early and late data are available, the aquifer properties T, S, K', and S' can be computed for *Case A* by combining the results of the early- and late-time solutions. For *Cases B* and *C*, the late-time data fall on the flat portions of the type curves, so a unique solution is not possible. However, the late-time equations can be used to compute expected drawdown.

10.4.1   Early-Time Solution

Match-Point Method

The theoretical basis of this method of analysis is described in Chapter 4. The specific steps are:

i Prepare a plot of $H(u, \beta)$ vs. $1/u$ on logarithmic paper. This plot is called the type curve (Figure 10.1). The type curve can be plotted using the values of $H(u, \beta)$ in Table 10.1 or values computed by the computer program TYPE5 (Appendix B).

ii Plot s vs. t (or s vs. t/r^2 for more than one observation well) using the same logarithmic scales used to prepare the type curve. This plot is called the data curve.

iii Overlay the data curve on the type curves. Shift the plots relative to each other, keeping the axes parallel, until a position of best fit is found between the data curve and one of the type curves. Try to use as much early drawdown data as possible. Unless a significant number of points fall within the period where leakage effects are insignificant, (i.e., where $t < 0.25t_i$ where t_i is the inflection point in the curve) a unique solution is impossible (Hantush, 1964).

iv From the best fit portion of the curves, select a match-point and record match-point values β^*, $H(u, \beta)^*$, $1/u^*$, s^*, and t^* or $t/r^2{}^*$.

v Substitute $H(u, \beta)^*$ and s^* into Equation 10.15 and solve for T.

vi Substitute the computed value of T and u^* and t^* or $t/r^2{}^*$ into Equation 10.16 and solve for S.

vii If simplifying assumptions can be made, such as $K"S" = 0$ or $K"S" = K'S'$, then β^* and the computed T and S values can be substituted into Equation 10.17 to solve for the product $K'S'$. For *Case A* conditions, providing that late-time data are available, S' can be computed as outlined in the next section, allowing for a direct solution for K'. For *Cases B and C*, equilibrium drawdown data from a minimum of three observation wells must be available to determine K' and S'.

10.4.2   Late-Time Solution - *Case A* only

Match-Point Method

The late-time solution procedure for *Case A* is outlined below. If early data are available (i.e., S has been computed from early data), this method can be used to compute T and S'. If early data are not available, this method can be used to compute T and the sum S + S'. The procedure is as follows:

i Prepare a plot of $W(\delta_1 u)$ vs. $1/u$ on logarithmic paper. The type curve can be plotted using the values of $W(u)$ in Table 8.1 (by replacing u with $\delta_1 u$) or values computed with the computer program TYPE3 (Appendix B).

ii Plot the data curve s vs. t (or s vs. t/r^2 for more than one observation well) using the same logarithmic scales used to prepare the type curve. This plot is called the data curve.

iii Overlay the data curve on the type curve. Shift the plots relative to each other, keeping the axes parallel until a best fit is found. This time, try to use as much of the late data as possible.

iv From the best fit portion of the curves, select a match-point and record match-point values $W(\delta_1 u)^*$, $1/\delta_1 u^*$, s^*, and t^* or t/r^2*.

v Substitute $W(\delta_1 u)^*$ and s^* into Equation 10.22 and solve for T. If early data were available, this value of T should be the same as that computed from the $H(u, \beta)$ curve.

vi Substitute $\delta_1 u^*$, t^*, and the computed value of T into the following equation and solve for $S(1 + S'/S)$

$$\delta_1 u = \frac{r^2 S(1+S'/S)}{4Tt} \tag{10.36}$$

If early drawdown data were available, use the computed value of S to solve for S'.

10.4.3 Equilibrium Solution

For *Cases B or C*, the procedures outlined in Section 9.4.2 can be used to compute K' or K" from equilibrium drawdown data. The procedure of Section 9.4.2 should be used with B computed from Equation 10.33.

Example 10.1

A pumping test was conducted in a silty sand aquifer (thickness = 20 ft) bounded below by a highly overconsolidated clay and above by a very soft, normally consolidated clay (thickness = 10 ft). The very soft clay is bounded above by a loose sand unconfined aquifer. Drawdown of the water table in the unconfined aquifer due to pumping is negligible. A 6-inch diameter well was pumped at 5 gal/min until equilibrium (constant drawdown) conditions were reached. Drawdown data from three observation wells and the pumping well are in Table 10.2. Find T, S, K', and S'.

Table 10.2 Time-drawdown data for Example 10.1.

| Time (min) | Drawdown (ft) | | | |
	r = 0.25 ft	r = 10 ft	r = 30 ft	r = 60 ft
0.1	6.97	0.47	0	0
0.2	7.53	0.84	0.02	0
0.3	7.85	1.07	0.06	0
0.4	8.06	1.25	0.11	0
0.5	8.23	1.39	0.15	0
0.6	8.36	1.50	0.20	0.01
0.7	8.47	1.59	0.24	0.01
0.8	8.56	1.68	0.28	0.01
0.9	8.64	1.75	0.31	0.02
1	8.72	1.81	0.35	0.02
2	9.18	2.23	0.60	0.09
3	9.43	2.47	0.76	0.15
4	9.61	2.64	0.88	0.21
5	9.74	2.76	0.98	0.26
6	9.85	2.87	1.06	0.30
7	9.94	2.95	1.12	0.34
8	10.02	3.02	1.18	0.37
9	10.08	3.09	1.23	0.40
10	10.14	3.15	1.28	0.43
20	10.52	3.51	1.59	0.64
30	10.74	3.72	1.77	0.78
40	10.89	3.87	1.91	0.88
50	11.01	3.98	2.01	0.95
60	11.10	4.08	2.09	1.02
70	11.18	4.15	2.16	1.08
80	11.25	4.22	2.22	1.18
90	11.31	4.28	2.28	1.17
100	11.36	4.33	2.33	1.21
200	11.71	4.68	2.65	1.48
300	11.88	4.85	2.79	1.58
400	12.07	5.04	2.98	1.76
500	12.21	5.17	3.11	1.88
600	12.31	5.27	3.21	1.98
700	12.38	5.34	3.28	2.05
800	12.44	5.40	3.34	2.10
900	12.48	5.45	3.38	2.15
1000	12.52	5.48	3.42	2.18
2000	12.67	5.63	3.57	2.33
3000	12.70	5.66	3.60	2.36
4000	12.71	5.67	3.61	2.37
5000	12.71	5.67	3.61	2.37
6000	12.71	5.67	3.61	2.37
7000	12.71	5.67	3.61	2.37
8000	12.71	5.67	3.61	2.37
9000	12.71	5.67	3.61	2.37
10000	12.71	5.67	3.61	2.37

If the storativity and hydraulic conductivity of the soft clay are assumed to be very small, the drawdown data can be interpreted using the *Case B* solution by setting K" and S" equal to zero. Since there is no analytical solution which is valid for all times, early and late data must be examined separately. It is necessary to estimate values for the hydraulic conductivity and storativity of the aquifer and upper aquitard to determine the appropriate data ranges to use with the early- and late-time solutions. Plausible ranges for these properties and the initial estimates used are listed below:

<u>Plausible range of values</u>   <u>Initial estimates</u>

$K = 1$ to $500 \dfrac{\text{gal}}{\text{day} \cdot \text{ft}^2}$ $K \approx 100 \dfrac{\text{gal}}{\text{day} \cdot \text{ft}^2}$

$S_s = 10^{-5}$ to 10^{-6} ft^{-1} $S_s \approx 5 \times 10^{-5}$ ft^{-1}

                                                                              $S \approx (5 \times 10^{-5} \text{ ft}^{-1}) (20 \text{ ft}) = 0.001$

$K' = 10^{-4}$ to $5 \dfrac{\text{gal}}{\text{day} \cdot \text{ft}^2}$ $K' \approx 0.05 \dfrac{\text{gal}}{\text{day} \cdot \text{ft}^2}$

$S_s' = 10^{-3}$ to 10^{-5} ft^{-1} $S_s' \approx 1 \times 10^{-4}$ ft^{-1}

                                                                              $S' \approx (1 \times 10^{-4} \text{ ft}^{-1}) (10 \text{ ft}) = 0.001$

The $H(u, \beta)$ type curve can be used when $t < m'S'/(10 K')$. Based on the initial estimates for aquifer and aquitard properties, this time can be computed

$$t = \frac{(10 \text{ ft}) (0.001)}{(10) (0.05 \dfrac{\text{gal}}{\text{day} \cdot \text{ft}^2}) (\dfrac{1 \text{ day}}{1440 \text{ min}}) (\dfrac{\text{ft}^3}{7.48 \text{ gal}})}$$

$$= 215 \text{ min}$$

Look for a change in the shape of the data curves near this time and only use earlier data in selecting the best fit position.

Next, the logarithmic plot of s vs. t/r^2 is placed over the $H(u, \beta)$ type curves. For only one observation point, s vs. t could have been plotted. Figure 10.1 shows the data curve for the observation well at $r = 30$ ft and type curve aligned in a best fit position (the drawdown data from each observation well will fall on separate curves since β is a function of r). Note that the drawdown values for large values of t/r^2 do not fall on the type curve. Values for the selected match-point for the $r = 30$ ft data curve are:

$$
\begin{aligned}
H(u, \beta)* &= 0.936 \\
1/u* &= 10 \\
\beta* &= 0.3 \\
s* &= 0.95 \text{ ft} \\
t/r^2* &= 0.0045 \text{ min/ft}^2
\end{aligned}
$$

Substituting known values into Equation 10.15 gives

$$T = \frac{Q H(u, \beta)*}{4\pi s*} = \frac{\left(5 \dfrac{\text{gal}}{\text{min}}\right) \left(\dfrac{1440 \text{ min}}{\text{day}}\right) (0.936)}{(4\pi) (0.95 \text{ ft})} = 565 \dfrac{\text{gal}}{\text{day} \cdot \text{ft}}$$

and

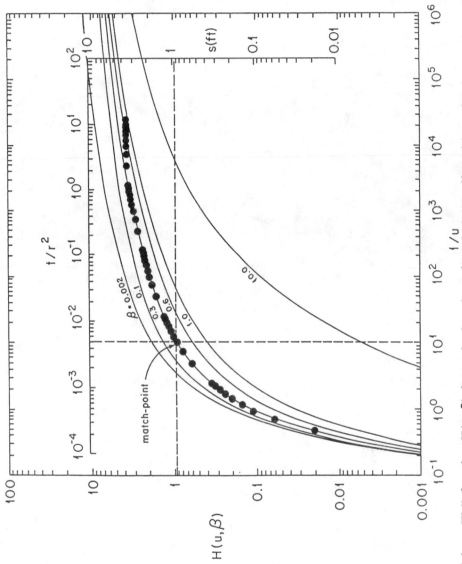

Figure 10.1   Well function H(u, β) for early drawdown in a leaky aquifer with aquitard storage overlain by time-drawdown data from Example 10.1.

$$K = \frac{T}{m} = \left(565 \frac{gal}{day \cdot ft}\right)\left(\frac{1}{20 \ ft}\right) = 28 \frac{gal}{day \cdot ft^2}$$

The coefficient of storage is computed using Equation 10.16

$$S = 4T\left(\frac{t}{r^2}\right)^* (u)^* = 4\left(565 \frac{gal}{day \cdot ft}\right)\left(0.0045 \frac{min}{ft}\right)\left(\frac{1}{10}\right)\left(\frac{ft^3}{7.48 \ gal}\right)\left(\frac{day}{1440 \ min}\right)$$
$$= 9.4 \times 10^{-5}$$

K'S' is computed using Equation 10.17 with K" = S" = 0

$$K'S' = \frac{16\beta^2 TSm'}{r^2}$$
$$= \frac{16 \ (0.3)^2 \ (565 \frac{gal}{day \cdot ft}) \ (9.4 \times 10^{-5}) \ (10 \ ft)}{(30 \ ft)^2}$$
$$= 8.5 \times 10^{-4} \frac{gal}{day \cdot ft^2}$$

For *Case B* conditions, late time-drawdown data fall on the flat portion of the $W(u\delta_3, r/B)$ curve so a unique fit is not possible. Therefore, the only way to directly compute K' or S' by this model is to follow the equilibrium distance-drawdown procedure outlined in Section 9.4.2.

Figure 10.2 shows the equilibrium distance-drawdown data curve matched with the $K_o(r/B)$ type curve. From the figure, the following match-point values were recorded

$$
\begin{array}{ll}
r/B^* & = 0.1 \\
K_o(r/B)^* & = 2.43 \\
s^* & = 4.6 \ ft \\
r^* & = 17.3 \ ft
\end{array}
$$

The value of T computed from early time-drawdown data should be checked and weighted or averaged if the two values differ. From Equation 9.19

$$T = \frac{QK_o(r/B)^*}{2\pi s^*} = \frac{\left(5 \frac{gal}{min}\right)(2.45)\left(\frac{1440 \ min}{day}\right)}{(2\pi) \ (4.6 \ ft)} = 610 \frac{gal}{day \cdot ft}$$

Averaging this value with the T computed previously gives

$$T = \frac{(610 + 565)}{2} = 588 \frac{gal}{day \cdot ft}$$

Recomputing S from the average T value gives S = 9.8 x 10^{-5}. K' can be computed from Equation 10.24

$$K' = \frac{Tm'}{B^2} = \frac{Tm'}{(\frac{r^*}{(r/B)^*})^2} = \frac{Tm'(r/B)^{*2}}{r^{*2}}$$

$$= \frac{(588 \frac{\text{gal}}{\text{day} \cdot \text{ft}}) (10 \text{ ft}) (0.1)^2}{(17.3 \text{ ft})^2}$$

$$= 0.2 \frac{\text{gal}}{\text{day} \cdot \text{ft}^2}$$

and

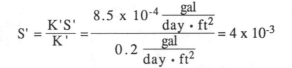

$$S' = \frac{K'S'}{K'} = \frac{8.5 \times 10^{-4} \frac{\text{gal}}{\text{day} \cdot \text{ft}^2}}{0.2 \frac{\text{gal}}{\text{day} \cdot \text{ft}^2}} = 4 \times 10^{-3}$$

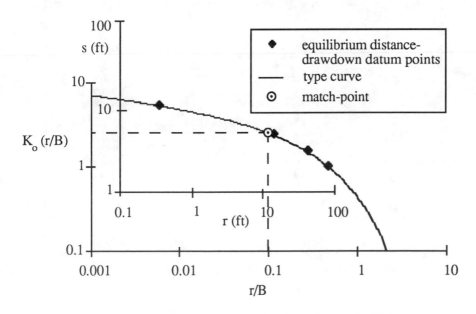

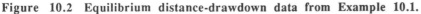

Figure 10.2 Equilibrium distance-drawdown data from Example 10.1.

Example 10.2

A pumping test is performed in a silty sand aquifer (thickness = 20 ft) corresponding to *Case A* conditions. The aquifer is bounded above by an aquiclude (thickness = 15 ft) consisting of soft clay. The aquifer is bounded below by an aquiclude (thickness = 22.5 ft) consisting of soft clay. The pumping rate is 18 gal/min. Drawdown data collected from an observation well 29 feet away from the pumping well are in Table 10.3. Compute T, S, K', K", S', and S".

Table 10.3   Time-drawdown data for Example 10.2.

Time (min)	Drawdown (ft) r = 29 ft	Time (min)	Drawdown (ft) r = 29 ft
0.1	0.01	90	2.37
0.2	0.03	100	2.5
0.3	0.05	200	3.4
0.4	0.06	300	4
0.5	0.08	400	4.4
0.6	0.10	1000	5.86
0.7	0.12	2000	7.04
0.8	0.14	3000	7.73
0.9	0.15	4000	8.22
1	0.17	5000	8.60
2	0.29	6000	8.91
3	0.39	7000	9.178
4	0.47	8000	9.41
5	0.54	9000	9.61
6	0.60	10000	9.79
7	0.65	20000	10.98
8	0.70	30000	11.68
9	0.75	40000	12.17
10	0.79	50000	12.55
20	1.15	60000	12.87
30	1.4	70000	13.13
40	1.6	80000	13.36
50	1.75	90000	13.56
60	1.95	100000	13.74
70	2.1	200000	14.94
80	2.2		

The time-drawdown data curve is placed over the $H(u, \beta)$ type curves and positioned so that early data align with one of the curves (later data should not be expected to fall on the same curve as the early data since the solution for drawdown is not a function of $H(u, \beta)$). From Figure 10.3, the match-points values are

$$
\begin{aligned}
1/u^* &= 100 \\
H(u, \beta)^* &= 0.267 \\
s^* &= 0.46 \text{ ft} \\
t^* &= 3.8 \text{ min} \\
\beta^* &= 5
\end{aligned}
$$

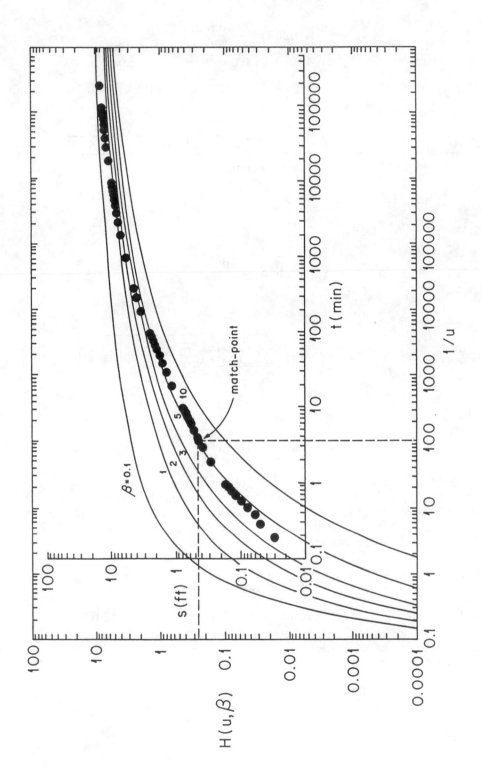

Figure 10.3 H(u, β) type curve for early data in a leaky aquifer with aquitard storage overlain by the time-drawdown data curve from Example 10.2.

Substituting match-point values into Equation 10.15 yields

$$T = \frac{QH(u, \beta)^*}{4\pi s^*} = \frac{\left(18 \frac{gal}{min}\right)(0.267)\left(\frac{1440 \ min}{day}\right)}{(4\pi)(0.46 \ ft)} = 1197 \frac{gal}{day \cdot ft}$$

Equation 10.16 yields

$$S = \frac{u^* 4Tt^*}{r^2} = \frac{\left(\frac{1}{100}\right)(4)\left(1197 \frac{gal}{day \cdot ft}\right)(3.8 \ min)\left(\frac{day}{1440 \ min}\right)\left(\frac{ft^3}{7.48 \ gal}\right)}{(29 \ ft)^2}$$

$$= 2 \times 10^{-5}$$

Rearranging Equation 10.17 gives

$$\sqrt{K'S'} + \sqrt{\frac{m'}{m}}\sqrt{K''S''} = \frac{4\beta\sqrt{TSm'}}{r}$$

Therefore,

$$\sqrt{K'S'} + \sqrt{\frac{15 \ ft}{10 \ ft}}\sqrt{K''S''} = \frac{4(5)\sqrt{1197 \frac{gal}{day \cdot ft}(2 \times 10^{-5})(15 \ ft)}}{29 \ ft}$$

and

$$\sqrt{K'S'} + \sqrt{1.5}\sqrt{K''S''} = 0.413\sqrt{\frac{gal}{day \cdot ft}}$$

At this point, a simplifying assumption must be made since there are too many unknown variables. The hydrogeologist studying this site estimated that the hydraulic conductivity and specific storage for the lower aquitard were about one-half and two-thirds, respectively, of the values for the upper aquitard. Since $S = S_s m$, then $S'' = S'$ and $K'S' = 2K''S''$. Substituting these values into the previous expression yields

$$K''S'' = 0.0246 \frac{gal}{day \cdot ft}$$

Now the late time data are analyzed to find S''. Figure 10.4 shows the time-drawdown data curve matched with the $W(u\delta_1)$ type curve of Model 3. This time, the late data are matched and early data fall below the curve. From Figure 10.4, the following match-point values were selected:

$$
\begin{aligned}
1/(u\delta_1)^* &= 100 \\
W(u\delta_1)^* &= 4.0 \\
s^* &= 6.9 \ ft \\
t^* &= 1890 \ min
\end{aligned}
$$

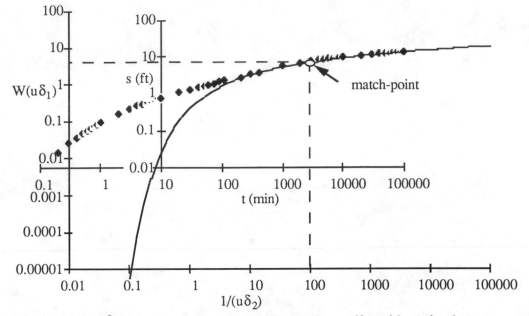

Figure 10.4 W(uδ_1) type curve for late data in a leaky aquifer with aquitard storage overlain by the time-drawdown data curve from Example 10.2.

As a check on early-time calculations, compute T from Equation 10.22,

$$T = \frac{QW(u\delta_1)^*}{4\pi s^*} = \frac{\left(18 \frac{gal}{min}\right)(4.0)\left(\frac{1440\ min}{day}\right)}{(4\pi)\ (6.9\ ft)} = 1196 \frac{gal}{day \cdot ft}$$

If the two computed transmissivities had been different, they could have been averaged.
    From early-data computations, S is known, so u can be computed for this match-point

$$u = \frac{r^2 S}{4Tt^*} = \frac{(29\ ft)^2\ (2 \times 10^{-5})}{4\left(1196 \frac{gal}{day \cdot ft}\right)(1890\ min)\left(\frac{ft^3}{7.48\ gal}\right)\left(\frac{day}{1440\ min}\right)}$$

$$= 2 \times 10^{-5}$$

From the match-point value for $1/(u\delta_1)^*$, δ_1 can be computed

$$\delta_1 = \frac{(u\delta_1)^*}{u} = \frac{\left(\frac{1}{100}\right)}{2 \times 10^{-5}} = 500$$

Equation 10.21 can be rewritten as

$$S' = S(\delta_1 - 1)$$

Thus,

$$S' = 2 \times 10^{-5} (500 - 1) = 0.01 = S''$$

Now the value of S" can be substituted into the expression developed for K"S" obtained from the analysis of the early drawdown data

$$K'' = \frac{(K''S'')}{S''} = \frac{0.0246 \frac{gal}{day \cdot ft}}{0.01} = 2.5 \frac{gal}{day \cdot ft^2}$$

and

$$K' = 2K'' = 5 \frac{gal}{day \cdot ft^2}$$

Chapter 11

MODEL 6: TRANSIENT, CONFINED, PARTIAL PENETRATION, ANISOTROPIC

11.1 CONCEPTUAL MODEL

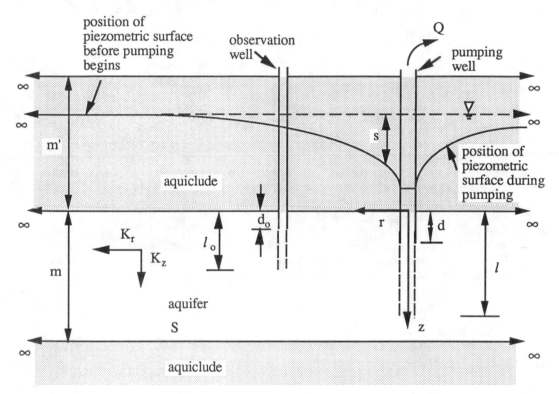

Definition of Terms

d = distance from aquifer top to top of pumping well screen (or uncased portion of hole), L

d_o = distance from aquifer top to top of observation well screen (or uncased portion of hole), L

K_r = aquifer horizontal hydraulic conductivity, LT^{-1}

K_z = aquifer vertical hydraulic conductivity, LT^{-1}

l = distance from aquifer top to bottom of pumping well screen, L

l_o = distance from aquifer top to bottom of observation well screen, L

m   = aquifer thickness, L

r   = radial distance from the pumping well to a point on the cone of depression
          (all distances are measured from the center of wells), L

s   = drawdown of piezometric surface during pumping, L

S   = aquifer storativity, dimensionless

z   = vertical distance from the aquifer top (positive z is downward), L

Assumptions

1.   The aquifer is bounded above and below by aquicludes.
2.   All layers are horizontal and extend infinitely in the radial direction.
3.   The initial piezometric surface (before pumping begins) is horizontal and extends
     infinitely in the radial direction.
4.   The aquifer is homogeneous and anisotropic. The horizontal and vertical hydraulic
     conductivities of the aquifer may be different.
5.   Groundwater density and viscosity are constant.
6.   Groundwater flow can be described by Darcy's Law.
7.   The pumping rate is constant.
8.   Head losses through the well screen and pump intake are negligible.
9.   The pumping well has an infinitesimal diameter.
10. The aquifer is compressible and completely elastic.

11.2 MATHEMATICAL MODEL

Governing Equation

The governing equation is derived by combining Darcy's Law with the principle of conservation of mass in a radial coordinate system (Hantush, 1964)

$$\left(\frac{K_r}{K_z}\right)^2\left[\frac{\partial^2 s}{\partial r^2} + \left(\frac{1}{r}\right)\frac{\partial s}{\partial r}\right] + \frac{\partial^2 s}{\partial z^2} = \left(\frac{S}{K_z m}\right)\frac{\partial s}{\partial t} \tag{11.1}$$

where K_r, K_z, and S are the aquifer horizontal and vertical hydraulic conductivities and storativity, respectively, s is drawdown, r is radial distance from the pumping well, z is vertical distance from the aquifer top, and t is time.

Initial Conditions

• Before pumping begins drawdown is zero everywhere

$$s(r, z, t = 0) = 0 \tag{11.2}$$

Boundary Conditions

- At an infinite distance from the pumping well, drawdown is zero

$$s(r = \infty, z, t) = 0 \qquad (11.3)$$

- There is no vertical flow at the top and bottom of the aquifer

$$\frac{\partial s(r, z = m, t)}{\partial z} = 0 \qquad (11.4)$$

$$\frac{\partial s(r, z = 0, t)}{\partial z} = 0 \qquad (11.5)$$

- Groundwater flow to the pumping well is constant and uniform over the screened interval (groundwater flow near the well is horizontal)

$$\lim_{r \to 0} \left[(l - d) r \frac{\partial s}{\partial r} \right] = 0 \qquad 0 < z < d$$

$$= -\frac{Q}{2\pi K_r} \qquad d < z < l$$

$$= 0 \qquad l < z < m \qquad (11.6)$$

11.3 ANALYTICAL SOLUTION

11.3.1 General Solution

The drawdown at any point in an observation well is (Hantush, 1964)

$$s = \left(\frac{Q}{4\pi K_r m} \right) [W(u) + f(u, x, d/m, l/m, z/m)] \qquad (11.7)$$

where

$$K_r = \frac{Q}{4\pi S m} [W(u) + f(u, x, d/m, l/m, z/m]$$

$$u = \frac{r^2 S}{4 K_r m t} \qquad (11.8)$$

W(u) is the Theis well function of Model 3 and

$$f = \left[\frac{2m}{\pi(l - d)} \right] \sum_{n=1}^{\infty} \left(\frac{1}{n} \right) \left[\sin\left(\frac{n\pi l}{m} \right) - \sin\left(\frac{n\pi d}{m} \right) \right] \cos\left(\frac{n\pi z}{m} \right) W(u, x) \qquad (11.9)$$

where

$$W(u, x) = \int_u^{\infty} \frac{e^{\left(\frac{-y - x^2}{4y} \right)}}{y} \, dy \qquad (11.10)$$

is the Hantush-Jacob leaky well function (Table 9.1) and

$$x = \frac{r}{m} \sqrt{\frac{K_z}{K_r}} \qquad (11.11)$$

To compute the average drawdown in an observation well, s_{ave}, Equation 11.7 must be integrated between the screened depths $z = d_o$ and $z = l_o$ and divided by the screen length $(l_o - d_o)$. The result is

$$s_{ave} = \left(\frac{Q}{4\pi K_r m}\right)[W(u) + f'(u, x, l/m, d/m, l_o/m, d_o/m)] \qquad (11.12)$$

where

$$f'(u, x, l/m, d/m, l_o/m, d_o/m) = \frac{2m^2}{\pi(l - d)(l_o - d_o)} \sum_{n = 1}^{\infty} \frac{1}{n^2}\left[\sin\left(\frac{n\pi l}{m}\right) - \sin\left(\frac{n\pi d}{m}\right)\right]$$

$$\cdot\left[\sin\left(\frac{n\pi l_o}{m}\right) - \sin\left(\frac{n\pi d_o}{m}\right)\right]W(u, n\pi x) \qquad (11.13)$$

where u , x, and $W(u, n\pi x)$ are as defined previously.

For convenience, Equation 11.12 can be written as

$$s_{ave} = \left(\frac{Q}{4\pi K_r m}\right)W(\text{partial penetration}) \qquad (11.14)$$

Values of the function W(partial penetration) can be computed with the computer program TYPE6; average drawdown can be computed with the computer program DRAW6 (Appendix B).

11.3.2   Special Case Solutions

Large Distance Solution

When partially penetrating observation wells are located a radial distance $r > 1.5m \sqrt{\frac{K_r}{K_z}}$ from the pumping well, the effects of partial penetration can be neglected and drawdown can be computed using

$$s_{ave} = \frac{Q}{4\pi K_r m} W(u) \qquad (11.15)$$

where $W(u)$ is the Theis well function of Model 3.

Early-Time or Deep Aquifer Solution

For relatively early times or very deep aquifers where $t < \dfrac{S}{20 \, m \, K_z} \left[2m - \dfrac{1}{2}(2l + l_o + d_o) \right]^2$ drawdown is given by (Hantush, 1964)

$$s = \frac{Q}{8\pi K_r(l - d)} \, E\left(u, \frac{l}{r}, \frac{d}{r}, \frac{z}{r} \right) \tag{11.16}$$

where

$$E\left(u, \frac{l}{r}, \frac{d}{r}, \frac{z}{r} \right) = M\left(u, \sqrt{\frac{K_r}{K_z}} \frac{l+z}{r} \right) + M\left(u, \sqrt{\frac{K_r}{K_z}} \frac{l-z}{r} \right)$$

$$- M\left(u, \sqrt{\frac{K_r}{K_z}} \frac{d+z}{r} \right) - M\left(u, \sqrt{\frac{K_r}{K_z}} \frac{d-z}{r} \right) \tag{11.17}$$

and

$$M(u, \beta) = \int_u^\infty \frac{e^{-y}}{y} \, \text{erf} \, [\beta(y)^{1/2}] \, dy \tag{11.18}$$

where erf(x) is the error function defined as (Hunt, 1983)

$$\text{erf}(x) = 1 - \frac{2}{(\pi)^{1/2}} \int_0^x e^{-y} \, dy \tag{11.19}$$

Selected values of $M(u, \beta)$ are in Table 11.1. Approximate expressions for $M(u, \beta)$ are:

$$\text{for } u < \frac{0.05}{\beta^2} < 0.01 \quad M(u, \beta) \approx z[\sinh^{-1} \beta - 2\beta \sqrt{u/\pi}] \tag{11.20}$$

$$u < \frac{0.05}{\beta^2} \quad M(u, \beta) \approx z[\sinh^{-1} \beta - \beta \text{erf} \sqrt{u}] \tag{11.21}$$

$$u > \frac{5}{\beta^2} \quad M(u, \beta) \approx W(u) \tag{11.22}$$

To compute the average drawdown Equation 11.16 must be integrated between the screened depths $z = d_o$ and $z = l_o$ and divided by the screen length (l_o- d_o).

Late-Time Solution

When $t > \dfrac{mS}{2K_z}$, the effect of partial penetration has reached its maximum and drawdown is given by (Hantush, 1961)

$$s_{ave} = \frac{Q}{4\pi T} \left[W(u) + f_s \left(\frac{r}{m}, \frac{l}{m}, \frac{d}{m}, \frac{l_o}{m}, \frac{d_o}{m} \right) \right] \tag{11.23}$$

where

$$f_s = \frac{4m^2}{\pi^2(l - d)(l_o - d_o)} \sum_{n = 1}^{\infty} \frac{1}{n^2} K_o\left(\frac{n\pi r}{m} \sqrt{\frac{K_z}{K_r}}\right)\left[\sin\left(\frac{n\pi l}{m}\right) - \sin\left(\frac{n\pi d}{m}\right)\right]$$

$$\left[\sin\left(\frac{n\pi l_o}{m}\right) - \sin\left(\frac{n\pi d_o}{m}\right)\right] \tag{11.24}$$

and

K_o = zero-order modified Bessel function of the second kind.

This approximation means that any two (or more) observation wells within the time range where $t > \frac{mS}{2K_z}$ will have identical slopes on their time-drawdown curves.

Table 11.1   Values of the function M(u, β) for use in early-time solutions (from Hantush, 1961).

u / β	0	0.1	0.2	0.3	0.4	0.5	0.6	0.7	0.8	0.9	1.0	1.2	1.4	1.6	1.8	2.0
0	0	0.1997	0.3974	0.5913	0.7801	0.9624	1.1376	1.3053	1.4653	1.6177	1.7627	2.0319	2.2759	2.4979	2.7009	2.8872
1		0.1994	0.3969	0.5907	0.7792	0.9613	1.1363	1.3037	1.4635	1.6157	1.7605	2.0292	2.2728	2.4943	2.6968	2.8827
2		0.1993	0.3967	0.5904	0.7788	0.9608	1.1357	1.3031	1.4628	1.6148	1.7595	2.0281	2.2715	2.4929	2.6929	2.8809
3		0.1993	0.3966	0.5902	0.7785	0.9605	1.1353	1.3026	1.4622	1.6142	1.7588	2.0272	2.2705	2.4907	2.6927	2.8794
4		0.1992	0.3965	0.5900	0.7783	0.9602	1.1349	1.3022	1.4617	1.6137	1.7582	2.0265	2.2696	2.4907	2.6927	2.8782
10^{-6} 5		0.1992	0.3964	0.5898	0.7780	0.9599	1.1346	1.3018	1.4613	1.6132	1.7577	2.0259	2.2689	2.4899	2.6918	2.8772
6		0.1991	0.3963	0.5897	0.7779	0.9596	1.1343	1.3014	1.4609	1.6127	1.7572	2.0253	2.2682	2.4891	2.6909	2.8762
7		0.1991	0.3962	0.5895	0.7777	0.9594	1.1341	1.3011	1.4605	1.6123	1.7568	2.0248	2.2676	2.4884	2.6901	2.8753
8		0.1990	0.3961	0.5894	0.7775	0.9592	1.1338	1.3009	1.4602	1.6120	1.7563	2.0243	2.2670	2.4877	2.6894	2.8745
9		0.1990	0.3960	0.5893	0.7774	0.9590	1.1336	1.3006	1.4599	1.6116	1.7560	2.0238	2.2665	2.4871	2.6887	2.8737
1		0.1989	0.3959	0.5892	0.7772	0.9588	1.1334	1.3003	1.4596	1.6113	1.7556	2.0234	2.2660	2.4865	2.6880	2.8730
2		0.1987	0.3954	0.5883	0.7760	0.9574	1.1316	1.2983	1.4572	1.6086	1.7526	2.0198	2.2618	2.4818	2.6827	2.8671
3		0.1984	0.3949	0.5876	0.7751	0.9562	1.1302	1.2967	1.4554	1.6066	1.7504	2.0171	2.2587	2.4782	2.6786	2.8625
4		0.1982	0.3945	0.5871	0.7744	0.9553	1.1291	1.2953	1.4539	1.6049	1.7485	2.0148	2.2560	2.4751	2.6752	2.8587
10^{-5} 5		0.1981	0.3942	0.5856	0.7737	0.9544	1.1281	1.2941	1.4526	1.6034	1.7468	2.0128	2.2536	2.4724	2.6721	2.8553
6		0.1979	0.3939	0.5861	0.7731	0.9537	1.1271	1.2931	1.4513	1.6020	1.7452	2.0110	2.2515	2.4700	2.6694	2.8523
7		0.1978	0.3936	0.5857	0.7725	0.9530	1.1263	1.2921	1.4502	1.6007	1.7438	2.0093	2.2495	2.4677	2.6669	2.8495
8		0.1976	0.3933	0.5853	0.7720	0.9523	1.1255	1.2912	1.4492	1.5996	1.7425	2.0077	2.2477	2.4657	2.6645	2.8469
9		0.1975	0.3931	0.5849	0.7715	0.9517	1.1248	1.2903	1.4482	1.5984	1.7413	2.0062	2.2460	2.4637	2.6623	2.8444
1		0.1974	0.3929	0.5846	0.7710	0.9511	1.1241	1.2895	1.4473	1.5974	1.7402	2.0049	2.2444	2.4619	2.6603	2.8421
2		0.1965	0.3910	0.5818	0.7673	0.9465	1.1185	1.2830	1.4398	1.5890	1.7308	1.9936	2.2313	2.4469	2.6434	2.8234
3		0.1958	0.3896	0.5796	0.7644	0.9429	1.1142	1.2780	1.4341	1.5825	1.7236	1.9850	2.2212	2.4354	2.6305	2.8091
4		0.1952	0.3883	0.5778	0.7620	0.9398	1.1106	1.2737	1.4292	1.5771	1.7176	1.9778	2.2128	2.4258	2.6197	2.7970
10^{-4} 5		0.1946	0.3873	0.5762	0.7599	0.9372	1.1074	1.2700	1.4250	1.5723	1.7123	1.9714	2.2053	2.4172	2.6101	2.7864
6		0.1941	0.3863	0.5748	0.7580	0.9348	1.1045	1.2666	1.4211	1.5680	1.7075	1.9658	2.1986	2.4095	2.6014	2.7768
7		0.1937	0.3854	0.5734	0.7562	0.9326	1.1018	1.2635	1.4176	1.5640	1.7030	1.9603	2.1924	2.4025	2.5934	2.7679
8		0.1933	0.3846	0.5722	0.7545	0.9305	1.0994	1.2607	1.4143	1.5603	1.6989	1.9554	2.1866	2.3959	2.5860	2.7597
9		0.1929	0.3838	0.5710	0.7530	0.9286	1.0970	1.2579	1.4112	1.5568	1.6951	1.9507	2.1812	2.3897	2.5791	2.7519

Table 11.1   (Continued).

u / β	0	0.1	0.2	0.3	0.4	0.5	0.6	0.7	0.8	0.9	1.0	1.2	1.4	1.6	1.8	2.0
0		0.1997	0.3974	0.5913	0.7801	0.9624	1.1376	1.3053	1.4653	1.6177	1.7627	2.0319	2.2759	2.4979	2.7009	2.8872
1		0.1925	0.3831	0.5699	0.7515	0.9267	1.0948	1.2554	1.4083	1.5535	1.6914	1.9463	2.1761	2.3838	2.5725	2.7446
2		0.1896	0.3772	0.5611	0.7397	0.9120	1.0771	1.2347	1.3846	1.5270	1.6619	1.9109	2.1348	2.3367	2.5195	2.6857
3		0.1873	0.3727	0.5543	0.7307	0.9007	1.0636	1.2189	1.3666	1.5066	1.6393	1.8838	2.1032	2.3006	2.4788	2.6406
4		0.1854	0.3689	0.5486	0.7231	0.8912	1.0521	1.2056	1.3513	1.4895	1.6203	1.8610	2.0766	2.2702	2.4447	2.6027
10^{-3} 5		0.1837	0.3655	0.5435	0.7163	0.8828	1.0421	1.1938	1.3379	1.4744	1.6035	1.8409	2.0532	2.2434	2.4146	2.5693
6		0.1822	0.3625	0.5390	0.7103	0.8752	1.0330	1.1832	1.3258	1.4608	1.5884	1.8228	2.0320	2.2193	2.3875	2.5393
7		0.1808	0.3597	0.5348	0.7047	0.8682	1.0246	1.1735	1.3147	1.4483	1.5745	1.8061	2.0126	2.1972	2.3626	2.5117
8		0.1795	0.3571	0.5310	0.6995	0.8618	1.0169	1.1645	1.3044	1.4367	1.5616	1.7907	1.9946	2.1766	2.3395	2.4861
9		0.1783	0.3547	0.5273	0.6947	0.8557	1.0096	1.1560	1.2947	1.4258	1.5495	1.7762	1.9777	2.1573	2.3179	2.4620
1		0.1772	0.3524	0.5239	0.6901	0.8500	1.0027	1.1480	1.2855	1.4155	1.5381	1.7625	1.9617	2.1391	2.2975	2.4394
2		0.1680	0.3340	0.4962	0.6533	0.8040	0.9476	1.0836	1.2121	1.3329	1.4464	1.6527	1.8340	1.9935	2.1342	2.2587
3		0.1610	0.3200	0.4753	0.6253	0.7691	0.9057	1.0349	1.1564	1.2703	1.3770	1.5697	1.7376	1.8839	2.0116	2.1233
4		0.1551	0.3083	0.4578	0.6020	0.7400	0.8708	0.9942	1.1100	1.2183	1.3193	1.5008	1.6577	1.7932	1.9103	2.0117
10^{-2} 5		0.1500	0.2981	0.4425	0.5817	0.7146	0.8404	0.9588	1.0596	1.1730	1.2691	1.4410	1.5884	1.7147	1.8229	1.9158
6		0.1455	0.2890	0.4289	0.5635	0.6919	0.8132	0.9272	1.0336	1.1326	1.2243	1.3877	1.5268	1.6450	1.7454	1.8307
7		0.1413	0.2807	0.4164	0.5470	0.6713	0.7885	0.8994	1.0008	1.0958	1.1837	1.3394	1.4711	1.5821	1.6756	1.7543
8		0.1375	0.2731	0.4050	0.5317	0.6522	0.7658	0.8720	0.9707	1.0621	1.1464	1.2951	1.4200	1.5246	1.6120	1.6848
9		0.1339	0.2660	0.3943	0.5176	0.6346	0.7447	0.8474	0.9428	1.0308	1.1118	1.2541	1.3729	1.4716	1.5534	1.6210
1		0.130	0.2593	0.3844	0.5043	0.6181	0.7249	0.8245	0.9167	1.0016	1.0795	1.2159	1.3290	1.4223	1.4991	1.5619
2		0.1051	0.2084	0.3081	0.4030	0.4920	0.5744	0.6500	0.7186	0.7806	0.8362	0.9297	1.0029	1.0595	1.1026	1.1352
3		8.74(-2)	0.1731	0.2554	0.3331	0.4053	0.4713	0.5309	0.5842	0.6313	0.6727	0.7400	0.7899	0.8261	0.8519	0.8699
4		7.39(-2)	0.1462	0.2153	0.2801	0.3397	0.3935	0.4415	0.4837	0.5203	0.5519	0.6015	0.6363	0.6602	0.6760	0.6863
10^{-1} 5		6.32(-2)	0.1248	0.1835	0.2381	0.2878	0.3323	0.3714	0.4052	0.4341	0.4584	0.4955	0.5203	0.5362	0.5462	0.5521
6		5.44(-2)	0.1074	0.1575	0.2039	0.2458	0.2828	0.3149	0.3423	0.3652	0.3842	0.4122	0.4300	0.4408	0.4471	0.4506
7		4.71(-2)	9.29(-2)	0.1360	0.1756	0.2111	0.2421	0.2686	0.2909	0.3093	0.3242	0.3455	0.3583	0.3657	0.3698	0.3719
8		4.10(-2)	8.06(-2)	0.1179	0.1519	0.1821	0.2082	0.2302	0.2484	0.2632	0.2750	0.2913	0.3007	0.3058	0.3084	0.3090
9		3.57(-2)	7.03(-2)	0.1026	0.1319	0.1576	0.1797	0.1980	0.2130	0.2250	0.2343	0.2468	0.2537	0.2572	0.2589	0.2597

Table 11.1   (Continued).

u / β	0.1	0.2	0.3	0.4	0.5	0.6	0.7	0.8	0.9	1.0	1.2	1.4	1.6	1.8	2.0
0	0.1997	0.3974	0.5913	0.7801	0.9624	1.1376	1.3053	1.4653	1.6177	1.7627	2.0319	2.2759	2.4979	2.7009	2.8872
1	3.13(-2)	6.14(-2)	8.95(-2)	0.1148	0.1369	0.1555	0.1709	0.1833	0.1929	0.2204	0.2101	0.2151	0.2175	0.2186	0.2191
2	9.01(-3)	1.72(-2)	2.51(-2)	3.16(-2)	3.67(-2)	4.07(-2)	4.35(-2)	4.55(-2)	4.69(-2)	4.77(-2)	4.85(-2)	4.88(-2)			4.88(-2)
3	2.82(-3)	5.44(-3)	7.68(-3)	9.47(-3)	1.08(-2)	1.17(-2)	1.23(-2)	1.26(-2)	1.28(-2)	1.30(-2)					1.30(-2)
4	9.20(-4)	1.76(-3)	2.44(-3)	2.96(-3)	3.31(-3)	3.53(-3)	3.66(-3)	3.72(-3)	3.76(-3)	3.77(-3)					3.77(-3)
5	3.07(-4)	5.80(-4)	7.96(-4)	9.49(-4)	1.05(-3)	1.10(-3)	1.13(-3)	1.14(-3)	1.15(-3)						1.15(-3)
6	1.04(-4)	1.95(-4)	2.64(-4)	3.10(-4)	3.36(-4)	3.50(-4)	3.56(-4)	3.59(-4)	3.60(-4)						3.60(-4)
7	3.56(-5)	6.61(-5)	8.84(-5)	1.02(-4)	1.10(-4)	1.13(-4)	1.15(-4)	1.15(-4)	1.15(-4)						1.15(-4)
8	1.23(-5)	2.26(-5)	2.99(-5)	3.42(-5)	3.63(-5)	3.72(-5)	3.75(-5)	3.76(-5)	3.77(-5)						3.77(-5)
9	4.28(-6)	7.79(-6)	1.02(-5)	1.15(-5)	1.21(-5)	1.23(-5)	1.24(-5)	1.24(-5)							1.24(-5)
10	1.49(-6)	2.70(-6)	3.48(-6)	3.90(-6)	4.07(-6)	4.13(-6)	4.15(-6)	4.16(-6)							4.16(-6)

Values of $M(u, \beta)$ are equal to $W(u)$ for u greater than 10 and all values of β

u / β	10	12	14	16	18	20	22	24	26	28	30
0	5.9964	6.3595	6.6668	6.9333	7.1684	7.3789	7.5692	7.7431	7.9030	8.0511	8.1890
1	5.9739	6.3325	6.6353	6.8973	7.1279	7.3339	7.5197	7.6891	7.8445	7.9881	8.1215
2	5.9645	6.3213	6.6223	6.8823	7.1111	7.3152	7.4992	7.6667	7.8202	7.9620	8.0935
3	5.9573	6.3127	6.6122	6.8709	7.0982	7.3008	7.4834	7.6495	7.8016	7.9419	8.0720
4	5.9513	6.3054	6.6038	6.8612	7.0873	7.2887	7.4701	7.6350	7.7859	7.9250	8.0539
5	5.9460	6.2990	6.5963	6.8527	7.0777	7.2781	7.4584	7.6222	7.7721	7.9101	8.0379
6	5.9412	6.2932	6.5896	6.8450	7.0691	7.2685	7.4478	7.6106	7.7596	7.8966	8.0235
7	5.9367	6.2879	6.5834	6.8379	7.0611	7.2596	7.4381	7.6000	7.7481	7.8842	8.0102
8	5.9326	6.2830	6.5776	6.8313	7.0537	7.2514	7.4290	7.5901	7.7374	7.8727	7.9979
9	5.9287	6.2783	6.5722	6.8251	7.0467	7.2436	7.4205	7.5809	7.7273	7.8619	7.9863

(u values in the lower table are multiplied by 10^{-6}.)

The numbers in parentheses are powers of 10 by which the other numbers are multiplied e.g., 8.74(-2) = 0.0874

Table 11.1   (Continued).

u / β	β	10	12	14	16	18	20	22	24	26	28	30
	0	5.9964	6.3595	6.6668	6.9333	7.1684	7.3789	7.5692	7.7431	7.9030	8.0511	8.1890
10^{-5}	1	5.9251	6.2739	6.5671	6.8193	7.0402	7.2363	7.4125	7.5721	7.7178	7.8517	7.9753
	2	5.8955	6.2385	6.5257	6.7720	6.9870	7.1773	7.3476	7.5013	7.6412	7.7692	7.8871
	3	5.8729	6.2113	6.4940	6.7358	6.9463	7.1321	7.2979	7.4472	7.5826	7.7061	7.8195
	4	5.8538	6.1884	6.4673	6.7053	6.1920	7.0940	7.2561	7.4016	7.5332	7.6531	7.7627
	5	5.8369	6.1682	6.4438	6.6785	6.8818	7.0606	7.2193	7.3515	7.4899	7.6065	7.7129
	6	5.8217	6.1500	6.4225	6.6542	6.8546	7.0303	7.1861	7.3253	7.4508	7.5644	7.6679
	7	5.8078	6.1332	6.4030	6.6319	6.8296	7.0025	7.1556	7.2921	7.4149	7.5259	7.6267
	8	5.7948	6.1177	6.3848	6.6112	6.8063	6.9767	7.1272	7.2613	7.3815	7.4901	7.5885
	9	5.7826	6.1030	6.3678	6.5917	6.7844	6.9525	7.1007	7.2324	7.3503	7.4565	7.5527
10^{-4}	1	5.7710	6.0892	6.3517	6.5734	6.7638	6.9295	7.0756	7.2051	7.3208	7.4249	7.5189
	2	5.6780	5.9778	6.2221	6.4257	6.5982	6.7463	6.8747	6.9869	7.0856	7.1729	7.2504
	3	5.6069	5.8928	6.1233	6.3133	6.4725	6.6075	6.7231	6.8227	6.9091	6.9844	7.0502
	4	5.5471	5.8214	6.0406	6.2194	6.3677	6.4920	6.5972	6.6868	6.7635	6.8294	6.8862
	5	5.4947	5.7589	5.9681	6.1374	6.2763	6.3915	6.4880	6.5692	6.6378	6.6961	6.7456
	6	5.4474	5.7026	5.9031	6.0638	6.1945	6.3019	6.3908	6.4648	6.5266	6.5784	6.6218
	7	5.4041	5.6511	5.8437	5.9967	6.1201	6.2205	6.3027	6.3704	6.4263	6.4726	6.5109
	8	5.3639	5.6034	5.7887	5.9348	6.0515	6.1456	6.2219	6.2841	6.3349	6.3763	6.4103
	9	5.3263	5.5588	5.7374	5.8771	5.9878	6.0762	6.1472	6.2044	6.2506	6.2879	6.3181
10^{-3}	1	5.2908	5.5168	5.6892	5.8230	5.9281	6.0113	6.0775	6.1303	6.1724	6.2061	6.2330
	2	5.0095	5.1861	5.3123	5.4037	5.4701	5.5184	5.5534	5.5788	5.5970	5.6101	5.6193
	3	4.8006	4.9437	5.0402	5.1056	5.1498	5.1795	5.1993	5.2124	5.2208	5.2263	5.2297
	4	4.6301	4.7481	4.8235	4.8714	4.9017	4.9205	5.9320	4.9390	4.9430	4.9454	4.9467
	5	4.4844	4.5830	4.6426	4.6783	4.6993	4.7114	4.7183	4.7220	4.7241	4.7251	4.7256
	6	4.3566	4.4396	4.4872	4.5140	4.5288	4.5367	4.5409	4.5429	4.5439	4.5444	4.5446
	7	4.2425	4.3128	4.3510	4.3714	4.3819	4.3871	4.3896	4.3908	4.3913	4.3915	4.3916
	8	4.1393	4.1991	4.2300	4.2455	4.2530	4.2565	4.2580	4.2587	4.2589	4.2590	4.2591
	9	4.0450	4.0961	4.1212	4.1331	4.1385	4.1408	4.1417	4.1421	4.1422	4.1423	

Table 11.1   (Continued).

u / β	0	10	12	14	16	18	20	22	24	26	28	30
0		5.9964	6.3595	6.6668	6.9333	7.1684	7.3789	7.5692	7.7431	7.9030	8.0511	8.1890
1		3.9382	4.0020	4.0224	4.0316	4.0355	4.0370	4.0378				
2		3.3403	3.3507	3.3537	3.3545	3.3547						
3		2.9558	2.9585	2.9590								
10^{-2} 4		2.6804	2.6812	2.6812								
5		2.4677	2.4679									
6		2.2952										

Values of M(u, β) are equal to W(u) for u greater than 6×10^{-2} and all values of β

u / β	0	32	34	36	38	40	42	44	46	48	50
0		8.3180	8.4392	8.5535	8.6615	8.7641	8.8616	8.9546	9.0435	9.1286	9.2102
1		8.2460	8.3627	8.4725	8.5761	8.6741	8.7671	8.8556	8.9400	9.0206	9.0977
2		8.2161	8.3309	8.4388	8.5406	8.6367	8.7279	8.8145	8.8971	8.9758	9.0510
3		8.1932	8.3066	8.4130	8.5133	8.6081	8.6978	8.7830	8.8641	8.9414	9.0152
4		8.1739	8.2861	8.3913	8.4904	8.5839	8.6725	8.7565	8.8364	8.9125	8.9851
10^{-6} 5		8.1569	8.2680	8.3722	8.4702	8.5627	8.6502	8.7331	8.8119	8.8870	8.9586
6		8.1415	8.2516	8.3549	8.4519	8.5435	8.6300	8.7120	8.7899	8.8640	8.9346
7		8.1273	8.2366	8.3390	8.4352	8.5258	8.6115	8.6926	8.7695	8.8428	8.9126
8		8.1142	8.2226	8.3242	8.4196	8.5094	8.5942	8.6745	8.7507	8.8231	8.8921
9		8.1018	8.2095	8.3103	8.4049	8.4940	8.5781	8.6576	8.7330	8.8047	8.8729
1		8.0901	8.1971	8.2972	8.3910	8.4794	8.5628	8.6416	8.7163	8.7872	8.8547
2		7.9960	8.0972	8.1914	8.2795	8.3621	8.4397	8.5127	8.5817	8.6469	8.7087
3		7.9240	8.0208	8.1106	8.1943	8.2725	8.3458	8.4145	8.4792	8.5401	8.5976
4		7.8636	7.9566	8.0428	8.1229	8.1975	8.2671	8.3322	8.3933	8.4507	8.5047
10^{-5} 5		7.8105	7.9003	7.9833	8.0602	8.1316	8.1982	8.2602	8.3182	8.3725	8.4235
6		7.7626	7.8496	7.9298	8.0038	8.0725	8.1362	8.1955	8.2508	8.3024	8.3507

Table 11.1   (Continued).

u / β	32	34	36	38	40	42	44	46	48	50
0	8.3180	8.4392	8.5535	8.6615	8.7641	8.8616	8.9546	9.0435	9.1286	9.2102
7	7.7188	7.8032	7.8807	7.9522	8.0183	8.0795	8.1364	8.1892	8.2384	8.2843
8	7.6781	7.7601	7.8353	7.9044	7.9682	8.0271	8.0817	8.1323	8.1792	8.2229
9	7.6401	7.7198	7.7928	7.8597	7.9214	7.9782	8.0306	8.0791	8.1241	8.1658
10^{-4} 1	7.6042	7.6818	7.7527	7.8177	7.8773	7.9321	7.9826	8.0292	8.0723	8.1122
2	7.3194	7.3811	7.4364	7.4861	7.5307	7.5709	7.6072	7.6399	7.6695	7.6962
3	7.1080	7.1588	7.2035	7.2429	7.2777	7.3084	7.3356	7.3597	7.3809	7.3997
4	6.9353	6.9778	7.0146	7.0465	7.0742	7.0982	7.1191	7.1371	7.1528	7.1663
5	6.7878	6.8237	6.8544	6.8805	6.9029	6.9219	6.9380	6.9518	6.9635	6.9734
6	6.6583	6.6890	6.7147	6.7363	6.7545	6.7696	6.7823	6.7929	6.8017	6.8090
7	6.5427	6.5690	6.5907	6.6087	6.6235	6.6357	6.6457	6.6539	6.6603	6.6661
8	6.4380	6.4607	6.4791	6.4942	6.5063	6.5162	6.5241	6.5305	6.5357	6.5397
9	6.3424	6.3620	6.3777	6.3903	6.4004	6.4084	6.4147	6.4197	6.4236	6.4267
10^{-3} 1	6.2543	6.2713	6.2848	6.2954	6.3037	6.3102	6.3153	6.3192	6.3222	6.3246
2	5.6257	5.6302	5.6333	5.6354	5.6377	5.6377	5.6383	5.6387	5.6390	5.6391
3	5.2318	5.2331	5.2339	5.2343	5.2346	5.2347	5.2348	5.2349		5.2349
4	4.9474	4.9478	4.9480	4.9481						4.9481
5	4.7259	4.7260	4.7260							4.7260
6	4.5447									4.5447
7										4.3916
8										4.2591
9										4.1423

Table 11.1   (Continued).

u / β	0	32	34	36	38	40	42	44	46	48	50
0	8.3180	8.4392	8.5535	8.6615	8.7641	8.8616	8.9546	9.0435	9.1286	9.2102	
10^{-2} 1											4.0378
2											3.3547
3											2.9590
4											2.6812
5											2.4679
6											2.2952

Values of M(u, β) are equal to W(u) for u greater than 6 x 10^{-2} and all values of β

u / β	0	50	52	54	56	58	60	62	64	66	68	70
0	9.2102	9.2886	9.3641	9.4368	9.5069	9.5747	9.6403	9.7037	9.7653	9.8249	9.8829	
10^{-6} 1	9.0977	9.1716	9.2426	9.3108	9.3765	9.4398	9.5008	9.5598	9.6168	9.6720	9.7255	
2	9.0510	9.1231	9.1922	9.2585	9.3223	9.3838	9.4430	9.5001	9.5553	9.6086	9.6602	
3	9.0152	9.0859	9.1535	9.2185	9.2809	9.3409	9.3986	9.4543	9.5081	9.5600	9.1602	
4	8.9851	9.0545	9.1210	9.1847	9.2459	9.3047	9.3613	9.4158	9.4684	9.5191	9.5681	
5	8.9586	9.0269	9.0924	9.1550	9.2152	9.2729	9.3285	9.3819	9.4335	9.4831	9.5311	
6	8.9346	9.0020	9.0665	9.1282	9.1874	9.2442	9.2988	9.3513	9.4019	9.4507	9.4977	
7	8.9126	8.9791	9.0427	9.1036	9.1619	9.1382	9.2716	9.3232	9.3730	9.4208	9.4670	
8	8.8921	8.9578	9.0206	9.0807	9.0592	9.1160	9.2463	9.2971	9.3460	9.3931	9.4385	
9	8.8729	8.9378	9.0000	9.0592	8.4940	9.1703	9.2225	9.2726	9.3208	9.3671	9.4118	
10^{-5} 1	8.8547	8.9190	8.9803	9.0389	9.0949	9.1486	9.2001	9.2495	9.2970	9.3426	9.3865	
2	8.7087	8.7673	8.8229	8.8759	8.9263	8.9743	9.0202	9.0640	9.1059	9.1461	9.1845	
3	8.5976	8.6519	8.7033	8.7521	8.7983	8.8422	8.8839	8.9237	8.9615	8.9975	9.0319	
4	8.5047	8.5555	8.6035	8.6488	8.6916	8.7321	8.7705	8.8069	8.8414	8.8742	8.9053	
5	8.4235	8.4713	8.5163	8.5587	8.5986	8.6362	8.6717	8.7053	8.7370	8.7671	8.7955	
6	8.3507	8.3959	8.4383	8.4780	8.5154	8.5505	8.5836	8.6147	8.6440	8.6716	8.6977	
7	8.2843	8.3271	8.3672	8.4046	8.4397	8.4726	8.5034	8.5324	8.5596	8.5851	8.6091	

Table 11.1 (Continued).

u / β	0	50	52	54	56	58	60	62	64	66	68	70
0		9.2102	9.2886	9.3641	9.4368	9.5069	9.5747	9.6403	9.7037	9.7653	9.8249	9.8829
8		8.2229	8.2636	8.3016	8.3370	8.3700	8.4009	8.4297	8.4568	8.4821	8.5057	8.5279
9		8.1658	8.2045	8.2405	8.2740	8.3052	8.3343	8.3614	8.3866	8.4102	8.4322	8.4528
10^{-4} 1		8.1122	8.1491	8.1833	8.2151	8.2446	8.2720	8.2974	8.3211	8.3431	8.3636	8.3827
2		7.6962	7.7203	7.7421	7.7618	7.7797	7.7958	7.8104	7.8236	7.8355	7.8463	7.8560
3		7.3997	7.4163	7.4309	7.4439	7.4553	7.4654	7.4742	7.4820	7.4889	7.4949	7.5002
4		7.1663	7.1780	7.1881	7.1968	7.2043	7.2108	7.2163	7.2211	7.2251	7.2286	7.2315
5		6.9734	6.9818	6.9888	6.9948	6.9998	7.0040	7.0075	7.0105	7.0129	7.0150	7.0167
6		6.8090	6.8151	6.8201	6.8242	6.8276	6.8304	6.8327	6.8345	6.8360	6.8372	6.8382
7		6.6661	6.6705	6.6741	6.6770	6.6793	6.6811	6.6826	6.6838	6.6847	6.6854	6.6860
8		6.5397	6.5430	6.5456	6.5476	6.5492	6.5504	6.5514	6.5521	6.5527	6.5531	6.5535
9		6.4267	6.4291	6.4310	6.4324	6.4335	6.4343	6.4350	6.4354	6.4358	6.4361	6.4363
10^{-3} 1		6.3246	6.3263	6.3277	6.3287	6.3294	6.3300	6.3304	6.3307	6.3310	6.3311	6.3312
2		5.6391	5.6392	5.6393	5.6393	5.6393						

Values of M(u, β) are equal to W(u) for u greater than 2×10^{-3} and all values of β

Table 11.1   (Continued).

u / β		72	74	76	78	80	82	84	86	88	90
	0	9.9392	9.9940	10.0473	10.0992	10.1498	10.1992	10.2474	10.2944	10.3404	10.3853
10^{-6}	1	9.7773	9.8276	9.8764	9.9239	9.9700	10.0148	10.0585	10.1011	10.1425	10.1830
	2	9.7102	9.7586	9.8056	9.8512	9.8955	9.9385	9.9803	10.0210	10.0606	10.0992
	3	9.6588	9.7058	9.7513	9.7955	9.8384	9.8800	9.9204	9.9597	9.9979	10.0351
	4	9.6155	9.6613	9.7057	9.7487	9.7904	9.8308	9.8700	9.9081	9.9452	9.9812
	5	9.5774	9.6222	9.6655	9.7075	9.7481	9.7875	9.8257	9.8628	9.8988	9.9938
	6	9.5431	9.5869	9.6293	9.6703	9.7101	9.7485	9.7858	9.8220	9.8571	9.8911
	7	9.5115	9.5545	9.5961	9.6362	9.6751	9.7127	9.7492	9.7845	9.8187	9.8519
	8	9.4822	9.5244	9.5652	9.6046	9.6426	9.6795	9.7151	9.7497	9.7831	9.8156
	9	9.4548	9.4962	9.5362	9.5749	9.6122	9.6483	9.6833	9.7171	9.7498	9.7815
10^{-5}	1	9.4288	9.4696	9.5089	9.5469	9.5835	9.6189	9.6532	9.6863	9.7183	9.7494
	2	9.2213	9.2566	9.2905	9.3230	9.3542	9.3843	9.4132	9.4410	9.4677	9.4935
	3	9.0647	9.0960	9.1260	9.1546	9.1819	9.2081	9.2332	9.2572	9.2802	9.3023
	4	8.9349	8.9630	8.9898	9.0153	9.0396	9.0628	9.0848	9.1059	9.1260	9.1451
	5	8.8224	8.8479	8.8721	8.8950	8.9167	8.9374	8.9570	8.9756	8.9932	9.0100
	6	8.7223	8.7455	8.7675	8.7882	8.8078	8.8263	8.8438	8.8603	8.8760	8.8908
	7	8.6317	8.6530	8.6730	8.6918	8.7095	8.7262	8.7419	8.7567	8.7706	8.7838
	8	8.5487	8.5682	8.5865	8.6036	8.6197	8.6348	8.6490	8.6623	8.6747	8.6865
	9	8.4720	8.4899	8.5067	8.5223	8.5370	8.5507	8.5635	8.5754	8.5866	8.5970
10^{-4}	1	8.4005	8.4170	8.4324	8.4468	8.4601	8.4726	8.4842	8.4949	8.5050	8.5143
	2	7.8648	7.8727	7.8798	7.8862	7.8920	7.8972	7.9019	7.9061	7.9098	7.9132
	3	7.5048	7.5088	7.5124	7.5154	7.5181	7.5204	7.5225	7.5242	7.5258	7.5271
	4	7.2341	7.2362	7.2380	7.2395	7.2408	7.2419	7.2428	7.2436	7.2442	7.2447
	5	7.0181	7.0192	7.0202	7.0209	7.0216	7.0221	7.0225	7.0228	7.0231	7.0233
	6	6.8390	6.8396	6.8401	6.8405	6.8408	6.8411	6.8413	6.8414	6.8416	6.8417
	7	6.6864	6.6868	6.6871	6.6873	6.6874	6.6875	6.6876	6.6877	6.6878	6.6878
	8	6.5537	6.5539	6.5541	6.5542	6.5543	6.5543	6.5544	6.5544	6.5544	6.5544
	9	6.4364	6.4365	6.4366	6.4366	6.4367	6.4367	6.4367	6.4368	6.4368	6.4368

Table 11.1   (Continued).

u / β	72	74	76	78	80	82	84	86	88	90
0	9.9392	9.9940	10.0473	10.0992	10.1498	10.1992	10.2474	10.2944	10.3404	10.3853
10^{-3} 1	6.3313	6.3314	6.3314	6.3315						6.3315
2										5.6393

Values of M(u, β) are equal to W(u) for u greater than 2×10^{-3} and all values of β

u / β	90	92	94	96	98	100	120	140	160	180	200
0	10.3853	10.4292	10.4722	10.5143	10.5555	10.5959	10.9604	11.2686	11.5355	11.7709	11.9815
1	10.1830	10.2224	10.2609	10.2986	10.3353	10.3712	10.6909	10.9543	11.1765	11.3674	11.5335
2	10.0992	10.1368	10.1735	10.2092	10.2441	10.2782	10.5795	10.8247	11.0288	11.2018	11.3502
3	10.0351	10.0713	10.1065	10.1409	10.1744	10.2070	10.4944	10.7258	10.9165	11.0760	11.2113
4	9.9812	10.0162	10.0503	10.0834	10.1158	10.1473	10.4231	10.6431	10.8225	10.9712	11.0958
10^{-6} 5	9.9338	9.9678	10.0008	10.0330	10.0643	10.0948	10.3605	10.5706	10.7404	10.8797	10.9953
6	9.8911	9.9242	9.9563	9.9876	10.0179	10.0475	10.3042	10.5055	10.6668	10.7979	10.9056
7	9.8519	9.8842	9.9155	9.9459	9.9754	10.0041	10.2526	10.4461	10.5997	10.7234	10.8241
8	9.8156	9.8470	9.8776	9.9072	9.9360	9.9640	10.2049	10.3911	10.5377	10.6548	10.7492
9	9.7815	9.8123	9.8421	9.8710	9.8991	9.9263	10.1602	10.3397	10.4800	10.5910	10.6797
1	9.7494	9.7794	9.8086	9.8368	9.8642	9.9808	10.1182	10.2914	10.4258	10.5313	10.6147
2	9.4935	9.5184	9.5423	9.5654	9.5876	9.6091	9.7870	9.9141	10.0059	10.0726	10.1211
3	9.3023	9.3234	9.3437	9.3631	9.3818	9.3957	9.5441	9.6413	9.7071	9.7516	9.7815
4	9.1451	9.1634	9.1808	9.1975	9.2134	9.2285	9.3478	9.4238	9.4721	9.5026	9.5216
10^{-5} 5	9.0100	9.0260	9.0411	9.0555	9.0692	9.0822	9.1819	9.2421	9.2781	9.2993	9.3116
6	8.8908	8.9048	8.9181	8.9306	8.9425	8.9538	9.0378	9.0859	9.1130	9.1279	9.1359
7	8.7838	8.7962	8.8078	8.8188	8.8292	8.8389	8.9101	8.9489	8.9694	8.9800	8.9853
8	8.6864	8.6974	8.7077	8.7174	8.7264	8.7349	8.7956	8.8269	8.8426	8.8502	8.8537
9	8.5970	8.6068	8.6159	8.6244	8.6324	8.6399	8.6917	8.7172	8.7293	8.7347	8.7370

Table 11.1   (Continued).

u / β	90	92	94	96	98	100	120	140	160	180	200
0	10.3853	10.4292	10.4722	10.5143	10.5555	10.5959	10.9604	11.2686	11.5355	11.7709	11.9815
1	8.5143	8.5231	8.5312	8.5387	8.5457	8.5522	8.5967	8.6175	8.6268	8.6307	8.6307
2	7.9132	7.9162	7.9189	7.9213	7.9235	7.9254	7.9361	7.9392	7.9399	7.9401	7.9402
3	7.5271	7.5282	7.5292	7.5300	7.5307	7.5314	7.5342	7.5347			
4	7.2447	7.2452	7.2456	7.2459	7.2461	7.2463	7.2471	7.2472			
10^{-4} 5	7.0233	7.0235	7.0237	7.0238	7.0239	7.0239	7.0242				
6	6.8417	6.8417	6.8418	6.8418	6.8419	6.8419	6.8419				
7	6.6878	6.6878	6.6878	6.6879							
8	6.5544	6.5544	6.5544								
9											

Values of M(u, β) are equal to W(u) for u greater than 8 x 10^{-4} and all values of β

u / β	220	240	260	280	300	320	340	360	380	400
0	12.1720	12.3459	12.5058	12.6539	12.7918	12.9207	13.0418	13.1560	13.2640	13.3664
1	11.6797	11.8093	11.9252	12.0294	12.1234	12.2088	12.2865	12.3574	12.4224	12.4821
2	11.4788	11.5912	11.6900	11.7773	11.8549	11.9240	11.9858	12.0411	12.0908	12.1355
3	11.3271	11.4269	11.5134	11.5888	11.6547	11.7125	11.7633	11.8081	11.8475	11.8824
4	11.2012	11.2909	11.3677	11.4337	11.4906	11.5398	11.5823	11.6192	11.6511	11.6788
10^{-6} 5	11.0919	11.1733	11.2420	11.3003	11.3499	11.3922	11.4282	11.4589	11.4850	11.5074
6	10.9945	11.0688	11.1307	11.1826	11.2261	11.2626	11.2933	11.3191	11.3408	11.3589
7	10.9065	10.9744	11.0304	11.0767	11.1151	11.1469	11.1732	11.1950	11.2130	11.2279
8	10.8257	10.8880	10.9388	10.9804	11.0144	11.0422	11.0649	11.0814	11.0984	11.1106
9	10.7509	10.8082	10.8545	10.8919	10.9221	10.9465	10.9661	10.9819	10.9945	11.0045
1	10.6811	10.7340	10.7763	10.8100	10.8369	10.8584	10.8754	10.8888	10.8994	10.9078
2	10.1563	10.1818	10.2001	10.2131	10.2224	10.2288	10.2333	10.2364	10.2385	10.2399
3	9.8014	9.8145	9.8230	9.8284	9.8319	9.8340	9.8353	9.8361	9.8365	9.8368
4	9.5332	9.5401	9.5442	9.5466	9.5479	9.5486	9.5490	9.5492	9.5493	9.5495

Table 11.1   (Continued).

u / β	220	240	260	280	300	320	340	360	380	400
0	12.1720	12.3459	12.5058	12.6539	12.7918	12.9207	13.0418	13.1560	13.2640	13.3664
10^{-5}										
5	9.3185	9.3222	9.3243	9.3253	9.3258	9.3261	9.3262	9.3263		9.3263
6	9.1401	9.1422	9.1432	9.1436	9.1438	9.1439	9.1440	9.1440		9.1440
7	8.9878	8.9890	8.9895	8.9897	8.9898	8.9898	8.9898			8.9899
8	8.8553	8.8559	8.8562	8.8563	8.8563					8.8502
9	8.7380	8.7384	8.7385	8.7385	8.7386					8.7386
10^{-4}										
1	8.6329	8.6331								8.6332
2	7.9402									7.9402
3	7.5348									7.5348
4	7.2472									7.2472
5	7.0242									7.0242
6	6.8420									6.8420
7	6.6879									6.6879
8	6.5545									6.5545

Values of M(u, β) are equal to W(u) for u greater than 8 x 10^{-4} and all values of β

u / β	2.2	2.4	2.6	2.8	3.0	3.2	3.4	3.6	3.8	4.0
0	3.0593	3.2188	3.3675	3.5064	3.6369	3.7597	3.8757	3.9856	4.0900	4.1894
10^{-6}										
1	3.0543	3.2134	3.3616	3.5001	3.6301	3.7525	3.8581	3.9775	4.0815	4.0804
2	3.0523	3.2112	3.3592	3.4975	3.6273	3.7495	3.8649	3.9742	4.0779	4.1766
3	3.0507	3.2095	3.3573	3.4955	3.6251	3.7472	3.8625	3.9716	4.0752	4.1738
4	3.0494	3.2080	3.3557	3.4938	3.6233	3.7453	3.8604	3.9694	4.0729	4.1714
5	3.0482	3.2067	3.3544	3.4923	3.6217	3.7436	3.8586	3.9675	4.0709	4.1592
6	3.0471	3.2056	3.3531	3.4910	3.6203	3.7420	3.8569	3.9658	4.0690	4.1673
7	3.0462	3.2045	3.3519	3.4897	3.6190	3.7406	3.8554	3.9642	4.0674	4.1655
8	3.0453	3.2035	3.3509	3.4886	3.6177	3.7393	3.8540	3.9627	4.0658	4.1639
9	3.0444	3.2026	3.3499	3.4875	3.6166	3.7380	3.8527	3.9613	4.0643	4.1623

Table 11.1   (Continued).

u / β		2.2	2.4	2.6	2.8	3.0	3.2	3.4	3.6	3.8	4.0
	0	3.0593	3.2188	3.3675	3.5064	3.6369	3.7597	3.8757	3.9856	4.0900	4.1894
10^{-5}	1	3.0436	3.2017	3.3489	3.4865	3.6155	3.7369	3.8515	3.9600	4.0629	4.1609
	2	3.0371	3.1946	3.3412	3.4782	3.6066	3.7274	3.8414	3.9493	4.0517	4.1490
	3	3.0321	3.1892	3.3352	3.4718	3.5998	3.7202	3.8337	3.9412	4.0431	4.1400
	4	3.0279	3.1846	3.3304	3.4665	3.5941	3.7140	3.8272	3.9343	4.0358	4.1323
	5	3.0242	3.1806	3.3260	3.4618	3.5890	3.7087	3.8215	3.9282	4.0294	4.1256
	6	3.0209	3.1769	3.3220	3.4575	3.5844	3.7038	3.8163	3.9227	4.0236	4.1195
	7	3.0178	3.1735	3.3184	3.4536	3.5802	3.6993	3.8116	3.9177	4.0183	4.1139
	8	3.0149	3.1704	3.3150	3.4499	3.5763	3.6951	3.8071	3.9130	4.0133	4.1087
	9	3.0122	3.1675	3.3118	3.4465	3.5727	3.6912	3.8030	3.9086	4.0087	4.1038
10^{-4}	1	3.0097	3.1647	3.3088	3.4433	3.5692	3.6875	3.7990	3.9044	4.0043	4.0992
	2	2.9891	3.1423	3.2845	3.4171	3.5412	3.6576	3.7673	3.8708	3.9688	4.0618
	3	2.9733	3.1251	3.2659	3.3970	3.5197	3.6347	3.7429	3.8450	3.9416	4.0332
	4	2.9600	3.1106	3.2502	3.3801	3.5015	3.6154	3.7224	3.8233	3.9187	4.0090
	5	2.9483	3.0978	3.2363	3.3652	3.4856	3.5984	3.7043	3.8042	3.8985	3.9878
	6	2.9378	3.0863	3.2238	3.3518	3.4712	3.5830	3.6880	3.7869	3.8802	3.9686
	7	2.9280	3.0757	3.2123	3.3394	3.4579	3.5688	3.6730	3.7710	3.8635	3.9509
	8	2.9190	3.0658	3.2017	3.3279	3.4456	3.5557	3.6590	3.7562	3.8479	3.9345
	9	2.9105	3.0565	3.1916	3.3171	3.4340	3.5434	3.6459	3.7423	3.8332	3.9191
10^{-3}	1	2.9024	3.0478	3.1821	3.3069	3.4231	3.5317	3.6335	3.7292	3.8194	3.9046
	2	2.8377	2.9771	3.1056	3.2245	3.3349	3.4377	3.5337	3.6236	3.7080	3.7874
	3	2.7881	2.9231	3.0471	3.1616	3.2675	3.3659	3.4575	3.5430	3.6230	3.6981
	4	2.7464	2.8776	2.9980	3.1087	3.2110	3.3056	3.3936	3.4754	3.5518	3.6233
	5	2.7098	2.8377	2.9548	3.0623	3.1613	3.2528	3.3375	3.4162	3.4894	3.5577
	6	2.6767	2.8018	2.9159	3.0205	3.1166	3.2052	3.2871	3.3629	3.4334	3.4989
	7	2.6464	2.7688	2.8802	2.9822	3.0757	3.1616	3.2409	3.3142	3.3821	3.4451
	8	2.6183	2.7382	2.8472	2.9466	3.0377	3.1213	3.1982	3.2691	3.3346	3.3953
	9	2.5920	2.7095	2.8162	2.9134	3.0022	3.0835	3.1582	3.2270	3.2903	3.3489

Table 11.1   (Continued).

u / β	0	2.2	2.4	2.6	2.8	3.0	3.2	3.4	3.6	3.8	4.0
0		3.0593	3.2188	3.3675	3.5064	3.6369	3.7597	3.8757	3.9856	4.0900	4.1894
1		2.5671	2.6825	2.7870	2.8820	2.9687	3.0480	3.1206	3.1873	3.2487	3.3052
2		2.3892	2.4675	2.5552	2.6337	2.7041	2.7673	2.8243	2.8756	2.9218	2.9637
3		2.2212	2.3072	2.3830	2.4498	2.5088	2.5610	2.6073	2.6482	2.6846	2.7168
4		2.0996	2.1759	2.2423	2.3000	2.3503	2.3942	2.4324	2.4658	2.4949	2.5202
10^{-2} 5		1.9951	2.0634	2.1221	2.1725	2.2158	2.2531	2.2851	2.3125	2.3361	2.3563
6		1.9031	1.9645	2.0167	2.0610	2.0986	2.1305	2.1574	2.1802	2.1995	2.2157
7		1.8205	1.8760	1.9227	1.9618	1.9946	2.0220	2.0449	2.0640	2.0798	2.0929
8		1.7455	1.7959	1.8378	1.8725	1.9012	1.9249	1.9444	1.9604	1.9734	1.9841
9		1.6767	1.7226	1.7603	1.7912	1.8164	1.8370	1.8536	1.8671	1.8780	1.8867
1		1.6133	1.6552	1.6892	1.7167	1.7389	1.7568	1.7825	1.7915	1.7915	1.7987
2		1.1596	1.1777	1.1909	1.2004	1.2073	1.2122	1.2179	1.2195	1.2195	1.2206
3		0.8823	0.8907	0.8962	0.8998	0.9021	0.9035	0.9049	0.9053	0.9053	0.9054
4		0.6928	0.6968	0.6992	0.7006	0.7014	0.7019	0.7023	0.7023	0.7023	0.7024
10^{-1} 5		0.5556	0.5576	0.5587	0.5592	0.5595	0.5597	0.5597	0.5597		
6		0.4525	0.4535	0.4540	0.4542	0.4543	0.4543				
7		0.3729	0.3734	0.3736	0.3737	0.3737					
8		0.3102	0.3104	0.3105							
9		0.2600	0.2601	0.2602							
1	0.2193	0.2194									

Values of M(u, β) are equal to W(u) for u greater than 1 and all values of β

Table 11.1   (Continued).

u / β	4.2	4.4	4.6	4.8	5.0	5.2	5.4	5.6	5.8	6.0
0	4.2842	4.3748	4.4516	4.5448	4.3248	4.7018	4.7760	4.8475	4.9167	4.9835
10^{-6}										
1	4.2747	4.3649	4.4512	4.5340	4.3136	4.6901	4.7638	4.8349	4.9036	4.9700
2	4.2708	4.3608	4.4469	4.5295	4.6089	4.6852	4.7588	4.8297	4.8982	4.9644
3	4.2678	4.3576	4.4436	4.5261	4.6053	4.6815	4.7549	4.8257	4.8940	4.9601
4	4.2653	4.3550	4.4408	4.5232	4.6023	4.6784	4.7516	4.8223	4.8905	4.9565
5	4.2630	4.3526	4.4384	4.5206	4.3996	4.6756	4.7488	4.8193	4.8874	4.9533
6	4.2610	4.3505	4.4362	4.5183	4.5972	4.6731	4.7462	4.8166	4.8846	4.9504
7	4.2591	4.3486	4.4341	4.5162	4.5950	4.6708	4.7438	4.8141	4.8821	4.9477
8	4.2574	4.3467	4.4323	4.5142	4.5929	4.6686	4.7415	4.8118	4.8797	4.9452
9	4.2558	4.3450	4.4305	4.5124	4.5910	4.6666	4.7394	4.8097	4.8774	4.9429
10^{-5}										
1	4.2542	4.3434	4.4288	4.5106	4.5892	4.6647	4.7375	4.8076	4.8753	4.9407
2	4.2418	4.3304	4.4152	4.4964	4.5744	4.6494	4.7215	4.7911	4.8582	4.9230
3	4.2323	4.3204	4.4048	4.4855	4.5631	4.6376	4.7093	4.7784	4.8450	4.9094
4	4.2243	4.3120	4.3960	4.4764	4.5535	4.6276	4.6989	4.7677	4.8339	4.8979
5	4.2172	4.3046	4.3882	4.4683	4.5451	4.6189	4.6899	4.7582	4.8242	4.8878
6	4.2108	4.2979	4.3812	4.4610	4.5375	4.6110	4.6816	4.7497	4.8153	4.8787
7	4.2049	4.2918	4.3748	4.4543	4.5305	4.6037	4.6741	4.7419	4.8072	4.8703
8	4.1994	4.2860	4.3688	4.4480	4.5240	4.5969	4.5670	4.7346	4.7997	4.8625
9	4.1943	4.2806	4.3632	4.4421	4.5178	4.5905	4.5604	4.7277	4.7926	4.8551
10^{-4}										
1	4.1894	4.2756	4.3578	4.4366	4.5121	4.5845	4.6542	4.7212	4.7859	4.8482
2	4.1502	4.2345	4.3149	4.3918	4.4654	4.5360	4.6038	4.6690	4.7317	4.7922
3	4.1202	4.2030	4.2820	4.3574	4.4296	4.4988	4.5652	4.6289	4.6903	4.7493
4	4.0948	4.1764	4.2542	4.3285	4.3995	4.4674	4.5326	4.5952	4.6553	4.7132
5	4.0725	4.1531	4.2298	4.3030	4.3729	4.4399	4.5040	4.5655	4.5246	4.6814
6	4.0524	4.1320	4.2077	4.2800	4.3490	4.4150	4.4781	4.5387	4.5969	4.6527
7	4.0338	4.1126	4.1875	4.2588	4.3270	4.3921	4.4544	4.5141	4.5714	4.6264
8	4.0166	4.0945	4.1686	4.2392	4.3065	4.3708	4.4323	4.4912	4.5477	4.6019
9	4.0004	4.0776	4.1509	4.2207	4.2873	4.3508	4.4116	4.4697	4.5255	4.5789

Table 11.1   (Continued).

u / β		4.2	4.4	4.6	4.8	5.0	5.2	5.4	5.6	5.8	6.0
	0	4.2842	4.3748	4.4516	4.5448	4.3248	4.7018	4.7760	4.8475	4.9167	4.9835
10^{-3}	1	3.9852	4.0616	4.1342	4.2033	4.2691	4.3320	4.3920	4.4494	4.5045	4.5572
	2	3.8623	3.9329	3.9998	4.0632	4.1233	4.1805	4.2349	4.2867	4.3360	4.3832
	3	3.7686	3.8350	3.8975	3.9566	4.0125	4.0654	4.1156	4.1332	4.2084	4.2514
	4	3.6902	3.7530	3.8120	3.8676	3.9199	3.9694	4.0161	4.0602	4.1020	4.1416
	5	3.6215	3.6813	3.7372	3.7898	3.8391	3.8856	3.9293	3.9705	4.0094	4.0462
	6	3.5599	3.6169	3.6702	3.7200	3.7667	3.8105	3.8517	3.8903	3.9267	3.9609
	7	3.5036	3.5581	3.6090	3.6564	3.7007	3.7422	3.7810	3.8174	3.8515	3.8835
	8	3.4516	3.5038	3.5524	3.5977	3.6398	3.6792	3.7159	3.7502	3.7823	3.8123
	9	3.4030	3.4532	3.4998	3.5430	3.5832	3.6206	3.6554	3.6878	3.7181	3.7463
10^{-2}	1	3.3574	3.4057	3.4503	3.4917	3.5300	3.5656	3.5987	3.6294	3.6580	3.6845
	2	3.0015	3.0357	3.0666	3.0946	3.1200	3.1430	3.1638	3.1827	3.1998	3.2153
	3	2.2753	2.7707	2.7932	2.1831	2.8307	2.8464	2.8602	2.8724	2.8832	2.8928
	4	2.5423	2.5615	2.5782	2.5927	2.6052	2.6161	2.5256	2.6337	2.6408	2.6468
	5	2.3735	2.3883	2.4009	2.4116	2.4207	2.4284	2.4350	2.4405	2.4452	2.4491
	6	2.2294	2.2408	2.2504	2.2584	2.2651	2.2705	2.2752	2.2790	2.2821	2.2846
	7	2.1038	2.1127	2.1201	2.1261	2.1310	2.1350	2.1382	2.1408	2.1429	2.1446
	8	1.9928	1.9998	2.0055	2.0101	2.0137	2.0166	2.0189	2.0207	2.0221	2.0233
	9	1.8937	1.8992	1.9036	1.9071	1.9098	1.9119	1.9135	1.9148	1.9158	1.9165
10^{-1}	1	1.8043	1.8087	1.8121	1.8147	1.8168	1.8183	1.8195	1.8204	1.8211	1.8216
	2	1.2213	1.2218	1.2221	1.2223	1.2224	1.2225	1.2226	1.2226	1.2226	1.2226
	3	0.9056	0.9056	0.9056	0.9057						0.9057
	4										0.7024
	5										0.5597
	6										0.4543
	7										0.3737
	8										0.3105
	9										0.2602

Table 11.1   (Continued).

u / β	4.2	4.4	4.6	4.8	5.0	5.2	5.4	5.6	5.8	6.0
0	4.2842	4.3748	4.4516	4.5448	4.3248	4.7018	4.7760	4.8475	4.9167	4.9835
1	0.2194									

Values of M(u, β) are equal to W(u) for u greater than 1 and all values of β

u / β	6.0	6.2	6.4	6.6	6.8	7.0	7.2	7.4	7.6	7.8	8.0
0	4.9835	5.0482	5.1109	5.1718	5.2308	5.2882	5.3440	5.3983	5.4511	5.5026	5.5529
10^{-6} 1	4.9700	5.0343	5.0965	5.1569	5.2155	5.2724	5.3278	5.3816	5.4340	5.4851	5.5349
2	4.9644	5.0523	5.0905	5.1507	5.2091	5.2659	5.3210	5.3747	5.4269	5.4778	5.5274
3	4.9601	5.0240	5.0860	5.1460	5.2043	5.2609	5.3159	5.3694	5.4215	5.4722	5.5217
4	4.9565	5.0203	5.0821	5.1420	5.2002	5.2566	5.3109	5.3649	5.4169	5.4675	5.5168
5	4.9533	5.0170	5.0787	5.1385	5.1965	5.2529	5.3071	5.3610	5.4128	5.4633	5.5126
6	4.9504	5.0140	5.0756	5.1353	5.1933	5.2495	5.3036	5.3574	5.4092	5.4596	5.5087
7	4.9477	5.0112	5.0728	5.1324	5.1902	5.2464	5.3003	5.3541	5.4058	5.4561	5.5052
8	4.9452	5.0087	5.0701	5.1297	5.1874	5.2435	5.2972	5.3511	5.4027	5.4529	5.5019
9	4.9429	5.0063	5.0676	5.1271	5.1848	5.2408	5.2944	5.3482	5.3997	5.4499	5.4988
10^{-5} 1	4.9407	5.0040	5.0653	5.1247	5.1823	5.2383	5.2926	5.3455	5.3969	5.4470	5.4958
2	4.9230	4.9857	5.0464	5.1052	5.1622	5.2176	5.2714	5.3236	5.3745	5.4240	5.4722
3	4.9094	4.9716	5.0319	5.0902	5.1468	5.2017	5.2550	5.3069	5.3573	5.4063	5.4541
4	4.8979	4.9598	5.0196	5.0776	5.1338	5.1883	5.2413	5.2927	5.3427	5.3914	5.4388
5	4.8878	4.9493	5.0089	5.0665	5.1224	5.1766	5.2292	5.2803	5.3299	5.3783	5.4253
6	4.8787	4.9399	4.9991	5.0565	5.1120	5.1659	5.2182	5.2690	5.3184	5.3664	5.4131
7	4.8703	4.9312	4.9902	5.0472	5.1025	5.1561	5.2081	5.2586	5.3077	5.3555	5.4020
8	4.8625	4.9232	4.9818	5.0386	5.0937	5.1470	5.1988	5.2490	5.2979	5.3453	5.3915
9	4.8551	4.9156	4.9740	5.0306	5.0853	5.1384	5.1900	5.2400	5.2886	5.3358	5.3818

Table 11.1 (Continued).

u/β		6.0	6.2	6.4	6.6	6.8	7.0	7.2	7.4	7.6	7.8	8.0
	0	4.9835	5.0482	5.1109	5.1718	5.2308	5.2882	5.3440	5.3983	5.4511	5.5026	5.5529
	1	4.8482	4.9084	4.9666	5.0229	5.0775	5.1303	5.1816	5.2314	5.2798	5.3268	5.3725
	2	4.7922	4.8506	4.9069	4.9614	5.0141	5.0651	5.1145	5.1624	5.2089	5.2541	5.2980
	3	4.7492	4.8062	4.8612	4.9142	4.9655	5.0151	5.0631	5.1096	5.1547	5.1985	5.2409
	4	4.7132	4.7689	4.8227	4.8745	4.9246	4.9730	5.0198	5.0652	5.1091	5.1516	5.1929
10^{-4}	5	4.6814	4.7361	4.7888	4.8396	4.8886	4.9360	4.9818	5.0261	5.0689	5.1105	5.1507
	6	4.6527	4.7065	4.7582	4.8081	4.8562	4.9026	4.9475	4.9908	5.0327	5.0733	5.1127
	7	4.6264	4.6793	4.7301	4.7791	4.8264	4.8719	4.9159	4.9584	4.9995	5.0393	5.0777
	8	4.6019	4.6540	4.7040	4.7522	4.7987	4.8435	4.8867	4.9284	4.9687	5.0076	5.0453
	9	4.5789	4.6302	4.6796	4.7270	4.7727	4.8168	4.8592	4.9002	4.9397	4.9780	5.0149
	1	4.5572	4.6078	4.6565	4.7032	4.7482	4.7915	4.8333	4.8736	4.9124	4.9500	4.9862
	2	4.3832	4.4282	4.4713	4.5125	4.5519	4.5898	4.6260	4.6609	4.6943	4.7266	4.7573
	3	4.2514	4.2923	4.3312	4.3684	4.4038	4.4376	4.4699	4.5007	4.5302	4.5584	4.5854
	4	4.1416	4.1792	4.2148	4.2487	4.2808	4.3114	4.3405	4.3682	4.3945	4.4197	4.4436
10^{-3}	5	4.0462	4.0809	4.1137	4.1448	4.1742	4.2021	4.2285	4.2535	4.2773	4.2998	4.3212
	6	3.9609	3.9932	4.0236	4.0523	4.0793	4.1048	4.1290	4.1517	4.1733	4.1936	4.2129
	7	3.8835	3.9136	3.9418	3.9684	3.9934	4.0168	4.0390	4.0598	4.0794	4.0978	4.1152
	8	3.8123	3.8404	3.8668	3.8914	3.9145	3.9360	3.9566	3.9756	3.9935	4.0103	4.0261
	9	3.7463	3.7726	3.7972	3.8202	3.8416	3.8617	3.8805	3.8980	3.9144	3.9297	3.9440
	1	3.6845	3.7093	3.7323	3.7537	3.7737	3.7923	3.8096	3.8258	3.8408	3.8548	3.8679
	2	3.2153	3.2293	3.2419	3.2534	3.2638	3.2731	3.2816	3.2892	3.2961	3.3023	3.3079
	3	2.8928	2.9012	2.9086	2.9151	2.9209	2.9259	2.9303	2.9342	2.9376	2.9405	2.9431
	4	2.6468	2.6520	2.6565	2.6603	2.6636	2.6664	2.6688	2.6708	2.6725	2.6740	2.6752
10^{-2}	5	2.4491	2.4523	2.4551	2.4574	2.4593	2.4609	2.4622	2.4632	2.4641	2.4648	2.4654
	6	2.2846	2.2867	2.2884	2.2898	2.2909	2.2918	2.2926	2.2931	2.2936	2.2940	2.2943
	7	2.1446	2.1460	2.1470	2.1479	2.1486	2.1491	2.1495	2.1498	2.1500	2.1502	2.1504
	8	2.0233	2.0241	2.0248	2.0253	2.0257	2.0260	2.0263	2.0264	2.0266	2.0257	2.0267
	9	1.9155	1.9171	1.9175	1.9178	1.9181	1.9183	1.9184	1.9185	1.9186	1.9186	1.9185

Table 11.1   (Continued).

u / β	6.0	6.2	6.4	6.6	6.8	7.0	7.2	7.4	7.6	7.8	8.0
0	4.9835	5.0482	5.1109	5.1718	5.2308	5.2882	5.3440	5.3983	5.4511	5.5026	5.5529
10^{-1} 1	1.8216	1.8219	1.8222	1.8224	1.8226	1.8227	1.8227	1.8228	1.8228	1.8229	
2	1.2226	1.2226	1.2226								

Values of M(u, β) are equal to W(u) for u greater than 2×10^{-1} and all values of β

u / β	8.2	8.4	8.6	8.8	9.0	9.2	9.4	9.6	9.8
0	5.6019	5.6497	5.6965	5.7421	5.7868	5.8305	5.8733	5.9151	5.9562
10^{-6} 1	5.5834	5.6308	5.8771	5.7223	5.7666	5.8098	5.8521	5.9935	5.9341
2	5.5758	5.6230	5.6691	5.7141	5.7581	5.8012	5.8433	5.8846	5.9250
3	5.5699	5.6170	5.6629	5.7078	5.7517	5.7946	5.8366	5.8777	5.9179
4	5.5649	5.6119	5.6577	5.7025	5.7463	5.7890	5.8309	5.8719	5.9120
5	5.5606	5.6074	5.6531	5.6978	5.7415	5.7841	5.8259	5.8668	5.9068
6	5.5566	5.6034	5.6490	5.6936	5.7371	5.7797	5.8214	5.8621	5.9021
7	5.5530	5.5996	5.6452	5.6897	5.7331	5.7756	5.8172	5.8579	5.8977
8	5.5496	5.5962	5.6416	5.6860	5.7294	5.7718	5.8133	5.8539	5.8937
9	5.5464	5.5929	5.6383	5.6826	5.7259	5.7683	5.8097	5.8502	5.8899
10^{-5} 1	5.5434	5.5898	5.6352	5.6794	5.7226	5.7649	5.8063	5.8467	5.8863
2	5.5192	5.5650	5.6097	5.6534	5.6961	5.7377	5.7785	5.8183	5.8573
3	5.5006	5.5460	5.5903	5.6335	5.6757	5.7169	5.7572	5.7966	5.8351
4	5.4849	5.5299	5.5738	5.6167	5.6585	5.6993	5.7392	5.7782	5.8164
5	5.4711	5.5158	5.5594	5.6018	5.6433	5.6838	5.7234	5.7621	5.7999
6	5.4587	5.5030	5.5463	5.5885	5.6296	5.6698	5.7091	5.7475	5.7850
7	5.4472	5.4913	5.5342	5.5762	5.6171	5.3570	5.6960	5.7341	5.7713
8	5.4365	5.4803	5.5231	5.5547	5.6053	5.6450	5.6837	5.7216	5.7586
9	5.4265	5.4701	5.5126	5.5540	5.5943	5.6338	5.6723	5.7099	5.7466

Table 11.1   (Continued).

u / β		8.2	8.4	8.6	8.8	9.0	9.2	9.4	9.6	9.8
	0	5.6019	5.6497	5.6965	5.7421	5.7868	5.8305	5.8733	5.9151	5.9562
10^{-4}	1	5.4170	5.4604	5.5026	5.5438	5.5840	5.6231	5.6614	5.6988	5.7353
	2	5.3406	5.3821	5.4225	5.4619	5.5002	5.5375	5.5739	5.6095	5.6441
	3	5.2822	5.3223	5.3612	5.3992	5.4361	5.4720	5.5070	5.5411	5.5744
	4	5.2330	5.2719	5.3097	5.3464	5.3822	5.4169	5.4508	5.4837	5.5158
	5	5.1898	5.2276	5.2644	5.3001	5.3348	5.3686	5.4014	5.4333	5.4644
	6	5.1508	5.1877	5.2236	5.2583	5.2921	5.3249	5.3568	5.3879	5.4180
	7	5.1150	5.1511	5.1861	5.2200	5.2530	5.2850	5.3160	5.3462	5.3755
	8	5.0818	5.1171	5.1513	5.1845	5.2166	5.2478	5.2781	5.3075	5.3361
	9	5.0508	5.0862	5.1187	5.1512	5.1826	5.2131	5.2427	5.2714	5.2992
10^{-3}	1	5.0213	5.0552	5.0880	5.1197	5.1505	5.1803	5.2092	5.2372	5.2644
	2	4.7870	4.8155	4.8430	4.8695	4.8950	4.9196	4.9433	4.9662	4.9882
	3	4.6113	4.5350	4.6597	4.6824	4.7042	4.7251	4.7452	4.7644	4.7829
	4	4.4664	4.4881	4.5089	4.5287	4.5476	4.5355	4.5829	4.6993	4.6150
	5	4.3415	4.3608	4.3792	4.3966	4.4131	4.4268	4.4438	4.4580	4.4716
	6	4.2311	4.2483	4.2646	4.2800	4.2946	4.3084	4.3214	4.3338	4.3455
	7	4.1316	4.1470	4.1616	4.1753	4.1882	4.2004	4.2119	4.2227	4.2329
	8	4.0409	4.0548	4.0678	4.0801	4.0916	4.1024	4.1125	4.1220	4.1309
	9	3.9575	3.9700	3.9817	3.9927	4.0029	4.0125	4.0215	4.0299	4.0377
10^{-2}	1	3.8801	3.8914	3.9020	3.9119	3.9210	3.9296	3.9375	3.9449	3.9518
	2	3.3130	3.3175	3.3215	3.3252	3.3285	3.3314	3.3340	3.3364	3.3385
	3	2.9453	2.9473	2.9489	2.9504	2.9517	2.9528	2.9537	2.9545	2.9552
	4	2.6762	2.6771	2.6778	2.6784	2.6789	2.6793	2.6797	2.6800	2.6802
	5	2.4659	2.4663	2.4666	2.4669	2.4671	2.4673	2.4674	2.4675	2.4676
	6	2.2945	2.2947	2.3948	2.2949	2.2950	2.2951	2.2951	2.2952	2.2952
	7	2.1505	2.1506	2.1506	2.1507	2.1507	2.1508			
	8	2.0268	2.0268	2.0269						
	9	1.9186								

Table 11.1   (Continued).

u / β	0	8.2	8.4	8.6	8.8	9.0	9.2	9.4	9.6	9.8
	0	5.6019	5.6497	5.6965	5.7421	5.7868	5.8305	5.8733	5.9151	5.9562
10^{-1}	1	1.8229								
	2	1.2226								

Values of $M(u, \beta)$ are equal to $W(u)$ for u greater than 2×10^{-1} and all values of β

11.4 METHODS OF ANALYSIS

11.4.1   General Solution

Match-Point Method

W(partial penetration) (Equation 11.14) is a function of six dimensionless parameters (u, x, l/m, d/m, l_o/m, d_o/m). If the aquifer thickness and screen locations of all wells are known, the only unknown parameters are u and x and the match-point method can be used to solve for K_r, K_z, and S directly. This requires a separate set of type curves for each observation well with different values of r, l_o, or d_o.

The theoretical basis for this method of analysis is described in Chapter 4. The specific steps are:

i   Prepare a plot of W(partial penetration) vs. 1/u on logarithmic paper. This plot is called the type curve (Figure 11.1). The type curve can be plotted using the program TYPE6 (Appendix B). A separate type curve will be required for each observation well with different values of r, l_o, or d_o. The type curves in Figure 11.1 were prepared for an observation well with l/m = 0.3, d/m = 0.05, l_o/m = 0.2, and d_o/m = 0.1.

ii   Plot s vs. t for each observation well using the same logarithmic scales used to prepare the type curve(s). This plot is called the data curve. Each observation well will have a separate data curve.

iii   Overlay the data curve for an observation well on the type curves developed for that well. Shift the plots relative to each other, keeping the respective axes parallel, until a position of best fit is found between the data curve and one of the type curves.

iv   From the best fit portion of the curves, select a match-point and record match-point values W(partial penetration)*, u*, x*, s_{ave}*, and t*.

v   Substitute W(partial penetration)* and s_{ave}* into Equation 11.14 and solve for K_r.

vi   Substitute the computed value of K_r and u* and t* into Equation 11.8 and solve for S.

vii   Compute K_z as

$$K_z = (x^*)^2 \frac{m^2 K}{r^2}$$

viii   Repeat steps iii through vii for each observation well. Average the values of K_r, K_z, and S from all observation wells to determine the effective properties of the aquifer.

Example 11.1

A pumping test was conducted on an aquifer with a saturated thickness of 100 feet. The pumping well was screened between the depths of 5 and 30 feet and the pumping rate was 50 gal/min. Drawdown data for an observation well, screened between the depths of 10 and 20 feet, located 10 feet from the pumping well are in Table 11.2. Compute K_r, K_z, and S.

Table 11.2   Time-drawdown data for Example 11.1

(min)	(ft)	(min)	(ft)
0.1	2.00	40	8.65
0.2	2.93	50	8.78
0.3	3.49	60	8.87
0.4	3.90	70	8.94
0.5	4.21	80	9.00
0.6	4.47	90	9.06
0.7	4.68	100	9.10
0.8	4.86	200	9.38
0.9	5.02	300	9.53
1	5.16	400	9.64
2	6.01	500	9.73
3	6.47	600	9.80
4	6.77	700	9.86
5	7.00	800	9.91
6	7.18	900	9.95
7	7.32	1000	9.99
8	7.45	2000	10.26
9	7.55	3000	10.41
10	7.65	4000	10.52
20	8.20	5000	10.61
30	8.48		

Application of the match-point method for this observation well is in Figure 11.1. The type curves were developed with the program TYPE6 with r = 10 ft, m = 100 ft, l = 30 ft, d = 5 ft, l_o = 20 ft, and d_o = 10 ft. The following match-point values were selected from the aligned curves in Figure 11.1:

$$s_{ave}^* = 2.6 \text{ ft}$$
$$t^* = 0.3 \text{ min}$$
$$W(\text{partial penetration})^* = 10$$
$$1/u^* = 20$$
$$x^* = 0.01$$

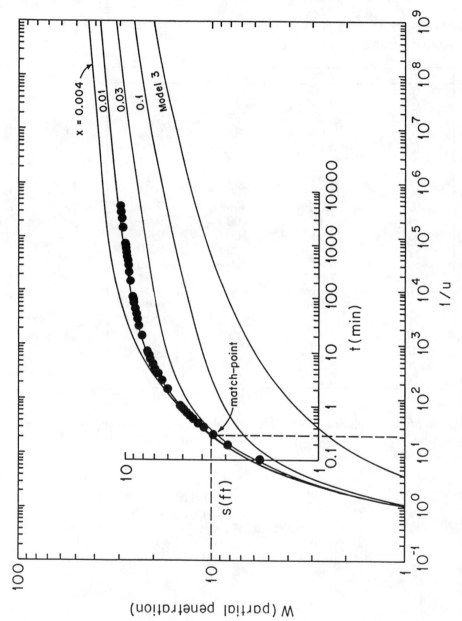

Figure 11.1   Data curve aligned with type curve for the partially-penetrating well of Example 11.1, where $l/m = 0.3$, $d/m = 0.05$, $l_o/m = 0.2$, and $d_o/m = 0.1$.

From Equation 11.14,

$$K_r = \frac{QW(\text{partial penetration})*}{s_{ave}* 4\pi m}$$

$$K_r = \frac{\left(50 \frac{gal}{min}\right) \left(\frac{ft^3}{7.48 \; gal}\right)(10)\left(\frac{1440 \; min}{1 day}\right)}{(2.6 \; ft)\,(4\pi)\,(100 \; ft)} = 29.5 \frac{ft}{day}$$

From Equation 11.8,

$$S = \frac{4\,K_r mt* u*}{r^2}$$

$$S = \frac{(4)\left(29.5 \frac{ft}{day}\right)(100 \; ft)\,(0.3 \; min)\left(\frac{1}{20}\right)\left(\frac{1 \; day}{1440 \; min}\right)}{(10 \; ft)^2} = 0.0012$$

Finally,

$$K_z = (x*)2\frac{m^2 \, K_r}{r^2}$$

$$K_z = \frac{(0.01)^2\,(100 \; ft)^2\left(29.5 \frac{ft}{day}\right)}{(10 \; ft)^2} = 0.30 \frac{ft}{day}$$

11.4.2   Early-Time or Deep Aquifer Solution

Inflection-Point Method

The *inflection-point method* can be used when the ratio of K_r to K_z can be reasonably estimated and when the well system geometry corresponds to the early-time or deep aquifer criteria i.e., when

$$t < \frac{S_s}{20 \, K_z}\left[2m - \frac{1}{2}(2l + l_o + d_o)\right]^2$$

In addition, the geometries of the pumping well and observation well must be such that Equation 11.17 will simplify to a single $M(u, \beta)$ term multiplied by a constant.

The method allows computation of K_r and S_s if only the early-time data are available. The aquifer thickness, m, may also be computed if early- and late-time data are obtained.

From Equation 11.22, $M(u, \beta) \approx W(u)$ when $u > 5/\beta^2$. Stated differently, during the period when $t < (r\beta)^2\,S/(20\,K_r)$, $s = (\text{constant})\,[W(u)]$. In this procedure, β is generally selected by simplifying the geometry of the system so that the four M( ) terms of

Equation 11.17 reduce to one M( ) term. For example, suppose there is a pumping well and an observation well where:

$$l \quad = 60 \text{ ft}$$
$$d \quad = 18 \text{ ft}$$
$$l_o \quad = 25 \text{ ft}$$
$$d_o \quad = 18 \text{ ft}$$
$$K_z \quad = K_r$$

Simplifying,

$$d \approx l/3$$

$$z = \frac{25 + 18}{2} \text{ ft} = 21.5 \text{ ft} \approx l/3$$

therefore, $l \approx 3z$ and $d \approx z$

Then

$$E\left(u, \frac{l}{r}, \frac{d}{r}, \frac{z}{r}\right) = M\left(u, \frac{(1)(3z+z)}{r}\right) + M\left(u, \frac{(1)(3z-z)}{r}\right)$$
$$- M\left(u, \frac{(1)(z+z)}{r}\right) - M\left(u, \frac{(1)(z-z)}{r}\right)$$
$$= M\left(u, \frac{4z}{r}\right)$$

and

$$\beta = \frac{4z}{r}$$

Thus, if early-time data are recorded, a plot of s vs. log(t) may show three distinct portions (Figure 11.2):

1. An initial portion where s ≈ (constant) [W(u)] which plots as a straight line except for the earliest times where the Cooper-Jacob straight line approximation is not valid.

2. An intermediate portion where the effects of partial penetration vary with time.

3. A final portion where the effect of partial penetration is at maximum.

The three portions may not always be visible. For example, if the screen length is large compared to the aquifer thickness, the duration of the first portion may be too short to be observed.

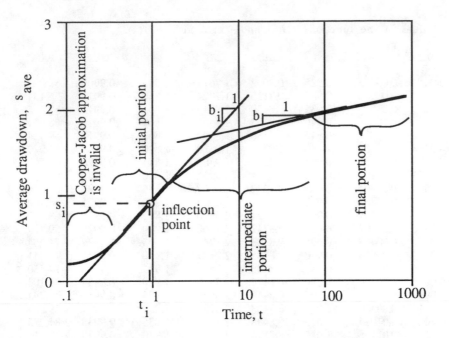

Figure 11.2   Idealized time-drawdown plot for a partially penetrating pumping well.

In general, there will be an inflection point in the plot between the initial and intermediate portions, usually within the period where $t < S_s(2m - r\beta)^2/(20K_r)$. The coordinates of the inflection point, (s_i, t_i), may be computed as follows:

$$t_i = \frac{r^2 S_s}{4 K_r u_i} \tag{11.25}$$

where

$$u_i = \left(\frac{x}{\beta}\right)^2 \tag{11.26}$$

and x is the argument of

$$f(x) = x e^{x^2} \, erf(x) = \frac{\beta^2}{\sqrt{\pi}} \tag{11.27}$$

Values of erf(x) and f(x) are in Table 11.3.

$$s_i = c M(u_i, \beta) \tag{11.28}$$

where

Table 11.3   Values of the functions erf(x) and $f(x) = xe^{x^2}$ erf(x).

x	erf(x)	f(x)	x	erf(x)	f(x)	x	erf(x)	f(x)
0.005	0.0056	0.00003	0.925	0.8092	1.7611	1.845	0.9909	55.002
0.010	0.0113	0.00011	0.930	0.8116	1.7924	1.850	0.9911	56.191
0.015	0.0169	0.00025	0.935	0.8139	1.8242	1.855	0.9913	57.405
0.020	0.0226	0.00045	0.940	0.8163	1.8565	1.860	0.9915	58.651
0.025	0.0282	0.00071	0.945	0.8186	1.8894	1.865	0.9917	59.924
0.030	0.0338	0.00102	0.950	0.8209	1.9229	1.870	0.9918	61.229
0.035	0.0395	0.00138	0.955	0.8232	1.9569	1.875	0.9920	62.563
0.040	0.0451	0.00181	0.960	0.8254	1.9915	1.880	0.9922	63.931
0.045	0.0507	0.00204	0.965	0.8277	2.0267	1.885	0.9923	65.328
0.050	0.0564	0.00283	0.970	0.8299	2.0625	1.890	0.9925	66.762
0.055	0.0620	0.00342	0.975	0.8321	2.0990	1.895	0.9926	68.226
0.060	0.0676	0.00407	0.980	0.8342	2.1360	1.900	0.9928	69.729
0.065	0.0732	0.00478	0.985	0.8368	2.1748	1.905	0.9929	71.264
0.070	0.0789	0.00558	0.990	0.8385	2.2121	1.910	0.9931	72.841
0.075	0.0845	0.00637	0.995	0.8406	2.2509	1.915	0.9932	74.450
0.080	0.0901	0.00725	1.000	0.8427	2.2907	1.920	0.9934	76.103
0.085	0.0957	0.00819	1.005	0.8448	2.3078	1.925	0.9935	77.791
0.090	0.1013	0.00919	1.010	0.8468	2.3721	1.930	0.9937	79.524
0.095	0.1069	0.01024	1.015	0.8488	2.4138	1.935	0.9938	81.295
0.100	0.1125	0.01136	1.020	0.8508	2.4811	1.940	0.9939	83.112
0.105	0.1181	0.01253	1.025	0.8528	2.4995	1.945	0.9941	84.970
0.110	0.1236	0.01384	1.030	0.8548	2.5435	1.950	0.9942	86.916
0.115	0.1292	0.01505	1.035	0.8567	2.5882	1.955	0.9943	88.827
0.120	0.1348	0.01640	1.040	0.8587	2.6338	1.960	0.9944	90.828
0.125	0.1403	0.01782	1.045	0.8606	2.6800	1.965	0.9946	92.873
0.130	0.1459	0.01928	1.050	0.8624	2.7264	1.970	0.9947	94.972
0.135	0.1514	0.02082	1.055	0.8643	2.7752	1.975	0.9948	97.118
0.140	0.1570	0.02241	1.060	0.8661	2.8240	1.980	0.9949	99.322
0.145	0.1625	0.02406	1.065	0.8680	2.8737	1.985	0.9950	101.58
0.150	0.1680	0.02577	1.070	0.8698	2.9242	1.990	0.9951	103.89
0.155	0.1735	0.02755	1.075	0.8716	2.9756	1.995	0.9952	106.26
0.160	0.1790	0.02938	1.080	0.8733	3.0280	2.000	0.9953	108.69
0.165	0.1845	0.03128	1.085	0.8751	3.0813	2.005	0.9954	111.17
0.170	0.1900	0.03325	1.090	0.8768	3.1355	2.010	0.9955	113.72
0.175	0.1955	0.03527	1.095	0.8785	3.1907	2.015	0.9956	116.33
0.180	0.2009	0.03736	1.100	0.8802	3.2470	2.020	0.9957	119.01
0.185	0.2064	0.03951	1.105	0.8819	3.3041	2.025	0.9958	121.75
0.190	0.2118	0.04171	1.110	0.8835	3.3623	2.030	0.9959	124.57
0.195	0.2173	0.04401	1.115	0.8852	3.4215	2.035	0.9960	127.45
0.200	0.2227	0.04636	1.120	0.8868	3.4819	2.040	0.9961	130.40
0.205	0.2281	0.04877	1.125	0.8884	3.5432	2.045	0.9962	133.43
0.210	0.2335	0.05125	1.130	0.8900	3.6058	2.050	0.9963	136.54
0.215	0.2389	0.05380	1.135	0.8915	3.6693	2.055	0.9963	139.72
0.220	0.2443	0.05642	1.140	0.8931	3.7342	2.060	0.9964	142.98
0.225	0.2497	0.05909	1.145	0.8946	3.8002	2.065	0.9965	146.32
0.230	0.2550	0.06183	1.150	0.8961	3.8674	2.070	0.9966	149.76
0.235	0.2604	0.06466	1.155	0.8976	3.9357	2.075	0.9967	153.27
0.240	0.2657	0.06755	1.160	0.8991	4.0055	2.080	0.9967	156.88
0.245	0.2710	0.07050	1.165	0.9006	4.0763	2.085	0.9968	160.58
0.250	0.2763	0.07354	1.170	0.9020	4.1485	2.090	0.9969	164.37
0.255	0.2816	0.07664	1.175	0.9034	4.2220	2.095	0.9970	168.26
0.260	0.2869	0.07981	1.180	0.9048	4.2970	2.100	0.9970	172.25
0.265	0.2922	0.08305	1.185	0.9062	4.3732	2.105	0.9971	176.34
0.270	0.2974	0.08637	1.190	0.9076	4.4509	2.110	0.9972	180.54

Table 11.3   (Continued).

x	erf(x)	f(x)	x	erf(x)	f(x)	x	erf(x)	f(x)
0.275	0.3027	0.08977	1.195	0.9090	4.5300	2.115	0.9972	184.84
0.280	0.3078	0.09324	1.200	0.9103	4.6106	2.120	0.9973	189.25
0.285	0.3131	0.09678	1.205	0.9116	4.6925	2.125	0.9973	193.78
0.290	0.3183	0.10039	1.210	0.9127	4.7748	2.130	0.9974	198.43
0.295	0.3235	0.10409	1.215	0.9143	4.8611	2.135	0.9975	203.17
0.300	0.3286	0.10788	1.220	0.9155	4.9481	2.140	0.9975	208.08
0.305	0.3338	0.11173	1.225	0.9168	5.0365	2.145	0.9976	213.09
0.310	0.3389	0.11566	1.230	0.9181	5.1261	2.150	0.9976	218.24
0.315	0.3440	0.11967	1.235	0.9193	5.2180	2.155	0.9977	223.52
0.320	0.3491	0.12376	1.240	0.9205	5.3115	2.160	0.9978	228.94
0.325	0.3542	0.12794	1.245	0.9217	5.4060	2.165	0.9978	234.49
0.330	0.3593	0.13221	1.250	0.9229	5.5033	2.170	0.9979	240.20
0.335	0.3643	0.13654	1.255	0.9241	5.6026	2.175	0.9979	246.05
0.340	0.3694	0.14098	1.260	0.9252	5.7031	2.180	0.9980	252.06
0.345	0.3744	0.14549	1.265	0.9264	5.8055	2.185	0.9980	258.22
0.350	0.3794	0.15009	1.270	0.9275	5.9103	2.190	0.9981	264.55
0.355	0.3844	0.15477	1.275	0.9286	6.0159	2.195	0.9981	271.04
0.360	0.3893	0.15956	1.280	0.9297	6.1252	2.200	0.9981	277.71
0.365	0.3943	0.16441	1.285	0.9308	6.2353	2.205	0.9982	284.55
0.370	0.3992	0.16938	1.290	0.9319	6.3481	2.210	0.9982	291.58
0.375	0.4041	0.17443	1.295	0.9330	6.4626	2.215	0.9983	298.78
0.380	0.4090	0.17956	1.300	0.9340	6.5798	2.220	0.9983	306.19
0.385	0.4139	0.18479	1.305	0.9350	6.6991	2.225	0.9984	313.78
0.390	0.4187	0.19014	1.310	0.9361	6.8211	2.230	0.9984	321.59
0.395	0.4236	0.19556	1.315	0.9371	6.9452	2.235	0.9984	330.12
0.400	0.4284	0.20109	1.320	0.9381	7.0717	2.240	0.9985	337.82
0.405	0.4332	0.20671	1.325	0.9391	7.1999	2.245	0.9985	346.27
0.410	0.4380	0.21243	1.330	0.9400	7.3318	2.250	0.9985	354.94
0.415	0.4427	0.21826	1.335	0.9410	7.4658	2.255	0.9986	363.85
0.420	0.4475	0.22419	1.340	0.9419	7.6015	2.260	0.9986	373.51
0.425	0.4522	0.23021	1.345	0.9428	7.7406	2.265	0.9986	382.38
0.430	0.4569	0.23636	1.350	0.9438	7.8827	2.270	0.9987	392.03
0.435	0.4616	0.24259	1.355	0.9447	8.0271	2.275	0.9987	401.94
0.440	0.4662	0.24896	1.360	0.9456	8.1751	2.280	0.9987	412.12
0.445	0.4709	0.25542	1.365	0.9464	8.3252	2.285	0.9988	422.57
0.450	0.4755	0.26201	1.370	0.9473	8.4790	2.290	0.9988	433.30
0.455	0.4801	0.26868	1.375	0.9482	8.6349	2.295	0.9988	444.33
0.460	0.4847	0.27546	1.380	0.9490	8.7948	2.300	0.9989	455.66
0.465	0.4892	0.28237	1.385	0.9499	8.9567	2.305	0.9989	467.31
0.470	0.4938	0.28943	1.390	0.9507	9.1228	2.310	0.9989	479.26
0.475	0.4983	0.29658	1.395	0.9515	9.2925	2.315	0.9989	491.55
0.480	0.5028	0.30385	1.400	0.9523	9.4645	2.320	0.9990	504.18
0.485	0.5072	0.31124	1.405	0.9531	9.6399	2.325	0.9990	517.15
0.490	0.5117	0.31874	1.410	0.9539	9.8203	2.330	0.9990	530.48
0.495	0.5161	0.32638	1.415	0.9546	10.003	2.335	0.9990	544.17
0.500	0.5205	0.33416	1.420	0.9554	10.190	2.340	0.9991	558.25
0.505	0.5249	0.34207	1.425	0.9561	10.380	2.345	0.9991	572.71
0.510	0.5292	0.35007	1.430	0.9569	10.575	2.350	0.9991	590.53
0.515	0.5336	0.35825	1.435	0.9576	10.773	2.355	0.9991	602.86
0.520	0.5379	0.36656	1.440	0.9583	10.976	2.360	0.9992	618.56
0.525	0.5422	0.37497	1.445	0.9590	11.182	2.365	0.9992	634.71
0.530	0.5465	0.38357	1.450	0.9597	11.393	2.370	0.9992	651.30
0.535	0.5507	0.39227	1.455	0.9604	11.606	2.375	0.9992	668.36
0.540	0.5549	0.40114	1.460	0.9611	11.826	2.380	0.9992	685.90

Table 11.3   (Continued).

x	erf(x)	f(x)	x	erf(x)	f(x)	x	erf(x)	f(x)
0.545	0.5591	0.41011	1.465	0.9617	12.050	2.385	0.9993	703.92
0.550	0.5633	0.41929	1.470	0.9624	12.278	2.390	0.9993	722.46
0.555	0.5675	0.42855	1.475	0.9630	12.510	2.395	0.9993	741.51
0.560	0.5716	0.43801	1.480	0.9637	12.748	2.400	0.9993	761.12
0.565	0.5757	0.44763	1.485	0.9643	12.990	2.405	0.9993	781.25
0.570	0.5798	0.45738	1.490	0.9649	13.238	2.410	0.9994	801.96
0.575	0.5839	0.46727	1.495	0.9655	13.490	2.415	0.9994	823.27
0.580	0.5879	0.47735	1.500	0.9661	13.750	2.420	0.9994	845.17
0.585	0.5919	0.48760	1.505	0.9667	14.012	2.425	0.9994	867.70
0.590	0.5959	0.49797	1.510	0.9673	14.282	2.430	0.9994	890.86
0.595	0.5999	0.50858	1.515	0.9679	14.554	2.435	0.9994	914.60
0.600	0.6039	0.51931	1.520	0.9684	14.835	2.440	0.9994	939.20
0.605	0.6078	0.53023	1.525	0.9690	15.121	2.445	0.9995	964.41
0.610	0.6117	0.54132	1.530	0.9695	15.414	2.455	0.9995	990.34
0.615	0.6156	0.55260	1.535	0.9701	15.711	2.460	0.9995	1017.0
0.620	0.6194	0.56402	1.540	0.9706	16.015	2.465	0.9995	1044.5
0.625	0.6232	0.57568	1.545	0.9712	16.326	2.465	0.9995	1072.7
0.630	0.6271	0.58754	1.550	0.9716	16.642	2.470	0.9995	1101.7
0.635	0.6308	0.59950	1.555	0.9721	16.965	2.475	0.9995	1131.6
0.640	0.6346	0.61173	1.560	0.9726	17.297	2.480	0.9996	1162.3
0.645	0.6383	0.62412	1.565	0.9731	17.634	2.485	0.9996	1194.0
0.650	0.6420	0.63675	1.570	0.9736	17.979	2.490	0.9996	1226.5
0.655	0.6457	0.64952	1.575	0.9741	18.330	2.495	0.9996	1260.0
0.660	0.6494	0.66256	1.580	0.9746	18.691	2.500	0.9996	1294.5
0.665	0.6530	0.67574	1.585	0.9750	19.058	2.505	0.9996	1330.0
0.670	0.6566	0.68921	1.590	0.9755	19.433	2.510	0.9996	1366.5
0.675	0.6602	0.70288	1.595	0.9759	19.815	2.515	0.9996	1404.0
0.680	0.6638	0.71669	1.600	0.9764	20.208	2.520	0.9996	1442.7
0.685	0.6673	0.73079	1.605	0.9768	20.606	2.525	0.9996	1482.5
0.690	0.6708	0.74514	1.610	0.9772	21.015	2.530	0.9997	1523.5
0.695	0.6743	0.75965	1.615	0.9776	21.433	2.535	0.9997	1565.7
0.700	0.6778	0.77446	1.620	0.9780	21.860	2.540	0.9997	1609.1
0.705	0.6813	0.78949	1.625	0.9784	22.293	2.545	0.9997	1653.8
0.710	0.6847	0.80477	1.630	0.9788	22.739	2.550	0.9997	1699.8
0.715	0.6881	0.82025	1.635	0.9792	23.193	2.555	0.9997	1747.2
0.720	0.6914	0.83606	1.640	0.9796	23.658	2.560	0.9997	1796.0
0.725	0.6948	0.85203	1.645	0.9800	24.132	2.565	0.9997	1846.2
0.730	0.6981	0.86827	1.650	0.9804	24.618	2.570	0.9997	1897.9
0.735	0.7014	0.88480	1.655	0.9808	25.113	2.575	0.9997	1951.2
0.740	0.7047	0.90166	1.660	0.9811	25.619	2.580	0.9997	2006.0
0.745	0.7079	0.91869	1.665	0.9815	26.137	2.585	0.9998	2062.5
0.750	0.7112	0.93612	1.670	0.9818	26.666	2.590	0.9998	2120.7
0.755	0.7144	0.95372	1.675	0.9822	27.205	2.595	0.9998	2180.6
0.760	0.7175	0.97161	1.680	0.9825	27.758	2.600	0.9998	2242.3
0.765	0.7207	0.98985	1.685	0.9828	28.322	2.605	0.9998	2305.9
0.770	0.7238	1.0083	1.690	0.9832	29.900	2.610	0.9998	2371.4
0.775	0.7269	1.0271	1.695	0.9835	29.489	2.615	0.9998	2438.8
0.780	0.7300	1.0464	1.700	0.9838	30.092	2.620	0.9998	2508.3
0.785	0.7331	1.0657	1.705	0.9841	30.709	2.625	0.9998	2579.9
0.790	0.7361	1.0855	1.710	0.9844	31.340	2.630	0.9998	2653.6
0.795	0.7391	1.1055	1.715	0.9847	31.983	2.635	0.9998	2713.8
0.800	0.7421	1.1259	1.720	0.9850	32.642	2.640	0.9998	2807.8
0.805	0.7451	1.1466	1.725	0.9853	33.315	2.645	0.9998	2888.5
0.810	0.7480	1.1677	1.730	0.9856	34.004	2.650	0.9998	2971.6

Table 11.3   (Continued).

x	erf(x)	f(x)	x	erf(x)	f(x)	x	erf(x)	f(x)
0.815	0.7509	1.1891	1.735	0.9859	34.707	2.655	0.9998	3057.3
0.820	0.7538	1.2108	1.740	0.9861	35.429	2.660	0.9998	3145.5
0.825	0.7567	1.2330	1.745	0.9864	36.164	2.665	0.9998	3236.5
0.830	0.7595	1.2554	1.750	0.9867	36.918	2.670	0.9998	3330.2
0.835	0.7623	1.2783	1.755	0.9870	37.688	2.675	0.9999	3426.9
0.840	0.7651	1.3015	1.760	0.9872	38.476	2.680	0.9999	3526.5
0.845	0.7679	1.3251	1.765	0.9874	39.279	2.685	0.9999	3629.1
0.850	0.7707	1.3492	1.770	0.9877	40.105	2.690	0.9999	3734.9
0.855	0.7734	1.3722	1.775	0.9879	40.946	2.695	0.9999	3844.0
0.860	0.7761	1.3970	1.780	0.9882	41.809	2.700	0.9999	3956.5
0.865	0.7788	1.4236	1.785	0.9884	42.689	2.705	0.9999	4072.4
0.870	0.7814	1.4492	1.790	0.9886	43.593	2.710	0.9999	4191.9
0.875	0.7841	1.4753	1.795	0.9889	44.514	2.715	0.9999	4315.1
0.880	0.7867	1.5018	1.800	0.9891	45.460	2.720	0.9999	4442.2
0.885	0.7893	1.5286	1.805	0.9893	46.423	2.725	0.9999	4573.2
0.890	0.7918	1.5561	1.810	0.9895	47.413	2.730	0.9999	4708.3
0.895	0.7944	1.5839	1.815	0.9897	48.422	2.735	0.9999	4847.6
0.900	0.7969	1.6122	1.820	0.9899	49.458	2.740	0.9999	4991.2
0.905	0.7994	1.6410	1.825	0.9901	50.514	2.745	0.9999	5139.4
0.910	0.8019	1.6702	1.830	0.9904	51.600	2.750	0.9999	5292.2
0.915	0.8043	1.7000	1.835	0.9906	52.706			
0.920	0.8068	1.7303	1.840	0.9907	53.843			

$$c = \frac{b_i}{2.3 e^{-u_i} \mathrm{erf}(x)} \tag{11.29}$$

and

$$b_i = \text{slope of the plot at the inflection point}$$
$$= \Delta s/\text{log cycle}$$

The procedure for computing K and S_s is:

i Plot s vs. log(t).

ii Draw a tangent to the curve in the region of the inflection point and measure the slope, b_i, as $\Delta s/\text{log cycle}$.

iii Use the known well system geometry to convert Equations 11.16 and 11.17 to the form

$$s = \frac{Q}{8\pi K_r(l - d)} E(u, \frac{l}{r}, \frac{d}{r}, \frac{z}{r}) = cM(u_i, \beta) \tag{11.30}$$

Some simplifying assumptions may be needed to reduce the four M(u, β) terms of Equation 11.17 to one M(u, β) term.

iv Compute $f(x) = \frac{\beta^2}{\sqrt{\pi}}$ (Equation 11.27).

v Compute x given f(x) from Table 11.2 or from Equation 11.27 ($f(x) = xe^{x^2} erf(x)$).

vi Compute $u_i = \left(\dfrac{x}{\beta}\right)^2$ (Equation 11.26).

vii Compute $c = \dfrac{b_i}{2.3 e^{-u_i} erf(x)}$ (Equation 11.29).

viii Compute K_r by combining Equations 11.29 and 11.30. For example, if c in Equation 11.30 is $c = \dfrac{Q}{8\pi K_r (l - d)}$, then combining Equations 11.29 and 11.30 yields

$$K_r = \dfrac{Q}{8\pi(l - d)c}$$

ix Compute $s_i = cM(u_i, \beta)$ (Equation 11.28). $M(u_i, \beta)$ may be obtained from Table 11.1.

x From the semilogarithmic plot, find the value of t_i corresponding to s_i. Compute S_s as $S_s = \dfrac{4Ku_i t_i}{r^2}$ (Equation 11.25).

xi At relatively late times $[t > 2m - r\beta)^2 S_s/(20K_z)]$, the plot becomes a straight line with slope to given by

$$b = \frac{\Delta s}{\log \text{ cycle}} = 2.3 \frac{Q}{4\pi K_r m} \qquad (11.31)$$

This relationship can be used to check the value of K computed in step viii, or it can be used to compute an unknown value of aquifer thickness m.

Example 11.2

A semilogarithmic plot of data from the pumping test of Example 11.1 is in Figure 11.3. The hydraulic conductivity and storativity of the aquifer are computed as follows:

i The slope of the tangent to the curve at the inflection point is $b_i = 3.21$ ft.

ii Observation well geometry gives

$$z = \frac{20 \text{ ft} + 10 \text{ ft}}{2} = 15 \text{ ft}$$

iii Well system geometry and the assumption of $\sqrt{\dfrac{K_r}{K_z}} = 7$. Substituted into Equation 11.17 yields

$$E\left(u, \frac{l}{r}, \frac{d}{r}, \frac{z}{r}\right) = M\left(u, \frac{(7)\,(30+15)}{10}\right) + M\left(u, \frac{(7)\,(30-15)}{10}\right)$$
$$- M\left(u, \frac{(7)\,(5+15)}{10}\right) - M\left(u, \frac{(7)\,(5-15)}{10}\right)$$

$$= M(u, 31.5) + M(u, 10.5) - M(u, 14) - M(u, -7)$$

Unfortunately, this is not of the form $E(u, \frac{l}{r}, \frac{d}{r}, \frac{z}{r}) = C, M(u, \beta)$, so the solution will only be approximate. To simplify conditions so that the above equation is of the correct form, set idealized well parameters as

$$\begin{aligned}
l &= 30 \text{ ft (unchanged)} \\
d &= 10 \text{ ft (instead of 5 ft)} \\
z &= 10 \text{ ft (instead of 15 ft)}
\end{aligned}$$

then

$$E\left(u, \frac{l}{r}, \frac{d}{r}, \frac{z}{r}\right) = M\left(u, \frac{(7)\,(30+10)}{10}\right) + M\left(u, \frac{(7)\,(30-10)}{10}\right)$$
$$- M\left(u, \frac{(7)\,(10+10)}{10}\right) - M\left(u, \frac{(7)\,(10-10)}{10}\right)$$

$$= M\left(u, \frac{(7)\,(30+10)}{10}\right) = M(u, 28)$$

Now the solution will have two sources of error - one from the assumption of $\sqrt{\frac{K_r}{K_z}}$ and one from the simplification of well geometry.

iv Compute

$$f(x) = \frac{\beta^2}{\sqrt{\pi}} = \frac{28^2}{\sqrt{\pi}} = 442.3$$

v Interpolating from Table 11.3,

$$x = 2.294 \text{ when } f(x) = 442.3$$

vi Compute

$$u_i = \left(\frac{x}{\beta}\right)^2 = \left(\frac{2.178}{28}\right)^2 = 0.0067$$

vii From Table 11.3

$$\text{erf}(x) = \text{erf}(2.294) = 0.9988$$

Then

$$C = \frac{b_i}{2.3e^{-u_i} \, erf(x)}$$

$$= \frac{3.21 \text{ ft}}{2.3(e^{-0.0067}) \, (0.9988)} = 1.4067 \text{ ft}$$

viii From Equation 11.16

$$C = \frac{Q}{8\pi K_r (l - d)}$$

Therefore

$$K_r = \frac{Q}{8\pi (l - d)C}$$

$$= \frac{50 \, \frac{\text{gal}}{\text{min}} \left(\frac{\text{ft}^3}{7.48 \text{ gal}}\right) \left(\frac{1440 \text{ min}}{\text{day}}\right)}{8\pi \, (30 \text{ ft} - 10 \text{ ft}) \, (1.4067 \text{ ft})}$$

$$= 13.6 \, \frac{\text{ft}}{\text{day}}$$

ix From Table 11.1,

$$M(0.0067, 28) = 4.544$$

Substituting,

$$s_i = cM(u_i, \beta) = 1.4069 \text{ ft} \, (4.544) = 6.39 \text{ ft}$$

x From Figure 11.3, when $s_i = 6.39$ ft, $t \approx 2$ min. Therefore,

$$S_s = \frac{4K_r u_i t_i}{r^2} = \frac{4 \left(13.6 \, \frac{\text{ft}}{\text{day}}\right) (0.0067) \, (2 \text{ min}) \left(\frac{\text{day}}{1440 \text{ min}}\right)}{(10 \text{ ft})^2}$$

$$= 4.6 \times 10^{-6}$$

xi From Figure 11.3, the slope, b, after the effect of partial penetration has reached a maximum is 0.90 ft. With this slope and Equation 11.30, either check the computed value of K_r, or determine the aquifer thickness if it was unknown. Assuming the value of m = 100 ft is correct, check K_r.

$$K_r = \frac{2.3Q}{4\pi mb}$$

$$= \frac{2.3 \left(50 \, \frac{\text{gal}}{\text{min}}\right) \left(\frac{\text{ft}^3}{7.48 \text{ gal}}\right) \left(\frac{1440 \text{ min}}{\text{day}}\right)}{4\pi \, (100 \text{ ft}) \, (0.90 \text{ ft})}$$

$$= 19.6 \frac{\text{ft}}{\text{day}}$$

This value of K_r might be more accurate than that calculated in step viii, since no simplifying assumptions have been made to arrive at this value.

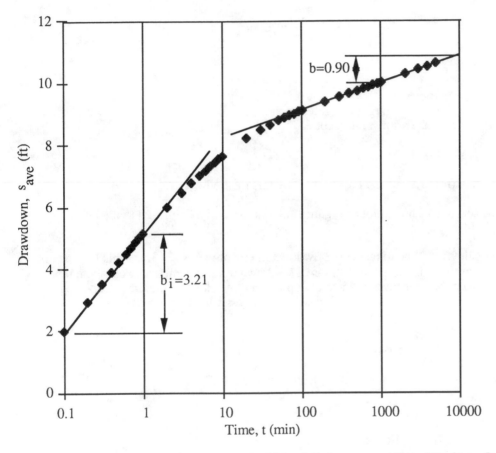

Figure 11.3   Drawdown data from Example 11.1 plotted on a semilogarithmic scale.

11.4.3   Late-Time Solution

Straight-Line Method

This method can be used with late-time data used to find T and S if the ratio of K_z to K_r can be estimated. When $t > \frac{mS}{2K_z}$, the effect of partial penetration is maximum and drawdown is given by Equation 11.23, or

$$s_{ave} = \frac{Q}{4\pi K_r m} [W(u) + f_s]$$

where f_s is defined in Equation 11.24.

Thus, T and S can be computed using a simple variation of the Cooper-Jacob straight-line method (Chapter 4),

$$T = \frac{2.30\,Q}{4\pi b} \tag{11.32}$$

$$S = \frac{2.25\,T}{r^2}\,t_o\,\exp(f_s) \tag{11.33}$$

where

b	= Δs/log cycle
	= slope of a straight line fitted to the data
t_o	= zero drawdown intercept

Example 11.3

Analyze the data from Example 11.1 and 11.2 by this method.

i From the semilogarithmic plot of Figure 11.3, the slope of a line fitted to the late-time data is 0.90 ft.

ii Extending the fitted line to the zero drawdown intercept yields t_o. Since this line cannot be conveniently constructed with Figure 11.3 drawn at its current scale, compute the intercept from the coordinates of a known point and the slope of the line. Choosing point 1 as $s_1 = 0$, $t_1 = t_o$ and point 2 as $s_2 = 10$ ft, $t_2 = 1000$ min,

$$t_o = \frac{t_2}{10^{(s_2/b)}} = \frac{1000\ \text{min}}{10^{(10/0.9)}} = 7.74 \times 10^{-9}\ \text{min}$$

iii Assume that $K_z/K_r = 50$. Compute f_s from Equation 11.24.

$$f_s = 15.23$$

iv Compute T and S from Equations 11.32 and 11.33

$$T = \frac{2.3\left(50\,\frac{\text{gal}}{\text{min}}\right)\left(\frac{\text{ft}^3}{7.48\ \text{gal}}\right)\left(\frac{1440\ \text{min}}{\text{day}}\right)}{4\pi\,(0.90\ \text{ft})} = 1.96 \times 10^3\,\frac{\text{ft}^2}{\text{day}}$$

$$K_r = \frac{T}{m} = \frac{1.96 \times 10^3\,\frac{\text{ft}^2}{\text{day}}}{100\ \text{ft}} = 19.6\ \text{ft/day}$$

$$S = \frac{2.25\left(1.96 \times 10^3\,\frac{\text{ft}^2}{\text{day}}\right)(7.74 \times 10^{-9}\ \text{min})\exp(15.23)\left(\frac{\text{day}}{1440\ \text{min}}\right)}{(10\ \text{ft})^2}$$

$$S = 0.001$$

Chapter 12

MODEL 7: TRANSIENT, CONFINED, LEAKY, PARTIAL PENETRATION, ANISOTROPIC

12.1 CONCEPTUAL MODEL

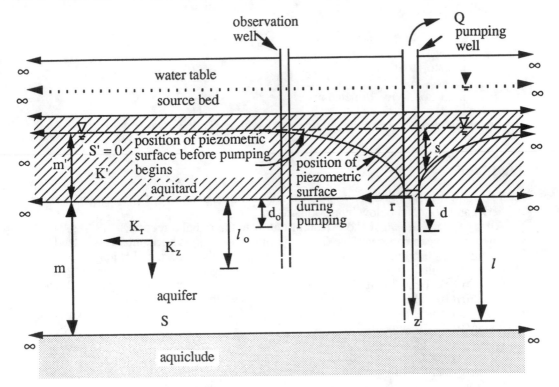

Definition of Terms

d   = distance from aquifer top to top of pumping well screen (or uncased portion of hole), L

d_o = distance from aquifer top to top of observation well screen (or uncased portion of hole), L

K'   = aquitard vertical hydraulic conductivity, LT^{-1}

K_r = aquifer horizontal hydraulic conductivity, LT^{-1}

K_z = aquifer vertical hydraulic conductivity, LT^{-1}

l = distance from aquifer top to bottom of pumping well screen, L

l_o = distance from aquifer top to bottom of observation well screen, L

m   = aquifer thickness, L

m'   = aquitard thickness, L

r   = radial distance from the pumping well to a point on the cone of depression
           (all distances are measured from the center of wells), L

s   = drawdown of piezometric surface during pumping, L

S   = aquifer storativity, dimensionless

S '   = aquitard storativity, dimensionless

z   = vertical distance from the aquifer top (positive z is downward), L

Assumptions

1. The aquifer is bounded above by an aquitard and an unconfined aquifer (the "source bed") and bounded below by an aquiclude.
2. All layers are horizontal and extend infinitely in the radial direction.
3. The initial piezometric surface (before pumping begins) is horizontal and extends infinitely in the radial direction. The water table in the source bed is horizontal, extends infinitely in the radial direction, and remains constant during pumping (zero drawdown). Drawdown of the water table in the source bed can be neglected when $t \leq S'm'/(10\ K')$ (Neuman and Witherspoon, 1969b).
4. The aquifer is homogeneous and anisotropic. The horizontal and vertical hydraulic conductivities may be different.
5. Groundwater density and viscosity are constant.
6. Groundwater flow can be described by Darcy's Law.
7. Groundwater flow in the aquitard is vertical. Groundwater flow in the aquifer is horizontal and directed radially toward the well. This assumption is valid when $K/K' > 100$ (Hantush, 1967).
8. The pumping rate is constant.
9. Head losses through the well screen and pump intake are negligible.
10. The pumping well has an infinitesimal diameter.
11. The aquifer is compressible and completely elastic. The aquitard is incompressible (i.e., no water is released from aquitard storage during pumping). This assumption is valid when $t > 0.036\ m'S'/K'$ (Hantush, 1960).

12.2 MATHEMATICAL MODEL

Governing Equation

The governing equation is derived by combining Darcy's Law for flow in the aquifer and aquitard with the principle of conservation of mass in a radial coordinate system (Hantush, 1964)

$$\left(\frac{K_r}{K_z}\right)^2 \left[\frac{\partial^2 s}{\partial r^2} + \left(\frac{1}{r}\right)\frac{\partial s}{\partial r}\right] + \frac{\partial^2 s}{\partial z^2} - \frac{sK'}{K_z mm'} = \frac{S}{K_z m}\frac{\partial s}{\partial t} \qquad (12.1)$$

where K_r, K_z, and S are the aquifer horizontal and vertical hydraulic conductivities and storativity, respectively, s is drawdown, r is radial distance from the pumping well, z is vertical distance from the aquifer top, t is time, and K' and m' are the aquitard vertical hydraulic conductivity and saturated thickness. The term sK'/m' is the rate of leakage through the aquitard.

Initial Conditions

• Before pumping begins drawdown is zero everywhere

$$s(r, z, t = 0) = 0 \qquad (12.2)$$

Boundary Conditions

• At an infinite distance from the pumping well, drawdown is zero

$$s(r = \infty, z, t) = 0 \qquad (12.3)$$

• There is no vertical flow at the top and bottom of the aquifer

$$\frac{\partial s(r, z = 0, t)}{\partial z} = 0 \qquad (12.4)$$

$$\frac{\partial s(r, z = m, t)}{\partial z} = 0 \qquad (12.5)$$

• Groundwater flow to the pumping well is constant and uniform over the screened interval (groundwater flow near the well is horizontal)

$$\lim_{r \to 0}\left[(l - d)r\frac{\partial s}{\partial r}\right] = 0 \qquad 0 < z < d$$

$$= -\frac{Q}{2\pi K_r} \qquad d < z < l$$

$$= 0 \qquad l < z < m \qquad (12.6)$$

12.3 ANALYTICAL SOLUTION

12.3.1 General Solution

The drawdown at any point in an observation well is (Hantush, 1964)

$$s = \left(\frac{Q}{4\pi K_r m}\right)\left[W(u_r, r/B_r) + f(u_r, x, r/B_r, d/m, l/m, z/m)\right] \qquad (12.7)$$

where

$$u_r = \frac{r^2 S}{4 K_r m t} \tag{12.8}$$

$$B_r^2 = \frac{K_r m m'}{K'} \tag{12.9}$$

$$x = \frac{r}{m} \sqrt{\frac{K_z}{K_r}} \tag{12.10}$$

$W(u_r, r/B_r)$ is the well function for leaky aquifer with fully penetrating wells (Equation 9.9), and

$$f = \left[\frac{2m}{\pi(l-d)} \right] \sum_{n=1}^{\infty} \left(\frac{1}{n} \right) \left[\sin\left(\frac{n\pi l}{m} \right) - \sin\left(\frac{n\pi d}{m} \right) \right] \cos\left(\frac{n\pi z}{m} \right)$$

$$W\left(u_r, \sqrt{(r/B_r)^2 + (n\pi x)^2} \right) \tag{12.11}$$

where

$$W\left(u_r, \sqrt{(r/B_r)^2 + (n\pi x)^2} \right) =$$

$$\int_{u_r}^{\infty} \frac{1}{y} \exp\left[-y - \frac{(r/B)^2 + (n\pi x)^2}{4y} \right] dy \tag{12.12}$$

To compute the average drawdown in an observation well, s_{ave}, Equation 12.7 must be integrated between the screened depths $z = d_o$ and $z = l_o$ and divided by the screen length $(l_o - d_o)$. The result is

$$s_{ave} = \frac{Q}{4\pi K_r m} \left[W(u_r, r/B_r) + F(u_r, x, r/B_r, l/m, d/m, l_o/m, d_o/m) \right] \tag{12.13}$$

where

$$F(u_r, x, r/B_r, l/m, d/m, l_o/m, d_o/m) =$$

$$\left[\frac{2m^2}{\pi^2 (l-d)(l_o - d_o)} \right] \sum_{n=1}^{\infty} \left(\frac{1}{n^2} \right) \left[\sin\left(\frac{n\pi l}{m} \right) - \sin\left(\frac{n\pi d}{m} \right) \right]$$

$$\left[\sin\left(\frac{n\pi l_o}{m} \right) - \sin\left(\frac{n\pi d_o}{m} \right) \right] W\left(u_r, \sqrt{(r/B_r)^2 + (n\pi x)^2} \right) \tag{12.14}$$

For convenience, Equation 12.13 can be written as

$$s_{ave} = \left(\frac{Q}{4\pi K_r m}\right) W(\text{partial penetration}) \tag{12.15}$$

Values of the function W(partial penetration) can be computed with the computer program TYPE7; average drawdown can be computed with the computer program DRAW7 (Appendix B).

12.3.2   Special Case Solutions

Large-Distance Solution

Drawdown data for an observation well located a distance $r > 1.5m \sqrt{\frac{K_r}{K_z}}$ from the pumping well can be interpreted using the methods of analysis for Model 4 (fully penetrating wells) (Hantush, 1964). Beyond this distance, the effects of partial penetration can be neglected as groundwater flow is essentially horizontal (see Section 4.3.2).

Late-Time Solution

When $t > \frac{mS}{2K_z}$ the effect of partial penetration has reached a maximum and is no longer a function of time (Hantush, 1964). Thus, the drawdown at a point is the sum of the leakage component (still time dependent) and constant component due to partial penetration

$$s = \frac{Q}{4\pi K_r m} [W(u_r, r/B_r) + f(x, r/B_r, d/m, l/m, z/m)] \tag{12.16}$$

where

$$f(x, r/B_r, d/m, l/m, z/m) = \frac{4m}{\pi(l - d)} \sum_{n=1}^{\infty} \left(\frac{1}{n}\right)\left[\sin\left(\frac{n\pi l}{m}\right) - \sin\left(\frac{n\pi d}{m}\right)\right]$$

$$\cdot \cos\left(\frac{n\pi z}{m}\right) K_0\left(\sqrt{\left(\frac{r}{B_r}\right)^2 + (n\pi x)^2}\right) \tag{12.17}$$

where $K_0(\ )$ is the zero-order modified Bessel Function of the second kind (Table 9.3).

Equilibrium Solution

As $t \to \infty$, $W(u_r, r/B_r) \to 2K_0(r/B_r)$, and all groundwater flow to the pumping well is derived from aquitard leakage. The solution for drawdown at a point is

$$s = \frac{Q}{2\pi K_r m} [K_0(r/B_r) + f(x, r/B_r, d/m, l/m, z/m)] \tag{12.18}$$

where

$$f = \frac{4m}{\pi} (l - d) \sum_{n = 1}^{\infty} \left(\frac{1}{n}\right) \left[\sin\left(\frac{n\pi l}{m}\right) - \sin\left(\frac{n\pi d}{m}\right) \right] \cos\left(\frac{n\pi z}{m}\right) K_o (n\pi x) \qquad (12.19)$$

12.4 METHODS OF ANALYSIS

The general solution for average drawdown (Equation 12.13) is a function of the pumping rate (Q), pumping and observation well geometries (r, m, l, d, m', l_o, and d_o), and aquifer and aquitard properties (K_r, K_z, S, and K'). The pumping rate is measured during the test and the well geometries are known from drilling and construction records. The aquifer properties are unknown and can be determined using a combination of match-point methods if the necessary drawdown data are available. In general, drawdown data from the pumping well and at least two observation wells are required. One observation well should be partially penetrating and should be located a distance r < 1.5 m $\sqrt{K_r/K_z}$ from the pumping well; the other may be fully penetrating (but this is not necessary) and should be located a distance r > 1.5 m $\sqrt{K_r/K_z}$ from the pumping well. Unless certain simplifying assumptions can be made (see below), the test should be continued long enough to reach equilibrium drawdown in all wells.

The well function for this model (Equation 12.14) is a function of seven dimensionless parameters. Of these four (l/m, d/m, l_o/m, and d_o/m) are known from the drilling and construction records for the pumping and observation wells. The remaining three (u_r, x, and r/B_r) are unknown. Since the match-point method can be used to determine only two of these parameters from drawdown data for a single well, analyses must be performed using drawdown data from two or more observation wells. However, it may be possible to assume the value of certain parameters, thus simplifying the analysis.

For example, it may be possible to estimate the value of $\sqrt{K_z/K_r}$ based on geologic descriptions of the aquifer material. Because r and m are known, this estimate removes the parameter x from the analysis and the parameters u_r and r/B_r can be determined with the match-point method. Similarly, if equilibrium drawdown data are available, the values of aquifer properties K_r, K', and S can be determined by the application of a simpler model (Model 3, 4, or 6). These values can be used to compute the parameter $u_r = r^2 S/(4K_r mt)$ thus eliminating it from the analysis of the transient drawdown data.

12.4.1 General Solution

The theoretical basis for the match-point method is described in Chapter 4. The specific steps are:

i Estimate the value of $\sqrt{K_z/K_r}$. Prepare several plots of W vs. $1/u_r$ on logarithmic paper, one plot for each value of r/B_r. The type curve can be plotted using values computed by the computer program TYPE7 (Appendix B). In most cases the observation well provides average drawdown data so the type curves produced by TYPE7 are computed using Equation 12.15. A separate type curve will be required for each observation well with different values of r, l_o, or d_o.

ii Plot s vs. t (the data curve) using the same logarithmic scales used to prepare the type curves.

iii Overlay the data curve on the type curves. Shift the plots relative to each other, keeping the respective axes parallel, until a position of best fit is found between the data curve and one of the type curves. Try to use as much early data as possible to improve the chances of obtaining a unique fit.

iv From the best fit portions of the curves select a match-point and record match-point values W^*, $1/u_r^*$, t^*, s^*, and r/B_r^*.

v Substitute s^* and W^* into Equation 12.15 and solve for K_r.

vi Substitute the computed value of K_r and u_r^* and t^* into Equation 12.8 and solve for S.

vii Substitute the computed value of K_r and r/B_r^* into Equation 12.9 and solve for K'.

The above procedure may be repeated with another family of type curves developed for a different assumed value of K_z/K_r. Theoretically, the data curves should have a unique fit with one curve only. Practically, real aquifers do not conform to the ideal model specifications and it is very difficult to discern the slight changes in curve shapes with differing aquifer properties. Therefore, it is often the best course to estimate certain properties (such as K_z/K_r) which have little effect on the computation of T or S or to use data from a time range or distance where the number of dimensionless parameters can be reduced.

Example 11.1 in the previous chapter illustrates a procedure very similar to that described above. In that example, $r/B_r = 0$, so instead of estimating K_z/K_r, different type curves are developed so the ratio can be determined directly.

12.4.2   Late-Time Solution

When $\dfrac{8m'S}{K'} > t > \dfrac{mS}{2K_z}$, the effects of partial penetration are constant, but flow is still derived from storage within the aquifer as well as aquitard leakage. Thus, the change in drawdown over time is determined from the sum of the leaky well function of Model 4 and a constant, as shown in Equation 12.16.

This method can be used to find K_r, S, and K' as long as K_z/K_r can be estimated. Since the effects of partial penetration are no longer time dependent K_z/K_r cannot be determined.

The method follows the general match-point procedure described in detail in Part I. The specific procedure is as follows:

i Prepare a family of type curves by plotting $W(u_r, r/B_r)$ vs. $1/u_r$ on a logarithmic scale. Plot curves for several different values of r/B_r. $W(u_r, r/B_r)$ is the leaky well function of Model 4.

ii Plot s vs. t (or s vs. t/r^2 for more than one well) on a scale identical to the type curve. This plot is called the data curve.

iii Overlay the data curve on the family of type curves. Shift the plots relative to each other, keeping the axes parallel, until a position of best fit is found between the data curve and one of the type curves. Remember to use only late-time data when choosing the best alignment.

iv With the curves aligned, select a match-point and record match-point values $W(u_r, r/B_r)^*$, $1/u_r^*$, s^*, t^* (or $(t/r^2)^*$), and $(r/B_r)^*$.

v Substitute $(r/B_r)^*$ and an estimated value of K_z/K_r into Equation 12.17 and compute f.

vi Substitute s^*, $W(u_r, r/B_r)^*$, and the computed value of f into Equation 12.16 and solve for K_r.

vii Substitute u_r^*, t^* (or $(t/r^2)^*$), and the computed value of K_r into Equation 12.8 and solve for S.

viii Solve for K' using Equation 12.9 rewritten as

$$K' = \frac{K_r m m' \ [(r/B_r)^*]^2}{r^2} \qquad (12.20)$$

In reality, it is very difficult to discern between the shapes of the different curves in this small time range and a unique fit is difficult to find. However, the solution is still valuable for predicting drawdown at late-times.

12.4.3 Equilibrium Solution

Equilibrium conditions can be assumed when $t > \frac{8m'S}{K'}$. If equilibrium drawdown data are available from at least three wells and α can be estimated, the following method can be used to compute K_r and K'.

i Prepare a plot of $K_o(r/B_r)$ vs. r/B_r on logarithmic paper. This plot is called the type curve and can be prepared using the values of $K_o(x)$ in Table 9.3.

ii Plot s vs. r (the data curve) using the same logarithmic scales used to prepare the type curve.

iii Overlay the data curve on the type curve. Shift the plots relative to each other, keeping respective axes parallel, until a position of best fit is found.

iv From the best fit portion of the curves, select match-point values for s^*, r^*, $K_o(r/B_r)^*$, r/B_r^*.

v Estimate a value for K_z/K_r and compute f using Equation 12.19.

vi Substitute s^*, $K_o(r/B_r)^*$, and the computed value of f into Equation 12.18 and solve for K_r.

vii Substitute r^* and $(r/B_r)^*$ into Equation 12.20 and solve for K'

Chapter 13

MODEL 8: TRANSIENT, CONFINED, WELL STORAGE

13.1 CONCEPTUAL MODEL

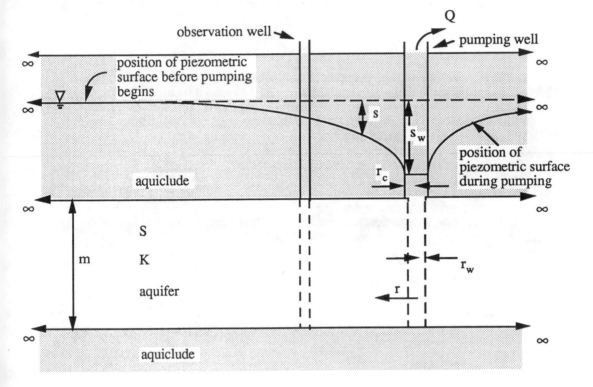

Definition of Terms

K = aquifer hydraulic conductivity, LT^{-1}

m = aquifer thickness, L

Q = constant pumping rate, L^3T^{-1}

r = radial distance from pumping well to a point on the cone of depression (all distances are measured from the center of wells), L

r_c = inside radius of pumping well casing within the range of water level fluctuation, L

r_w = effective radius of the well bore or open hole, L
        = borehole radius if the filter is much more permeable than the aquifer
        = screen radius if no filter is used or if the filter has a hydraulic conductivity similar
          to that of the aquifer

s   = drawdown of piezometric surface during pumping, L

s_w = drawdown in the pumping well, L

S   = aquifer storativity, dimensionless

Assumptions

1. The aquifer is bounded above and below by aquicludes.
2. All layers are horizontal and extend infinitely in the radial direction.
3. The initial piezometric surface (before pumping begins) is horizontal and extends
   infinitely in the radial direction.
4. The aquifer is homogeneous and isotropic.
5. Groundwater density and viscosity are constant.
6. Groundwater flow can be described by Darcy's Law.
7. Groundwater flow is horizontal and is directed radially toward the well.
8. The pumping and observation wells are screened over the entire aquifer thickness.
9. The pumping rate is constant.
10. Head losses through the well screen and pump intake are negligible.
11. The aquifer is compressible and completely elastic.

13.2 MATHEMATICAL MODEL

Governing Equation

The governing equation is derived by combining Darcy's Law with the principle of
conservation of mass (for the aquifer and well casing) in a radial coordinate system
(Papadopulos and Cooper, 1967)

$$\frac{\partial^2 s}{\partial r^2} + \left(\frac{1}{r}\right)\frac{\partial s}{\partial r} = \left(\frac{S}{T}\right)\frac{\partial s}{\partial t}, \quad r \geq r_w \tag{13.1}$$

where s is drawdown, r is radial distance from the pumping well, S and T = Km are the
storativity and transmissivity of the aquifer, t is time, and r_w is the effective radius of the
pumping well.

Initial Conditions

• Before pumping begins, drawdown is zero in the aquifer and pumping well

$$s(r, t = 0) = 0, \quad r \geq r_w \tag{13.2}$$

$$s_w(t = 0) = 0 \tag{13.3}$$

where s_w is the drawdown in the pumping well.

Boundary Conditions

- The drawdown in the aquifer at the face of the well is equal to the drawdown in the well

$$s(r = r_w, t) = s_w(t) \tag{13.4}$$

- At an infinite distance from the pumping well drawdown is zero

$$s(r = \infty, t) = 0 \tag{13.5}$$

- The pumping rate, Q, is the sum of groundwater entering the pumping well from the aquifer and the rate of decrease of water stored in the well casing

$$Q = -2\pi r_w T \frac{\partial s(r_w,t)}{\partial r} + \pi r_c^2 \frac{\partial s_w(t)}{\partial t}, \quad t > 0 \tag{13.6}$$

The first term on the right hand side of Equation 13.6 will be positive for flow from the aquifer to the well $\left(\dfrac{\partial s(r_w,t)}{\partial r} < 0 \right)$ and the second term will be positive for a decrease in water stored in the well casing $\left(\dfrac{\partial s_w(t)}{\partial t} > 0 \right)$.

13.3 ANALYTICAL SOLUTION

13.3.1 General Solution

The solution for drawdown in the pumping well, s_w, is (Papadopulos and Cooper, 1967)

$$s_w = \left(\frac{Q}{4\pi T} \right) F(u_w, \alpha) \tag{13.7}$$

where

$$u_w = \frac{r_w^2 S}{4Tt} \tag{13.8}$$

$$\alpha = \frac{r_w^2 S}{r_c^2} \tag{13.9}$$

and $F(u_w, \alpha)$ is a well function

$$F(u_w, \alpha) = \frac{32\alpha^2}{\pi^2} \int_0^\infty \frac{(1 - \exp(-\beta^2/(4u_w)))}{\beta^3 \, \Delta(\beta)} \, d\beta \tag{13.10}$$

where

$$\Delta(\beta) = [\beta J_o(\beta) - 2\alpha J_1(\beta)]^2 + [\beta Y_o(\beta) - 2\alpha Y_1(\beta)]^2 \qquad (13.11)$$

J_o = Bessel function of the first kind and zero order
J_1 = Bessel function of the first kind and first order
Y_o = Bessel function of the second kind and zero order
Y_1 = Bessel function of the second kind and first order
β = dummy variable for integration

Selected values of the function $F(u_w, \alpha)$ are in Table 13.1.

Table 13.1. Values of the well function $F(u_w, \alpha)$ for an isotropic, fully-penetrated nonleaky aquifer with well storage (after Papadopulos and Cooper, 1967, p. 243).

u_w	$\alpha = 10^{-1}$	$\alpha = 10^{-2}$	$\alpha = 10^{-3}$	$\alpha = 10^{-4}$	$\alpha = 10^{-5}$
10	9.755×10^{-3}	9.976×10^{-4}	9.998×10^{-5}	1.000×10^{-5}	1.000×10^{-6}
1	9.192×10^{-2}	9.914×10^{-3}	9.991×10^{-4}	1.000×10^{-4}	1.000×10^{-5}
5×10^{-1}	1.767×10^{-1}	1.974×10^{-2}	1.997×10^{-3}	2.000×10^{-4}	2.000×10^{-5}
2	4.062×10^{-1}	4.890×10^{-2}	4.989×10^{-3}	4.999×10^{-4}	5.000×10^{-5}
1	7.336×10^{-1}	9.665×10^{-2}	9.966×10^{-3}	9.997×10^{-4}	1.000×10^{-4}
5×10^{-2}	1.260×10^{0}	1.896×10^{-1}	1.989×10^{-2}	1.999×10^{-3}	2.000×10^{-4}
2	2.303×10^{0}	4.529×10^{-1}	4.949×10^{-2}	4.995×10^{-3}	5.000×10^{-4}
1	3.276×10^{0}	8.520×10^{-1}	9.834×10^{-2}	9.984×10^{-3}	1.000×10^{-3}
5×10^{-3}	4.255×10^{0}	1.540×10^{0}	1.945×10^{-1}	1.994×10^{-2}	2.000×10^{-3}
2	5.420×10^{0}	3.043×10^{0}	4.725×10^{-1}	4.972×10^{-2}	4.998×10^{-3}
1	6.212×10^{0}	4.545×10^{0}	9.069×10^{-1}	9.901×10^{-2}	9.992×10^{-3}
5×10^{-4}	6.960×10^{0}	6.031×10^{0}	1.688×10^{0}	1.965×10^{-1}	1.997×10^{-2}
2	7.866×10^{0}	7.557×10^{0}	3.523×10^{0}	4.814×10^{-1}	4.982×10^{-2}
1	8.572×10^{0}	8.443×10^{0}	5.526×10^{0}	9.340×10^{-1}	9.932×10^{-2}
5×10^{-5}	9.318×10^{0}	9.229×10^{0}	7.631×10^{0}	1.768×10^{0}	1.975×10^{-1}
2	1.024×10^{1}	1.020×10^{1}	9.676×10^{0}	3.828×10^{0}	4.861×10^{-1}
1	1.093×10^{1}	1.087×10^{1}	1.068×10^{1}	6.245×10^{0}	9.493×10^{-1}
5×10^{-6}	1.163×10^{1}	1.162×10^{1}	1.150×10^{1}	8.991×10^{0}	1.817×10^{0}
2	1.255×10^{1}	1.254×10^{1}	1.249×10^{1}	1.174×10^{1}	4.033×10^{0}
1	1.324×10^{1}	1.324×10^{1}	1.321×10^{1}	1.291×10^{1}	6.779×10^{0}
5×10^{-7}	1.393×10^{1}	1.393×10^{1}	1.392×10^{1}	1.378×10^{1}	1.013×10^{1}
2	1.485×10^{1}	1.485×10^{1}	1.484×10^{1}	1.479×10^{1}	1.371×10^{1}
1	1.554×10^{1}	1.554×10^{1}	1.554×10^{1}	1.551×10^{1}	1.513×10^{1}
5×10^{-8}	1.623×10^{1}	1.623×10^{1}	1.623×10^{1}	1.622×10^{1}	1.605×10^{1}
2	1.705×10^{1}	1.705×10^{1}	1.705×10^{1}	1.714×10^{1}	1.708×10^{1}
1	1.784×10^{1}	1.784×10^{1}	1.784×10^{1}	1.784×10^{1}	1.781×10^{1}
5×10^{-9}	1.854×10^{1}	1.854×10^{1}	1.854×10^{1}	1.854×10^{1}	1.851×10^{1}
2	1.945×10^{1}	1.945×10^{1}	1.945×10^{1}	1.945×10^{1}	1.940×10^{1}
1	2.015×10^{1}	2.015×10^{1}	2.015×10^{1}	2.015×10^{1}	2.015×10^{1}

The solution for drawdown in the aquifer is (Papadopulos, 1967; Papadopulos and Cooper, 1967)

$$s = \frac{Q}{4\pi T} F(u, \alpha, \rho) \tag{13.12}$$

where

$$u = \frac{r^2 S}{4Tt} \tag{13.13}$$

and

$$\rho = r/r_w \tag{13.14}$$

$$F(u, \alpha, \rho) = \frac{8\alpha}{\pi} \int_o^\infty [1 - \exp{(-\beta^2 \rho^2/(4u))}]\{J_o(\beta\rho)\,[\beta Y_o(\beta) - 2\alpha Y_1(\beta)]$$
$$- Y_o(\beta\rho)\,[\beta J_o(\beta) - 2\alpha J_1(\beta)]\} \frac{d\beta}{\beta^2 \Delta(\beta)} \tag{13.15}$$

and α, J_o, J_1, Y_o, and Y_1 are as defined earlier. It should be noted that for $\rho = 1$

$$F(u_w, \alpha) = F(u, \alpha, \rho = 1) \tag{13.16}$$

Selected values of the function $F(u, \alpha, \rho)$ are in Table 13.2. Values of $F(u, \alpha, \rho)$ can also be computed with the computer program TYPE8; drawdown can be computed with the computer program DRAW8 (Appendix B).

13.3.2   Special Case Solutions

The effects of well storage on drawdown can be neglected if either $t > \dfrac{2.5 \times 10^3\, r_c^2}{T}$ or $r > 300 r_w$ (Papadopulos and Cooper, 1967). In this case the Model 3 solution can be used.

For a short period of time after pumping begins, all water pumped is derived from well storage. When u_w is sufficiently large, $F(u_w, \alpha)$ vs. u_w plots as a straight line and

$$s_w = \frac{Q\alpha}{4\pi T u_w} \tag{13.17}$$

$$= \frac{Q\,t}{\pi r_c^2} = \frac{\text{Volume of water discharged}}{\text{Area of well casing}} \tag{13.18}$$

Table 13.2   Values of the confined aquifer with well storage function, $F(u, \alpha, \rho)$, for various values of α (from Papadopulos, 1967).

Values of the function $F(u, \alpha, \rho)$ for $\alpha = 10^{-1}$

u	$\rho = 1$	$\rho = 2$	$\rho = 5$	$\rho = 10$	$\rho = 20$	$\rho = 50$	$\rho = 100$	$\rho = 200$
2×10^0	4.88×10^{-2}	1.96×10^{-2}	1.75×10^{-2}	2.41×10^{-2}	3.48×10^{-2}	4.24×10^{-2}	4.48×10^{-2}	4.50×10^{-2}
1	9.19×10^{-2}	7.01×10^{-2}	9.55×10^{-2}	1.41×10^{-1}	1.85×10^{-1}	2.09×10^{-1}	2.14×10^{-1}	2.15×10^{-1}
5×10^{-1}	1.77×10^{-1}	1.95×10^{-1}	3.21×10^{-1}	4.44×10^{-1}	5.20×10^{-1}	5.49×10^{-1}	5.55×10^{-1}	5.59×10^{-1}
2	4.06×10^{-1}	5.78×10^{-1}	9.42×10^{-1}	1.13×10^0	1.19×10^0	1.22×10^0		
1	7.34×10^{-1}	1.11×10^0	1.60×10^0	1.76×10^0	1.80×10^0			
5×10^{-2}	1.26×10^0	1.84×10^0	2.33×10^0	2.43×10^0	2.46×10^0			
2	2.30×10^0	2.97×10^0	3.28×10^0	3.34×10^0	3.35×10^0			
1	3.28×10^0	3.81×10^0	4.00×10^0	4.03×10^0				
5×10^{-3}	4.26×10^0	4.60×10^0	4.70×10^0	4.72×10^0				
2	5.42×10^0	5.58×10^0	5.63×10^0	5.64×10^0				
1	6.21×10^0	6.30×10^0	6.33×10^0					
5×10^{-4}	6.96×10^0	7.01×10^0						
2	7.87×10^0	7.93×10^0						
1	8.57×10^0	8.63×10^0						
5×10^{-5}	9.32×10^0							
2	10.24×10^0							

Table 13.2 (Continued).

Values of the function $F(u, \alpha, \rho)$ for $\alpha = 10^{-2}$

u	$\rho = 1$	$\rho = 2$	$\rho = 5$	$\rho = 10$	$\rho = 20$	$\rho = 50$	$\rho = 100$	$\rho = 200$
2×10^0	4.99×10^{-3}	2.13×10^{-3}	2.11×10^{-3}	3.52×10^{-3}	7.47×10^{-3}	2.03×10^{-2}	3.44×10^{-2}	4.35×10^{-2}
1	9.91×10^{-3}	7.99×10^{-3}	1.32×10^{-2}	2.69×10^{-2}	6.12×10^{-2}	1.42×10^{-1}	1.91×10^{-1}	2.11×10^{-1}
5×10^{-1}	1.97×10^{-2}	2.40×10^{-2}	5.40×10^{-2}	1.21×10^{-1}	2.63×10^{-1}	4.65×10^{-1}	5.31×10^{-1}	5.51×10^{-1}
2	4.89×10^{-2}	8.34×10^{-2}	2.33×10^{-1}	5.12×10^{-1}	9.15×10^{-1}	1.16×10^0	1.20×10^0	1.22×10^0
1	9.67×10^{-2}	1.93×10^{-1}	5.67×10^0	1.12×10^0	1.58×10^0	1.78×10^0	1.81×10^0	
5×10^{-2}	1.90×10^{-1}	4.16×10^{-1}	1.18×10^0	1.95×10^0	2.32×10^0	2.44×10^0	2.46×10^0	
2	4.53×10^{-1}	1.03×10^0	2.42×10^0	3.11×10^0	3.29×10^0	3.34×10^0	3.35×10^0	
1	8.52×10^{-1}	1.87×10^0	3.48×10^0	3.90×10^0	4.00×10^0	4.03×10^0		
5×10^{-3}	1.54×10^0	3.05×10^0	4.43×10^0	4.65×10^0	4.71×10^0	4.72×10^0		
2	3.04×10^0	4.78×10^0	5.52×10^0	5.61×10^0	5.63×10^0	5.64×10^0		
1	4.55×10^0	5.90×10^0	6.27×10^0	6.31×10^0	6.33×10^0			
5×10^{-4}	6.03×10^0	6.81×10^0	6.99×10^0	7.01×10^0				
2	7.56×10^0	7.85×10^0	7.92×10^0	7.94×10^0				
1	8.44×10^0	8.59×10^0	8.63×10^0					
5×10^{-5}	9.23×10^0	9.30×10^0						
2	10.20×10^0	10.23×10^0						
1	10.87×10^0	10.93×10^0						
5×10^{-6}	11.62×10^0	11.63×10^0						
2	12.54×10^0	10.23×10^0						
1	13.24×10^0	10.93×10^0						

Table 13.2 (Continued).

Values of the function $F(u, \alpha, \rho)$ for $\alpha = 10^{-3}$

u	$\rho = 1$	$\rho = 2$	$\rho = 5$	$\rho = 10$	$\rho = 20$	$\rho = 50$	$\rho = 100$	$\rho = 200$
2×10^0	5.00×10^{-4}	2.15×10^{-4}	2.15×10^{-4}	3.70×10^{-4}	8.35×10^{-4}	3.05×10^{-3}	8.38×10^{-3}	1.50×10^{-2}
1	9.99×10^{-4}	8.11×10^{-4}	1.37×10^{-3}	2.95×10^{-3}	7.58×10^{-3}	2.81×10^{-2}	7.56×10^{-2}	1.47×10^{-1}
5×10^{-1}	2.00×10^{-3}	2.45×10^{-3}	5.77×10^{-3}	1.42×10^{-2}	3.90×10^{-2}	1.54×10^{-1}	3.23×10^{-1}	4.78×10^{-1}
2	4.99×10^{-3}	8.71×10^{-3}	2.67×10^{-2}	7.24×10^{-2}	2.03×10^{-1}	6.59×10^0	1.02×10^0	1.17×10^0
1	9.97×10^{-3}	2.07×10^{-2}	7.16×10^{-2}	2.01×10^{-1}	5.41×10^0	1.38×10^0	1.70×10^0	1.79×10^0
5×10^{-2}	1.99×10^{-2}	4.66×10^{-2}	1.74×10^{-1}	4.87×10^{-1}	1.19×10^0	2.27×10^0	2.40×10^0	2.45×10^0
2	4.95×10^{-2}	1.29×10^{-1}	5.05×10^{-1}	1.31×10^0	2.52×10^0	3.22×10^0	3.32×10^0	3.35×10^0
1	9.83×10^{-2}	2.70×10^{-1}	1.04×10^0	2.38×10^0	3.59×10^0	3.96×10^0	4.02×10^0	
5×10^{-3}	1.95×10^{-1}	5.47×10^{-1}	1.96×10^0	3.68×10^0	4.50×10^0	4.69×10^0	4.72×10^0	
2	4.73×10^{-1}	1.31×10^0	3.81×10^0	5.23×10^0	5.55×10^0	5.63×10^0	5.64×10^0	
1	9.07×10^{-1}	2.39×10^0	5.34×10^0	6.13×10^0	6.28×10^0	6.32×10^0		
5×10^{-4}	1.69×10^0	3.98×10^0	6.57×10^0	6.92×10^0	7.00×10^0	7.02×10^0		
2	3.52×10^0	6.44×10^0	7.77×10^0	7.90×10^0	7.93×10^0			
1	5.53×10^0	7.95×10^0	8.55×10^0	8.61×10^0	8.63×10^0			
5×10^{-5}	7.63×10^0	9.02×10^0	9.08×10^0	9.31×10^0				
2	9.68×10^0	10.12×10^0	10.22×10^0	10.24×10^0				
1	10.68×10^0	10.88×10^0	10.93×10^0					
5×10^{-6}	11.50×10^0	11.95×10^0	11.62×10^0					
2	12.49×10^0	12.53×10^0	12.54×10^0					
1	13.21×10^0	13.23×10^0	13.24×10^0					
5×10^{-7}	13.92×10^0	13.93×10^0						
2	14.84×10^0							
1	15.54×10^0							

Table 13.2 (Continued).

Values of the function $F(u, \alpha, \rho)$ for $\alpha = 10^{-4}$

u	$\rho = 1$	$\rho = 2$	$\rho = 5$	$\rho = 10$	$\rho = 20$	$\rho = 50$	$\rho = 100$	$\rho = 200$
2×10^0	5.00×10^{-5}	2.17×10^{-5}	2.18×10^{-5}	3.73×10^{-5}	8.46×10^{-5}	3.16×10^{-4}	9.56×10^{-4}	3.83×10^{-3}
1	1.00×10^{-4}	8.15×10^{-5}	1.38×10^{-4}	2.98×10^{-4}	7.77×10^{-4}	3.23×10^{-3}	1.01×10^{-2}	3.42×10^{-2}
5×10^{-1}	2.00×10^{-4}	2.47×10^{-4}	5.81×10^{-4}	1.45×10^{-3}	4.10×10^{-3}	1.80×10^{-2}	5.62×10^{-2}	1.75×10^{-1}
2	5.00×10^{-4}	8.76×10^{-3}	2.71×10^{-3}	7.54×10^{-3}	2.27×10^{-2}	1.03×10^{-1}	3.04×10^{-1}	7.10×10^{-1}
1	1.00×10^{-3}	2.09×10^{-3}	7.34×10^{-3}	2.16×10^{-2}	6.69×10^{-2}	2.97×10^{-1}	7.92×10^{-1}	1.43×10^0
5×10^{-2}	2.00×10^{-3}	4.72×10^{-3}	1.82×10^{-2}	5.55×10^{-2}	1.74×10^{-1}	7.30×10^{-1}	1.62×10^0	2.24×10^0
2	5.00×10^{-3}	1.32×10^{-2}	5.56×10^{-2}	1.74×10^{-1}	5.36×10^{-1}	1.87×10^0	2.95×10^0	3.28×10^0
1	9.98×10^{-3}	2.81×10^{-2}	1.23×10^{-1}	3.86×10^{-1}	1.14×10^0	3.08×10^0	3.84×10^0	4.02×10^0
5×10^{-3}	1.99×10^{-2}	5.88×10^{-2}	2.64×10^{-1}	8.13×10^{-1}	2.17×10^0	4.25×10^0	4.63×10^0	4.71×10^0
2	4.97×10^{-2}	1.53×10^{-1}	6.89×10^{-1}	1.97×10^0	4.14×10^0	5.47×10^0	5.60×10^0	5.63×10^0
1	9.90×10^{-2}	3.10×10^0	1.36×10^0	3.44×10^0	5.61×10^0	6.24×10^0	6.31×10^0	6.33×10^0
5×10^{-4}	1.97×10^{-1}	6.18×10^0	2.53×10^0	5.26×10^0	6.71×10^0	6.98×10^0	7.01×10^0	
2	4.81×10^{-1}	1.48×10^0	4.95×10^0	7.33×10^0	7.82×10^0	7.92×10^0	7.94×10^0	
1	9.34×10^{-1}	2.72×10^0	7.03×10^0	8.37×10^0	8.57×10^0	8.62×10^0		
5×10^{-5}	1.77×10^0	4.65×10^0	8.65×10^0	9.20×10^0	9.29×10^0	9.32×10^0		
2	3.83×10^0	7.87×10^0	10.02×10^0	10.19×10^0	10.23×10^0	10.24×10^0		
1	6.25×10^0	9.92×10^0	10.83×10^0	10.91×10^0	10.93×10^0			
5×10^{-6}	8.99×10^0	11.23×10^0	11.57×10^0	11.62×10^0	11.63×10^0			
2	11.74×10^0	12.40×10^0	12.52×10^0	12.54×10^0				
1	12.91×10^0	13.17×10^0	13.23×10^0	13.24×10^0				
5×10^{-7}	13.78×10^0	13.90×10^0	13.93×10^0					
2	14.79×10^0	14.83×10^0						
1	15.51×10^0	15.53×10^0						
5×10^{-8}	16.22×10^0	16.23×10^0						
2	17.14×10^0							
1	17.48×10^0							

Table 13.2 (Continued).

Values of the function $F(u, \alpha, \rho)$ for $\alpha = 10^{-5}$

u	$\rho = 1$	$\rho = 2$	$\rho = 5$	$\rho = 10$	$\rho = 20$	$\rho = 50$	$\rho = 100$	$\rho = 200$
2×10^0	5.00×10^{-6}	2.27×10^{-6}	2.48×10^{-6}	4.19×10^{-6}	9.00×10^{-6}	3.21×10^{-5}	9.77×10^{-5}	3.15×10^{-4}
1	1.00×10^{-5}	8.36×10^{-6}	1.44×10^{-5}	3.07×10^{-5}	7.89×10^{-5}	3.27×10^{-4}	1.04×10^{-3}	3.44×10^{-3}
5×10^{-1}	2.00×10^{-5}	2.51×10^{-5}	5.94×10^{-5}	1.47×10^{-4}	4.14×10^{-4}	1.84×10^{-3}	6.02×10^{-3}	2.00×10^{-2}
2	5.00×10^{-5}	8.87×10^{-5}	2.74×10^{-4}	7.61×10^{-4}	2.31×10^{-3}	1.08×10^{-2}	3.61×10^{-2}	1.19×10^{-1}
1	1.00×10^{-4}	2.11×10^{-4}	7.42×10^{-4}	2.18×10^{-3}	6.85×10^{-3}	3.30×10^{-2}	1.10×10^{-1}	3.50×10^{-1}
5×10^{-2}	2.00×10^{-4}	4.77×10^{-4}	1.84×10^{-3}	5.56×10^{-3}	1.82×10^{-2}	8.90×10^{-2}	2.92×10^{-1}	8.57×10^{-1}
2	5.00×10^{-4}	1.34×10^{-3}	5.64×10^{-3}	1.80×10^{-2}	5.92×10^{-2}	2.89×10^{-1}	8.91×10^{-1}	2.12×10^0
1	1.00×10^{-3}	2.84×10^{-3}	1.26×10^{-2}	4.09×10^{-2}	1.36×10^{-1}	6.49×10^{-1}	1.80×10^0	3.34×10^0
5×10^{-3}	2.00×10^{-3}	5.96×10^{-3}	2.74×10^{-2}	9.03×10^{-2}	3.01×10^{-1}	1.35×10^0	3.14×10^0	4.40×10^0
2	5.00×10^{-3}	1.56×10^{-2}	7.43×10^{-2}	2.47×10^{-1}	8.06×10^{-1}	3.03×10^0	5.01×10^0	5.52×10^0
1	9.99×10^{-3}	3.20×10^{-2}	1.55×10^{-1}	5.15×10^{-1}	1.60×10^0	4.75×10^0	6.06×10^0	6.27×10^0
5×10^{-4}	2.00×10^{-2}	6.54×10^{-2}	3.20×10^{-1}	1.04×10^0	2.96×10^0	6.31×10^0	6.90×10^0	6.99×10^0
2	4.98×10^{-2}	1.66×10^{-1}	8.08×10^{-1}	2.45×10^0	5.58×10^0	7.71×10^0	7.89×10^0	7.93×10^0
1	9.93×10^{-2}	3.34×10^{-1}	1.58×10^0	4.28×10^0	7.54×10^0	8.52×10^0	8.61×10^0	8.63×10^0
5×10^{-5}	1.98×10^{-1}	6.62×10^{-1}	2.93×10^0	6.63×10^0	8.90×10^0	9.21×10^0	9.31×10^0	
2	4.86×10^{-1}	1.59×10^0	5.86×10^0	9.36×10^0	10.10×10^0	10.22×10^0	10.24×10^0	
1	9.49×10^{-1}	2.95×10^0	8.53×10^0	10.60×10^0	10.86×10^0	10.92×10^0		
5×10^{-6}	1.82×10^0	5.15×10^0	10.67×10^0	11.48×10^0	11.59×10^0	11.62×10^0		
2	4.03×10^0	9.08×10^0	12.28×10^0	12.49×10^0	12.53×10^0	12.54×10^0		
1	6.78×10^0	11.76×10^0	13.12×10^0	13.21×10^0	13.23×10^0	13.24×10^0		
5×10^{-7}	10.13×10^0	13.41×10^0	13.88×10^0	13.92×10^0	13.93×10^0			
2	13.71×10^0	14.68×10^0	14.83×10^0	14.85×10^0				
1	15.13×10^0	15.46×10^0	15.54×10^0					
5×10^{-8}	16.05×10^0	16.20×10^0						
2	17.08×10^0	17.14×10^0						
1	17.81×10^0	17.84×10^0						
5×10^{-9}	18.51×10^0							
2	19.40×10^0							
1	20.15×10^0							

13.4 METHODS OF ANALYSIS

13.4.1   General Solution

Match-Point Method

The theoretical basis for this method is described in Chapter 4. The specific steps are:

i   Prepare a plot of $F(u, \alpha, \rho)$ vs. $1/u$ on logarithmic paper. This plot is called the type curve (Figure 13.1). A separate type curve must be prepared for each well with a different value of ρ. The type curve(s) can be plotted using the values of $F(u, \alpha, \rho)$ in Table 13.2 (for pumping well only use the values of $F(u, \alpha)$ in Table 13.1) or values computed by the computer program TYPE8 (Appendix B).

ii   Plot s vs. t using the same logarithmic scales used to prepare the type curve(s). This plot is called the data curve.

iii   Overlay the data curve on the type curve. Shift the plots relative to each other, keeping the respective axes parallel, until a position of best fit is found between the data curve and one of the type curves. Use as much of the late data as possible; initially almost all of the flow is derived from well storage and a unique solution is not possible.

iv   From the best fit portion of the curves, select a match-point and record match-point values $F(u, \alpha, \rho)^*$, $1/u^*$, t^*, s^* or $t/r^2{}^*$, and α^*.

v   Substitute $F(u, \alpha, \rho)^*$ and s^* into Equation 13.12 and solve for T.

vi   Substitute u^* and t^*, and the computed value of T into Equation 13.13 and solve for S. This value should agree with S computed by substituting α^* into Equation 13.9.

It can be very difficult to select the correct value of α curve since the curves differ in shape only slightly. When α varies by an order of magnitude, the validity of the computed S value is questionable. Therefore it is often more reliable to estimate S from geologic and hydraulic conditions (Papadopulos, 1967). The value of T is much less sensitive to the choice of α used to develop the type curve(s).

Straight-Line Method

This method cannot be used to interpret drawdown data if the pumping well has significant well storage until sufficient time has elapsed that drawdown conforms to the assumptions of the Theis solution (Model 3).

Example 13.1

A two-foot diameter well which fully penetrates a 75-foot thick confined silty sand aquifer is pumped at 80 gal/min. The casing and screen diameter are equal. Drawdown data are listed below for an observation well located 10 feet from the pumping well. Compute T and S.

t (min)	s (ft)	t (min)	s (ft)
0.1	0.008	60	8.91
0.2	0.04	70	9.19
0.3	0.09	80	9.42
0.4	0.15	90	9.63
0.5	0.21	100	9.81
0.6	0.28	200	10.98
0.7	0.36	300	11.65
0.8	0.43	400	12.12
0.9	0.51	500	12.48
1	0.59	600	12.77
2	1.35	700	13.02
3	2.05	800	13.24
4	2.66	900	13.43
5	3.19	1000	13.6
6	3.66	2000	14.71
7	4.06	3000	15.35
8	4.41	4000	15.81
9	4.73	5000	16.17
10	5.01	6000	16.46
20	6.72	7000	16.71
30	7.59	8000	16.92
40	8.16	9000	17.11
50	8.58	10000	17.28

The drawdown data are plotted in Figure 13.1. The data curve has been placed over a series of type curves developed using the program TYPE8 with $\rho = 10$ ft/1 ft = 10. From hydrogeologic information, the storativity was estimated to be in the range of 0.01 to 0.0001, so α-curves were developed for this range.

With a type curve and data curve aligned, the following match-point values were selected:

$$1/u^* \qquad = 1.0$$
$$t^* \qquad\qquad = 0.1 \text{ min}$$
$$F(u, \alpha, \rho)^* = 5.8 \times 10^{-3}$$
$$s^* \qquad\qquad = 0.01 \text{ ft}$$
$$\alpha^* \qquad\qquad = 0.002$$

From Equation 13.12,

$$T = \frac{Q\, F(u, \alpha, \rho)^*}{s^*\, 4\pi} = \frac{\left(80\, \frac{gal}{min}\right)(5.8 \times 10^{-3})\left(\frac{1440\, min}{day}\right)}{(0.01\, ft)\,(4\pi)}$$

$$= 5317\, \frac{gal}{day \cdot ft}$$

From Equation 13.13,

$$S = \frac{u^*\, 4Tt^*}{r^2} = \frac{1.0\,(4)\left(5317\, \frac{gal}{day \cdot ft}\right)(0.1\, min)\left(\frac{ft^3}{7.48\, gal}\right)\left(\frac{day}{1440\, min}\right)}{(10\, ft)^2}$$

$$= 0.002$$

This value for S matches with the one computed from Equation 13.9, where

$$S = \frac{\alpha^*\, r_c^2}{r_w^2} = 0.002\, \frac{(1.0\, ft)^2}{(1.0\, ft)^2} = 0.002$$

Well storage must be considered to properly interpret drawdown data in this example. If data collected while well storage was influencing drawdown are ignored, the remaining drawdown data fall on the relatively flat portion of the Model 3 type curve where a unique solution is not possible (Figure 13.1).

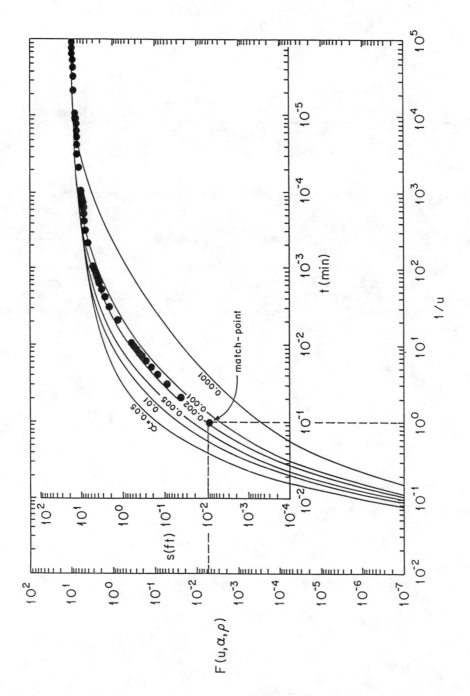

Figure 13.1 Data curve from Example 13.1 matched with a family of type curves where $\rho = 10$.

Chapter 14

MODEL 9: TRANSIENT, CONFINED, LEAKY, WELL STORAGE

14.1 CONCEPTUAL MODEL

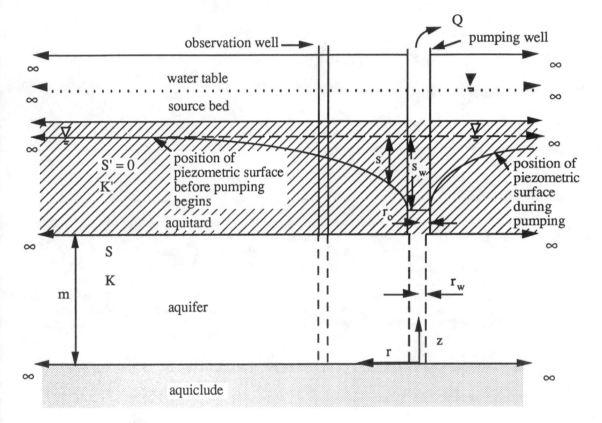

Definition of Terms

K $\quad$ = aquifer hydraulic conductivity, LT^{-1}

K' $\quad$ = aquitard vertical hydraulic conductivity, LT^{-1}

m $\quad$ = aquifer thickness, L

m' $\quad$ = aquitard thickness, L

Q $\quad$ = constant pumping rate, L^3T^{-1}

r   = radial distance from pumping well to a point on the cone of depression
        (all distances are measured from the center of wells), L

r_c = inside radius of pumping well casing within the range of the water level
        fluctuation, L

r_w = effective radius of the well bore or open hole, L
        = borehole radius if the filter is much more permeable than the aquifer
        = screen radius if no filter is used or if the filter has a hydraulic conductivity similar
          to that of the aquifer

s   = drawdown of piezometric surface during pumping, L

s_w = drawdown in the pumping well, L

S   = aquifer storativity, dimensionless

S '   = aquitard storativity, dimensionless

z   = vertical distance from the aquifer base, L

Assumptions

1. The aquifer is bounded above by an aquitard and an unconfined aquifer (the "source
    bed") and bounded below by an aquiclude.
2. All layers are horizontal and extend infinitely in the radial direction.
3. The initial piezometric surface (before pumping begins) is horizontal and extends
    infinitely in the radial direction. The water table in the source bed is horizontal,
    extends infinitely in the radial direction, and remains constant during pumping (zero
    drawdown). Drawdown of the water table in the source bed can be neglected when
    $t \leq S'm'/(10K')$ (Neuman and Witherspoon, 1969b).
4. The aquifer and aquitard are homogeneous and isotropic.
5. Groundwater density and viscosity are constant.
6. Groundwater flow can be described by Darcy's Law.
7. Groundwater flow in the aquitard is vertical. Groundwater flow in the aquifer is
    horizontal and directed radially toward the well. This assumption is valid when
    $K/K' > 100$ m/m' (Hantush, 1967).
8. The pumping and observation wells are screened over the entire aquifer thickness.
9. The pumping rate is constant.
10. Head losses through the well screen and pump intake are negligible.
11. The aquifer is compressible and completely elastic. The aquitard is incompressible
    (i.e., no water is released from storage during pumping). This assumption is valid
    when $t > 0.036$ m'S'/K' (Hantush, 1960).

14.2 MATHEMATICAL MODEL

Governing Equation

        The governing equation is derived by combining Darcy's Law with the principle of
conservation of mass (for the aquifer <u>and</u> well casing) in a radial coordinate system (Lai
and Su, 1974)

$$\frac{\partial^2 s}{\partial r^2} + \left(\frac{1}{r}\right)\frac{\partial s}{\partial r} - \frac{sK'}{Kmm'} = \left(\frac{S}{Km}\right)\frac{\partial s}{\partial t}, \qquad r \geq r_w \tag{14.1}$$

where s is drawdown, r is radial distance from the pumping well, K' and m' are the hydraulic conductivity and saturated thickness of the aquitard, K and m are the hydraulic conductivity and saturated thickness of the aquifer, S is the storativity of the aquifer, and r_w is the radius of the pumping well. Recalling Darcy's Law it is easy to see that the term sK'/m represents the rate of aquitard leakage.

Initial Conditions

• Before pumping begins, drawdown is zero in the aquifer and pumping well

$$s(r, t = 0) = 0, \qquad r \geq r_w \tag{14.2}$$

$$s_w(t = 0) = 0 \tag{14.3}$$

where s_w is the drawdown in the pumping well

Boundary Conditions

• The drawdown in the aquifer at the face of the well is equal to the drawdown in the well

$$s(r = r_w, t) = s_w(t) \tag{14.4}$$

• At an infinite distance from the pumping well drawdown is zero

$$s(r = \infty, t) = 0 \tag{14.5}$$

• The pumping rate, Q, is the sum of groundwater entering the pumping well from the aquifer and the rate of decrease of water stored in the well casing

$$Q = -2\pi r_w Km \frac{\partial s(r_w,t)}{\partial r} + \pi r_c^2 \frac{\partial s_w(t)}{\partial t}, \quad t > 0 \tag{14.6}$$

The first term on the right hand side of Equation 14.6 will be positive for flow from the aquifer to the well $\left(\frac{\partial s(r_w,t)}{\partial r} < 0\right)$ and the second term will be positive for a decrease in water stored in the well casing $\left(\frac{\partial s_w(t)}{\partial t} > 0\right)$.

14.3 ANALYTICAL SOLUTION

14.3.1 General Solution

The solution for drawdown in the pumping well, s_w, is (Lai and Su, 1974)

$$s_w = \frac{Q}{4\pi T} F(u_w, \alpha, r_w/B) \tag{14.7}$$

where

$$u_w = \frac{r_w^2 S}{4Tt} \qquad (14.8)$$

$$\alpha = \frac{r_w^2 S}{r_c^2} \qquad (14.9)$$

$$\frac{r_w}{B} = \frac{r_w}{\sqrt{\dfrac{Kmm'}{K'}}} \qquad (14.10)$$

where $T = Km$ and $F(u_w, \alpha, r_w/B)$ is a well function

$$F(u_w, \alpha, r_w/B) = \frac{32\alpha^2}{\pi^2} \int_0^\infty \frac{(1 - \exp(-(\beta^2 + (r_w/B)^2 (1/4u))}{(\beta^2 + (r_w/B)^2)\, \Delta\,(\beta)} \beta\, d\beta \qquad (14.11)$$

where

$$\Delta(\beta) = [(\beta^2 + (r_w/B)^2)\, J_o(\beta) - 2\alpha\beta\, J_1(\beta)]^2$$
$$+ [(\beta^2 + (r_w/B)^2)\, Y_o(\beta) - 2\alpha\beta\, Y_1(\beta)]^2 \qquad (14.12)$$

$$
\begin{aligned}
J_o &= \text{Bessel function of the first kind and zero order} \\
J_1 &= \text{Bessel function of the first kind and first order} \\
Y_o &= \text{Bessel function of the second kind and zero order} \\
Y_1 &= \text{Bessel function of the second kind and first order} \\
\beta &= \text{dummy variable for integration}
\end{aligned}
$$

The solution for drawdown in the aquifer is (Lai and Su, 1974)

$$s = \frac{Q}{4\pi T} F(u, \alpha, r_w/B, \rho) \qquad (14.13)$$

where

$$u = \frac{r^2 S}{4Tt} \qquad (14.14)$$

$$\rho = \frac{r}{r_w} \qquad (14.15)$$

and $F(u, \alpha, r_w/B, \rho)$ is a well function

$$F(u, \alpha, r_w/B, \rho) = \frac{8\alpha}{\pi} \int_0^\infty \frac{[1 - \exp(-(\beta^2 + (r/B)^2)(1/4u))]}{(\beta^2 + (r/B)^2 \Delta(\beta))} \{J_o(\beta\rho)[(\beta^2 +$$

$$(r/B)^2)Y_o(\beta) - 2\alpha\beta Y_1(\beta)] - Y_o(\beta\rho)[(\beta^2 +$$

$$(r/B)^2)J_o(\beta) - 2\alpha\beta J_1(\beta)]\} \beta \, d\beta \qquad (14.16)$$

and α, J_o, J_1, Y_o, and Y_1 are as defined earlier. For $\rho = 1$

$$F(u, \alpha, r_w/B, \rho) = F(u_w, \alpha, r_w/B) \qquad (14.17)$$

Values of $F(u, \alpha, r_w/B, \rho)$ may be computed with the computer program TYPE9; drawdown can be computed with the computer program DRAW9 (Appendix B).

14.3.2   Special Case Solutions

If the aquitard is impermeable ($K' = 0$), $r_w/B = 0$ and Equations 14.11 and 14.16 reduce to Equations 13.10 and 13.15, the solutions for Model 8.

14.4   METHODS OF ANALYSIS

14.4.1   General Solution

Match-Point Method

The well function for this model (Equation 14.16) is a function of four dimensionless variables (u, α, r_w/B, and ρ). ρ is known from the well configuration. To proceed the analyst must assume a value for either u, α, or r_w/B. The remaining two parameters can then be determined by the match-point method. For example, if the aquitard hydraulic conductivity, K', is of most interest S should be estimated and used to compute α. The values of u and r_w/B can then be determined using the match-point method. Similarly, if the storativity of the aquifer is of most interest the ratio K/K' should be estimated and used to compute r_w/B. The values of u and α can then be determined.

The theoretical basis of the match-point method is described in Chapter 4. The steps listed below are for the case where S can be estimated with reasonable accuracy. The procedure can be easily modified for situations where K/K' or $T = Km$ can be estimated.

i   Estimate S and compute α using Equation 14.9. For each observation well compute ρ using Equation 14.15 and prepare a plot of $F(u, \alpha, r_w/B, \rho)$ vs. $1/u$ on logarithmic paper. This plot is called the type curve. A separate type curve must be prepared for each well with a different value of ρ. The type curves can be plotted using the values of $F(u, \alpha, r_w/B, \rho)$ computed with the computer program TYPE9 (Appendix B).

ii Plot s vs. t using the same logarithmic scales used to prepare the type curve(s). This plot is called the data curve.

iii Overlay the data curve on the type curve(s). Shift the plots relative to each other, keeping the respective axes parallel, until a position of best fit is found between the data curve and one of the type curves. Use as much late data as possible; initially almost all of the flow is derived from well storage and a unique solution is not possible.

iv From the best fit portions of the curves, select a match-point and record match-point values $F(u, \alpha, r_w/B, \rho)^*$, $1/u^*$, t^*, s^*, and r_w/B^*.

v Substitute $F(u, \alpha, r_w/B, \rho)^*$ and s^* into Equation 14.13 and solve for $T = Km$.

vi Substitute the computed value of T and r_w/B^* into Equation 14.10 and solve for K'.

vii Substitute the computed value of T and t^* and u^* into Equation 14.8 and solve for S. This serves as a check on the initial estimate for S. If necessary the procedure can be repeated with the computed value of S.

Example 14.1

A two-foot diameter well which fully penetrates a 75-foot thick confined silty sand aquifer is pumped at 50 gal/min. The casing and screen diameter are equal. Overlying the aquifer is a 20-foot thick silty aquitard. Drawdown data are listed below for an observation well located 10 feet from the pumping well. Assume the storativity of the aquifer is 0.002. Compute T, K, and K'.

t (min)	s (ft)	t (min)	s (ft)
5	1.51	200	42.03
10	3.99	225	43.38
15	6.55	250	44.51
20	9.03	275	45.46
25	11.37	300	46.27
30	13.56	325	46.97
35	15.60	350	47.58
40	17.50	375	48.12
45	19.26	400	48.59
50	20.90	450	49.39
55	22.42	500	50.03
60	23.83	550	50.56
65	25.16	600	51.00
70	26.37	650	51.36
75	27.51	700	51.67
80	28.58	750	51.93
85	29.58	800	52.15
90	30.52	900	52.50
95	32.00	1000	52.77
100	32.22	1100	53.00
125	35.69	1200	53.12
150	38.32	1300	53.24
175	40.38	1400	53.33

The drawdown data are plotted in Figure 14.1. Substituting $S = 0.002$ and $r_w = r_c$ = 1 foot into Equation 14.9 gives $\alpha = 0.002$. From Equation 14.15 $\rho = 10$ ft/1 ft = 10. A set of type curves was prepared using program TYPE9 (Appendix B). With the type curves and data curve aligned, the following match-point values were selected:

$$
\begin{aligned}
1/u^* &= 100 \\
t^* &= 100 \text{ min} \\
F(u, \alpha, r_w/B, \rho)^* &= 3.0 \\
s^* &= 31 \text{ ft} \\
r_w/B^* &= 0.008
\end{aligned}
$$

Substituting into Equation 14.13 gives

$$
T = \frac{Q\ F(u, \alpha, r_w/B, \rho)^*}{s^*\ 4\pi} = \frac{\left(50\ \dfrac{\text{gal}}{\text{min}}\right)(3.0)\left(\dfrac{1440\ \text{min}}{\text{day}}\right)}{(31\ \text{ft})\ (4\pi)}
$$

$$
= 554\ \frac{\text{gal}}{\text{day} \cdot \text{ft}}
$$

$$
K = \frac{T}{m} = \frac{\left(554\ \dfrac{\text{gal}}{\text{day} \cdot \text{ft}}\right)\left(\dfrac{\text{ft}^3}{7.48\ \text{gal}}\right)}{(75\ \text{ft})} = 1\ \frac{\text{ft}}{\text{day}}
$$

From Equation 14.10,

$$
K' = \frac{Kmm'(r_w/B^*)^2}{r_w{}^2} = \frac{\left(1\ \dfrac{\text{ft}}{\text{day}}\right)(75\ \text{ft})(20\ \text{ft})(0.008)^2}{(1\ \text{ft})^2} = 0.1\ \frac{\text{ft}}{\text{day}}
$$

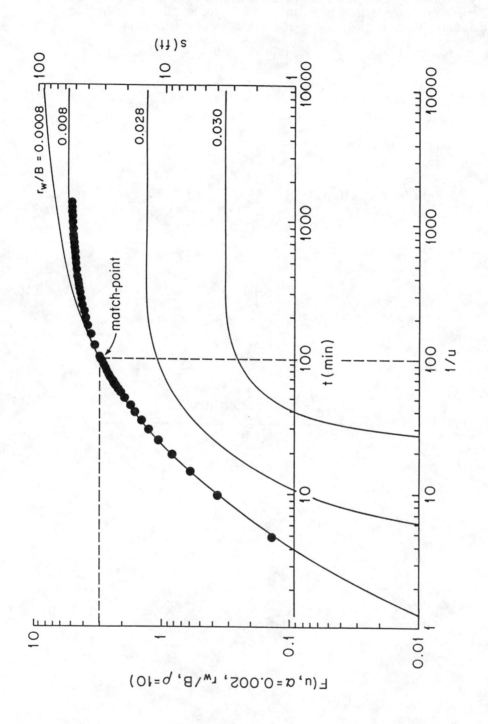

Figure 14.1 Application of match-point method for Example 14.1.

Chapter 15

MODEL 10: TRANSIENT, CONFINED, FISSURE - BLOCK SYSTEM

15.1 CONCEPTUAL MODEL

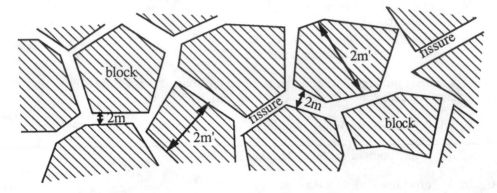

The "real" system of fissures and blocks is idealized as shown below (Boulton and Streltsova, 1977).

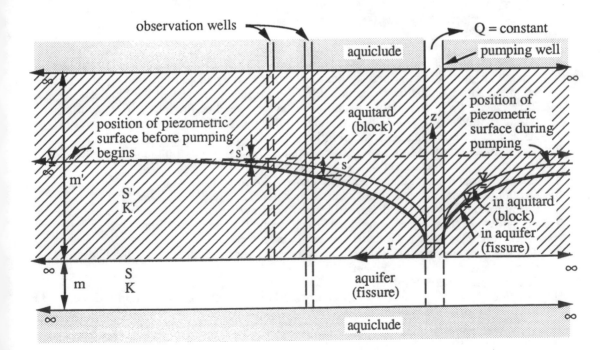

Definition of Terms

K   = hydraulic conductivity of the fissure (aquifer), LT^{-1}

K'   = hydraulic conductivity of the block (aquitard), LT^{-1}

m   = half the thickness of the average fissure, L
        = the model aquifer thickness

m'   = half the thickness of the average block, L
        = the model aquitard thickness

Q   = constant rate of pumping from the fissure, L^3T^{-1}

r   = radial distance from pumping well to a point on the cone of depression
        (all distances are measured from the center of wells), L

s   = average drawdown in the fissure (aquifer), L

s'   = drawdown at any point in the block (aquitard), L

S   = fissure (aquifer) storativity, dimensionless

S'   = block (aquitard) storativity, dimensionless

z   = vertical distance from the base of the block (aquitard), L

Assumptions

1. A large number of porous blocks separated by randomly distributed and arbitrarily oriented fissures in a confined aquifer can be modeled by an aquitard and aquifer bounded above and below by aquicludes. The average thickness of the blocks is 2m'. Due to symmetry, only one half of the system need be modeled. The "aquifer" (fissure) and "aquitard" (block) are assigned thicknesses of m and m', respectively.
2. All layers are horizontal and extend infinitely in the radial direction.
3. The initial piezometric surface (before pumping begins) is horizontal and extends infinitely in the radial direction.
4. The aquifer and aquitard are homogeneous and isotropic.
5. Groundwater density and viscosity are constant.
6. Groundwater flow can be described by Darcy's Law.
7. Groundwater flow in the aquitard is vertical, while flow in the aquifer is horizontal and directed radially toward the well.
8. The pumping and observation wells are screened over the entire aquifer thickness and are impermeable over the entire aquitard thickness.
9. The pumping rate is constant.
10. The thickness of the aquifer is small compared to the thickness of the aquitard (i.e., m/m' is small).
11. Head losses through the well screen and pump intake are negligible.
12. The aquifer and aquitard are compressible and completely elastic.

15.2 MATHEMATICAL MODEL

15.2.1 In the Aquitard (Block)

Governing Equation

The governing equation is derived by combining Darcy's Law with the principle of conservation of mass in a radial coordinate system

$$\frac{\partial^2 s'}{\partial z^2} = \left(\frac{S'}{T'}\right)\frac{\partial s'}{\partial t} \tag{15.1}$$

where s' is drawdown in the aquitard, z is the distance from the base of the aquitard, S' and T' = K'm' are the storativity and transmissivity of the aquitard, and t is time.

Initial Conditions

- Before pumping begins drawdown is zero everywhere

$$s'(r, z, t = 0) = 0 \tag{15.2}$$

Boundary Conditions

- The top of the aquitard represents a "groundwater divide" ; no groundwater flows across this boundary due to the symmetry of the system

$$\frac{\partial s' \ (r, \ z = m', \ t)}{\partial z} = 0 \tag{15.3}$$

- At the bottom of the aquitard, the drawdown is equal to the drawdown in the aquifer

$$s'(r, z = 0, t) = s(r, z = 0, t) \tag{15.4}$$

15.2.2 In the Aquifer (Fissure)

Governing Equation

The governing equation is derived by combining Darcy's Law with the principle of conservation of mass in a radial coordinate system.

$$\frac{\partial^2 s}{\partial r^2} + \frac{1}{r}\frac{\partial s}{\partial r} - \frac{s'}{B^2} = \frac{S}{Km}\frac{\partial s}{\partial t} \tag{15.5}$$

where s is drawdown , r is radial distance from the pumping well, S, K, and m are the storativity, hydraulic conductivity, and thickness of the aquifer, t is time, and

$$B^2 = \frac{Kmm'}{K'} \tag{15.6}$$

Referring to the definition sketch for the conceptual model and recalling Darcy's Law, it is easy to see that s'K'/m' represents the rate of vertical groundwater flow from the aquitard to the aquifer.

Initial Conditions

- Before pumping begins drawdown is zero everywhere

$$s(r, t = 0) = s'(r, t = 0) = 0 \tag{15.7}$$

Boundary Conditions

- At an infinite distance from the pumping well drawdown is zero

$$s(r = \infty, t) = s'(r = \infty, t) = 0 \quad t \geq 0 \tag{15.8}$$

- Groundwater flow to the pumping well is constant and uniform over the aquifer thickness (which is a result of the assumption of horizontal groundwater flow in the aquifer)

$$\lim_{r \to 0} r\frac{\partial s}{\partial r} = -\frac{Q}{2\pi T} \tag{15.9}$$

where $T = Km$.

15.3 ANALYTICAL SOLUTION

15.3.1 General Solution

The solution to Equation 15.1 for drawdown in the aquitard is (Boulton and Streltsova, 1977)

$$s' = \frac{Q}{2\pi T} W'(u, r/m', b, c) \tag{15.10}$$

where

$$u = \frac{r^2 S}{4Tt} \tag{15.11}$$

$$b = \frac{T'S}{TS'} \tag{15.12}$$

$$c = \frac{T'}{T} \tag{15.13}$$

and $W'(u, r/m', b, c)$ is a well function

$$W'(u, r/m', b, c) = \int_0^\infty \beta \, J_o\left(\frac{r\beta}{m'}\right)\left(\sum_{n=1}^\infty \varphi_n \phi_n\right) d\beta \tag{15.14}$$

where

β = dummy variable for integration
J_o = zero-order Bessel function of the first kind

$$\varphi_n = \frac{1 - \exp\left(\dfrac{-\gamma_n^2 T't}{S'm'^2}\right)}{\left(\dfrac{T'S\gamma_n^2}{TS'}\right) + 0.5 \dfrac{T'}{T} \gamma_n^2 (\tan\gamma_n + \gamma_n \sec^2\gamma_n)} \tag{15.15}$$

γ_n is the positive root of

$$\gamma_n \left(\frac{T'S}{TS'} \gamma_n + \frac{T'}{T} \tan \gamma_n\right) = \beta^2 \tag{15.16}$$

and

$$\phi_n = \cos(\gamma_n \, z/m') + \tan\gamma_n \, \sin(\gamma_n \, z/m') \tag{15.17}$$

The solution to Equation 15.5 for drawdown in the aquifer is

$$s = \frac{Q}{2\pi T} \, W(u, \, r/m', \, b, \, c) \tag{15.18}$$

where $W(u, r/m', b, c)$ is a well function

$$W(u, \, r/m', \, b, \, c) = \int_0^\infty \beta \, J_o\left(\frac{r\beta}{m'}\right)\left(\sum_{n=1}^\infty \varphi_n\right) d\beta \tag{15.19}$$

Values of the function $W(u, r/m', b, c)$ can be computed with the computer program TYPE10; average drawdown can be computed with the computer program DRAW10 (Appendix B).

15.3.2   Special Case Solution

Incompressible Fissure

If the aquifer (fissure) can be assumed to be incompressible ($S = b = 0$) the drawdown equations can be simplified. For the aquitard (block)

$$s' = \frac{Q}{2\pi T} \int_0^\infty \beta \, J_o\left(\frac{r\beta}{m'}\right)\left(\sum_{n=1}^\infty \varphi_n \, \phi_n\right) d\beta \tag{15.20}$$

and for the aquifer (fissure)

$$s = \frac{Q}{2\pi T} \int_0^\infty \beta \, J_o\left(\frac{r\beta}{m'}\right)\left(\sum_{n=1}^\infty \varphi_n\right) d\beta \tag{15.21}$$

where

$$\varphi_n = \frac{2\left[1 - \exp\left(\dfrac{-\gamma_n^2 \, T't}{S'm'^2}\right)\right]}{\beta^2\left(1 + \beta^2 \dfrac{T}{T'}\right) + \dfrac{T'}{T} \gamma_n^2} \tag{15.22}$$

$$\phi_n = \cos\left(\gamma_n \, z/m'\right) + \tan\gamma_n \sin\left(\gamma_n \, z/m'\right) \tag{15.23}$$

and γ_n is the positive root of

$$\gamma_n \tan\gamma_n = \beta^2 \, T/T' \tag{15.24}$$

15.4 METHODS OF ANALYSIS

15.4.1 General Solution

The well functions used in the match-point method are functions of three dimensionless parameters (u, b, and c) assuming r/m' is known. To proceed it is necessary to estimate one of these parameters (usually c = T'/T); the remaining two can then be determined by application of the match-point method using drawdown data for an observation well screened in either the aquifer or aquitard. The steps for application of the match-point method are described below assuming that aquifer drawdown data are used:

i Estimate c = T'/T using the geologic descriptions of the aquifer and aquitard materials obtained from drilling records.

ii Prepare a plot of W(u, r/m', b, c) vs. 1/u on logarithmic paper. This plot is called the type curve. The type curve can be plotted using values of W(u, r/m', b, c) computed with the computer program TYPE10 (Appendix B).

iii Plot s vs. t (or s vs. t/r^2 for more than one observation well) using the same logarithmic scales used to prepare the type curve. This plot is called the data curve.

iv Overlay the data curve on the type curve(s). Shift the plots relative to each other, keeping respective axes parallel, until a position of best fit is found between the data curve and one of the type curves.

v From the best fit portion of the curves, select a match-point and record match-point values W(u, r/m', b, c)*, 1/u*, b*, s*, and t*.

vi Substitute W(u, r/m', b, c)* and s* into Equation 15.18 and solve for T.

vii Substitute the computed value of T and 1/u* and t* (or t/r^2*) into Equation 15.11 and solve for S.

viii Substitute the estimated value of c and b* into Equations 15.12 and 15.13 and solve for S'.

ix Substitute the computed value of T into Equation 15.13 and solve for T'. Compare the ratio T'/T with the original estimate. If the result is different repeat steps ii to ix with the computed value of c.

Example 15.1

A pumping test was conducted in a fractured, columnar basalt aquifer. The average block size (2m') is about 80 ft. The pumping well was screened over a fracture; the pumping rate was 1925 gal/day. Drawdown data are listed below for an observation well located 10 feet from the pumping well. Assume the ratio T'/T is about 0.033 and the ratio S'/S is about 0.01. Compute T and S.

t (min)	s (ft)	t (min)	s (ft)
1	0.01	55	0.35
2	0.02	60	0.51
3	0.04	120	0.46
4	0.06	180	0.51
5	0.08	240	0.55
6	0.10	300	0.58
7	0.11	360	0.61
8	0.12	420	0.63
9	0.14	480	0.64
10	0.15	540	0.66
15	0.19	600	0.67
20	0.23	720	0.70
25	0.25	840	0.72
30	0.28	960	0.74
35	0.30	1080	0.75
40	0.31	1200	0.77
45	0.33	1320	0.78
50	0.34	1440	0.79

The drawdown data are plotted in Figure 15.1. With T'/T = c = 0.033 and with S'/S = 0.01, b = 3.33; and r/m' = 10 ft/ 40 ft = 0.25. Using these parameters, a set of type curves was prepared using program TYPE10 (Appendix B). With the type curves and data curve aligned, the following match-point values were selected:

$$
\begin{aligned}
1/u^* &= 12 \\
t^* &= 30 \text{ min} \\
W(u, r/m', b, c)^* &= 2 \\
s^* &= 1 \text{ ft} \\
b^* &= 3.333
\end{aligned}
$$

Substituting into Equation 15.18 gives

$$ T = \frac{Q}{2\pi s^*} W(u, r/m', b, c)^* $$

$$ = \left(\frac{1925 \text{ gal}}{\text{day}}\right)\left(\frac{ft^3}{7.48 \text{ gal}}\right)\left(\frac{2}{2\pi \ 1ft}\right) $$

$$= 82 \frac{ft^2}{day}$$

$$K = \frac{T}{m} = \left(82\frac{ft^2}{day}\right)\left(\frac{1}{40\ ft}\right) = 2\frac{ft}{day}$$

Substituting into Equation 15.11 gives

$$S = \frac{u*4Tt*}{r^2}$$

$$= \frac{\left(\frac{1}{12}\right)(4)\left(82\frac{ft}{day}\right)(30\ min)\left(\frac{day}{1440\ min}\right)}{(10\ ft)^2}$$

$$= 0.0057$$

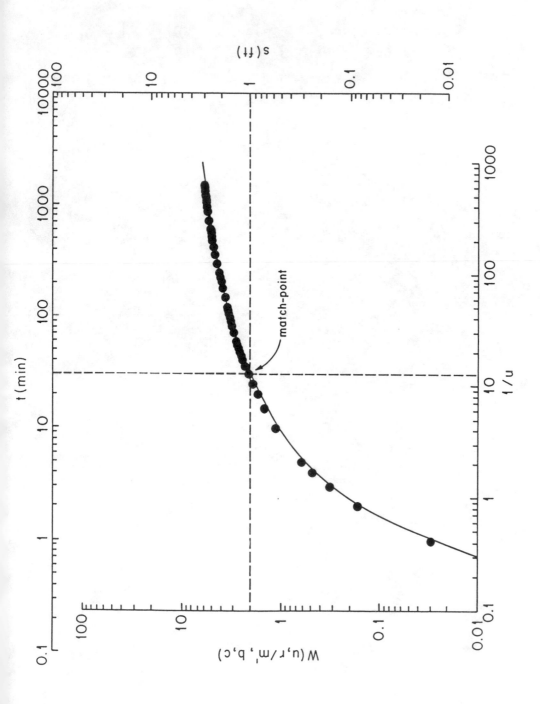

Figure 15.1 Application of match-point method for Example 15.1.

MODEL 11: TRANSIENT, CONFINED, ANISOTROPIC IN TWO HORIZONTAL DIRECTIONS

16.1 CONCEPTUAL MODEL

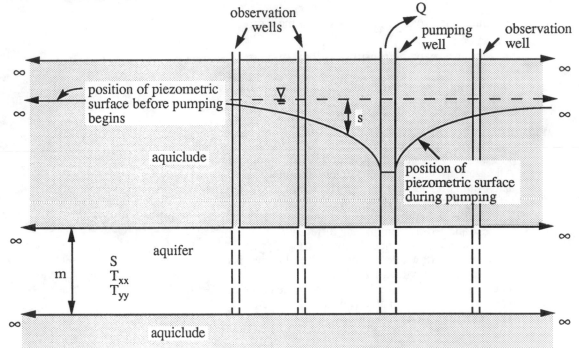

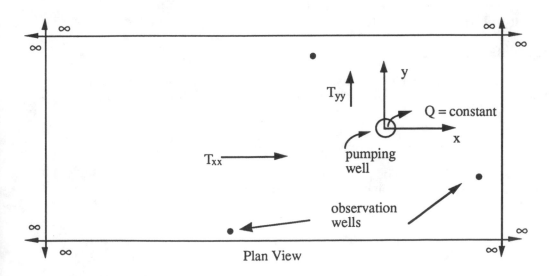

Plan View

Definition of Terms

m   = aquifer thickness, L

Q   = constant pumping rate, L^3T^{-1}

s   = drawdown of piezometric surface during pumping, L

S   = aquifer storativity, dimensionless

T_{xx} = aquifer horizontal transmissivity in the x direction, L^2T^{-1}

T_{yy} = aquifer horizontal transmissivity in the y direction, L^2T^{-1}

x, y   = horizontal coordinates with origin at the pumping well, L

Assumptions

1. The aquifer is bounded above and below by aquicludes.
2. All layers are horizontal and extend infinitely in the radial direction.
3. The initial piezometric surface (before pumping begins) is horizontal and extends infinitely in the radial direction.
4. The aquifer is homogeneous and anisotropic. The transmissivities of the aquifer in the x and y direction may be different.
5. Groundwater density and viscosity are constant.
6. Groundwater flow can be described by Darcy's Law.
7. Groundwater flow is horizontal.
8. The well is screened (or uncased) over the entire aquifer thickness.
9. The pumping well has an infinitesimal diameter.
10. The pumping rate is constant.
11. The aquifer is compressible and completely elastic.

16.2 MATHEMATICAL MODEL

Governing Equation

The governing equation is derived by combining Darcy's Law with the principle of conservation of mass in a two-dimensional coordinate system (Papadopulos, 1965)

$$T_{xx}\frac{\partial^2 s}{\partial x^2} + 2T_{xy}\frac{\partial^2 s}{\partial x \partial y} + T_{yy}\frac{\partial^2 s}{\partial y^2} + Q\,\delta(x)\,\delta(y) = S\frac{\partial s}{\partial t} \qquad (16.1)$$

where T_{xx}, T_{xy}, and T_{yy} are components of aquifer transmissivity, s is drawdown, x and y are coordinates, Q is the pumping rate, $\delta(x)$ is the Dirac delta function, S is aquifer storativity, and t is time.

Initial Conditions

• Before pumping begins drawdown is zero everywhere

$$s(x, y, t = 0) = 0 \qquad x, y \geq 0 \tag{16.2}$$

Boundary Conditions

• At an infinite distance from the pumping well, drawdown is zero

$$s(x = \pm\infty, y, t) = 0 \tag{16.3}$$

$$s(x, y = \pm\infty, t) = 0 \tag{16.4}$$

16.3 ANALYTICAL SOLUTION

16.3.1 General Solution

The solution is (Papadopulos, 1965)

$$s = \frac{Q}{4\pi\sqrt{D'}} W(u_{xy}) \tag{16.5}$$

where

$$W(u_{xy}) = \int_{u_{xy}}^{\infty} \frac{e^y}{y} dy$$

is the well function of Model 3 with u_{xy} replacing u, and

$$u_{xy} = \frac{S}{4t} \frac{[T_{xx}(y^2) + T_{yy}(x^2) - 2T_{xy}(xy)]}{D'} \tag{16.6}$$

$$D' = T_{xx}T_{yy} - T_{xy}^2 \tag{16.7}$$

Values of the function $W(u_{xy})$ are in Table 8.1. Values of the function $W(u_{xy})$ can also be computed with the computer program TYPE11; average drawdown can be computed with the computer program DRAW11 (Appendix B).

16.4 METHODS OF ANALYSIS

Match-Point Method

A minimum of three observation wells is required. The locations of the observation wells are arbitrary as long as no two observation wells are radially aligned with the pumping well. However, the best results are achieved by maximizing radial separation of the observation wells (e.g., by placing the three observation wells at the corners of an equilateral triangle centered about the pumping well) (Maslia and Randolph, 1987).

The theoretical basis for this method of analysis is described in Chapter 4. The specific steps are (Papadopulos, 1965; Maslia and Randolph, 1987):

i Prepare a plot of $W(u_{xy})$ vs. $1/u_{xy}$ on logarithmic paper. This plot is called the type curve (Figure 8.1). The type curve can be plotted using the values of $W(u)$ in Table 8.1 or values computed by the computer program TYPE11 (Appendix B).

ii Prepare a separate plot of s vs. t for each observation well using the same logarithmic scales used to prepare the type curves. These plots are called the data curves. Note that composite data curves (s vs. r^2/t) and distance-drawdown data curves (s vs. r) cannot be used with Model 11 because, in an anisotropic aquifer, groundwater flow to the pumping well is not radially symmetric.

iii Overlay the data curve for one observation well on the type curve. Shift the plots relative to each other, keeping the respective axes parallel until a position of best fit is found between the data curve and type curve.

iv From the best fit portion of the curves, select a match-point and record match-point values s^*, t^*, $W(u_{xy})^*$, and u_{xy}.

v Repeat steps iii and iv for each observation well.

vi Substitute the values of s^* and $W(u_{xy})^*$ for each observation well into Equation 16.5 and solve for D'. The value of D' should be approximately the same for each well, therefore, average the values (if appropriate) from all of the wells.

vii Substitute the observation well coordinates x and y, u_{xy}^*, and the computed average value of D' into Equation 16.6 rearranged into the form

$$ST_{xx}(y^2) + ST_{yy}(x^2) - 2T_{xy}(xy) = 4tu_{xy}D' \tag{16.8}$$

This will result in a system of three or more simultaneous equations of the general form

$$\begin{bmatrix} y_1^2 & x_1^2 & -2x_1y_1 \\ y_2^2 & x_2^2 & -2x_2y_2 \\ : & : & : \\ y_n^2 & x_n^2 & -2x_ny_n \end{bmatrix} \begin{Bmatrix} ST_{xx} \\ ST_{yy} \\ ST_{xy} \end{Bmatrix} = \begin{Bmatrix} 4t_1^*u_{xy_1}^*D' \\ 4t_2^*u_{xy_2}^*D' \\ : \\ 4t_n^*u_{xy_n}^*D' \end{Bmatrix} \tag{16.9}$$

where n is the number of observation wells. Solve Equation 16.9 to find ST_{xx}, ST_{yy}, and ST_{xy}. If more than three wells are used, a least squares optimization procedure is required to solve the system of equations. A description of this procedure and a computer program for performing the necessary calculations are in Maslia and Randolph (1987).

viii Multiplying Equation 16.7 by S^2 and rearranging yields

$$S = \sqrt{\frac{(ST_{xx})(ST_{yy}) - (ST_{xy})^2}{D'}} \tag{16.10}$$

Substitute the computed values of ST_{xx}, ST_{yy}, and ST_{xy} and D' into Equation 16.10 and solve for S then compute T_{xx}, T_{yy}, and T_{xy}.

ix Compute the principal values of transmissivity, $T_{\varepsilon\varepsilon}$ and $T_{\eta\eta}$, using

$$T_{\varepsilon\varepsilon} = \frac{1}{2}\left[(T_{xx} + T_{yy}) + \sqrt{(T_{xx} - T_{yy})^2 + 4\,(T_{xy})^2}\,\right] \tag{16.11}$$

$$T_{\eta\eta} = \frac{1}{2}\left[(T_{xx} + T_{yy}) - \sqrt{(T_{xx} - T_{yy})^2 + 4\,(T_{xy})^2}\,\right] \tag{16.12}$$

The axis of the maximum principal value, is directed at an angle θ from the x-axis, where

$$\theta = \tan^{-1}\left[\frac{T_{\varepsilon\varepsilon} - T_{xx}}{T_{xy}}\right] \tag{16.13}$$

x   Write the equation for the theoretical transmissivity ellipse

$$\frac{\varepsilon^2}{T_{\varepsilon\varepsilon}} + \frac{\eta^2}{T_{\eta\eta}} = 1 \tag{16.14}$$

where

ε, η = the axes of the principal coordinate system rotated by θ degrees from the arbitrary x-y coordinate system.

$\sqrt{T_{\varepsilon\varepsilon}}$ = the major axis of the transmissivity ellipse

$\sqrt{T_{\eta\eta}}$ = the minor axis of the transmissivity ellipse

xi If necessary the value of transmissivity can be computed for an arbitrary direction ρ. This can be done by substituting the computed values of $T_{\varepsilon\varepsilon}$ and $T_{\eta\eta}$ into

$$\frac{1}{T_\rho} = \left(\frac{1}{T_{\varepsilon\varepsilon}}\right)\cos^2\beta + \left(\frac{1}{T_{\eta\eta}}\right)\sin^2\beta \tag{16.15}$$

where T_ρ is the transmissivity in the ρ direction and β is the angle of ρ from the ε axis, with β positive in the counter-clockwise direction

xii Compute the directional transmissivity, T_d, for each observation well using Equation 16.16 and compare it with the transmissivity computed for that direction using Equation 16.15 as a check of model accuracy

$$T_d = \frac{Sr^2}{4u_{xy}^* t^*} \tag{16.16}$$

where

T_d = the directional transmissivity at a particular observation well
S = the storage coefficient computed from Equation 16.10

r   = the radial distance from the pumping well to the observation well
t^* = the match-point value of time for the observation well
u_{xy}^*= the match-point value of the well function variable at that
          observation well

Values of $\sqrt{T_d}$ for several observation wells, plotted at the correct angular location should fall on the transmissivity ellipse defined by Equation 16.14.

Straight-Line Method

This method, developed by Papadopulos (1965) can be used only under conditions where u_{xy} is small enough so that the Cooper and Jacob (1946) approximation for $W(u_{xy})$ is valid for all observation wells. Substituting u_{xy} into the Cooper and Jacob approximation (Appendix A)

$$W(u_{xy}) = 2.303 \log \left(\frac{2.25}{4u_{xy}}\right) \tag{16.17}$$

Substituting Equations 16.17 and 16.6 into Equation 16.5 yields

$$s = \frac{2.3.3Q}{4\pi \sqrt{D'}} \log \left\{\frac{2.25t}{S} \left[\frac{D'}{T_{xx}(y^2) + T_{yy}(x^2) - 2T_{xy}(xy)}\right]\right\} \tag{16.18}$$

which is the basis for this method.

The step-by-step procedure for computing transmissivity and storativity is:

i   For each observation well plot s vs. log(t). From Equation 16.18, the plot should be a straight line with slope Δs and zero drawdown intercept t_o, where

$$\Delta s = \frac{2.303Q}{4\pi\sqrt{D'}} \tag{16.19}$$

and

$$t_o = \frac{S}{2.25}\left[\frac{T_{xx}(y^2) + T_{yy}(x^2) - 2T_{xy}(xy)}{D'}\right] \tag{16.20}$$

From the slope of each plot, a value for D' can be computed. Average (if appropriate) the values for D' from each observation well to find D' for the aquifer.

ii Equation 16.20 can be rearranged as

$$ST_{xx}(y^2) + ST_{yy}(x^2) - 2T_{xy}(xy) = 2.25t_oD' \tag{16.21}$$

By substituting the computed value of D' into this equation, three or more linear equations of the form

$$[A]\{x\} = \{B\} \tag{16.22}$$

can be written, where

$$[A] = \begin{bmatrix} y_1^2 & x_1^2 & -2x_1y_1 \\ y_2^2 & x_2^2 & -2x_2y_2 \\ : & : & : \\ y_n^2 & x_n^2 & -2x_ny_n \end{bmatrix}, \quad \{x\} = \begin{Bmatrix} ST_{xx} \\ ST_{yy} \\ ST_{xy} \end{Bmatrix}, \quad \{B\} = \begin{Bmatrix} 2.25(t_o)_1 D' \\ 2.25(t_o)_2 D' \\ : \\ 2.25(t_o)_n D' \end{Bmatrix}$$

As in the match-point method, when more than three observation wells are used, a least squares optimization procedure, such as described in Maslia and Randolph (1987), is required to solve the system of equations. Substitute values of $(t_o)_n$ and D' into Equation 16.22 and solve for ST_{xx}, ST_{yy}, and ST_{xy}.

iii Substitute the computed values of ST_{xx}, ST_{yy}, and ST_{xy} into Equation 16.10 and solve for the storativity and transmissivity components T_{xx}, T_{yy}, and T_{xy}.

iv If desired compute the magnitude and direction of principal transmissivities and the value of transmissivity for an arbitrary direction using steps vii to x in Section 16.4.

Example 16.1

This example is from Papadopulos (1965). A pumping test was conducted to determine the storage coefficient and principal transmissivities for an anisotropic, confined aquifer. The pumping rate for the test was 200 gal/min. The relative locations of the pumping and observation wells are in Figure 16.1. The drawdown data are in Table 16.1. Compute T_{xx}, T_{yy}, and S.

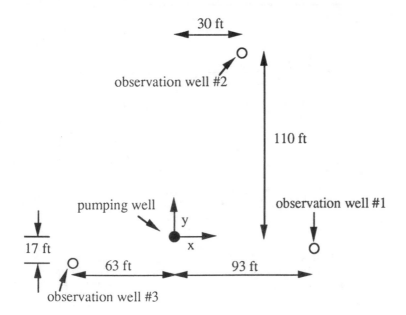

Figure 16.1   Locations of pumping and observation wells for Example 16.1.

Table 16.1   Drawdown data for observation wells in Example 16.1.

Time, t (min)	Drawdown, s (ft)		
	Well #1	Well #2	Well #3
0.5	1.10	0.50	1.61
1	1.94	1.13	2.50
2	2.99	2.00	3.57
3	3.55	2.50	4.21
4	3.99	2.99	4.65
6	4.61	3.57	5.28
8	5.08	4.02	5.76
10	5.42	4.36	6.08
15	6.08	5.02	6.79
20	6.62	5.55	7.25
30	7.23	6.08	7.92
40	7.69	6.62	8.38
50	8.04	6.96	8.76
60	8.33	7.25	9.02
90	9.02	7.92	9.72
120	9.52	8.38	10.23
150	9.83	8.76	10.56
180	10.09	9.02	10.86
240	10.61	9.52	11.33
300	10.99	9.83	11.69
360	11.28	10.23	11.97
480	11.77	10.65	12.47
720	12.41	11.33	13.11

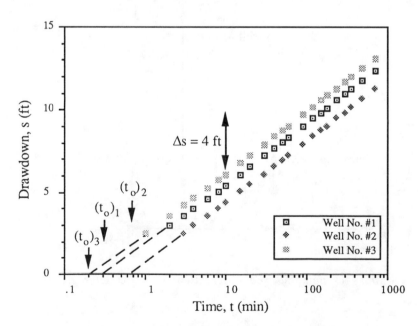

Figure 16.2   Semilogarithmic plot of drawdown data for Example 16.1.

The location of the coordinate system origin is arbitrary. In Figure 16.1, the origin is at the pumping well

observation well	x (ft)	y (ft)
1	93	0
2	30	110
3	-63	-17

A semilogarithmic plot of the data (Figure 16.2) shows that the straight-line method of analysis is applicable. The slope of the three lines is the same, 4 ft per log cycle. The zero drawdown intercepts are

observation well	t_o (min)	t_o (day)
1	0.35	2.43×10^{-4}
2	0.70	4.86×10^{-4}
3	0.20	1.39×10^{-4}

From Equation 16.19

$$D' = \left[\frac{2.303\ Q}{4\pi\ \Delta s} \right]^2$$

$$= \left[\frac{(2.303) \left(200\ \frac{gal}{min} \right) \left(\frac{ft}{7.48\ gal} \right) \left(\frac{1440\ min}{day} \right)}{4\pi\ (4\ ft)} \right]^2$$

$$= 3.112 \times 10^6\ \frac{ft^4}{day}$$

Substituting into Equation 16.21 yields

$$\begin{bmatrix} (0)^2 & (93)^2 & -2(93)\,(0) \\ (110)^2 & (30)^2 & -2(30)\,(110) \\ (-17)^2 & (-63)^2 & -2(-63)\,(-17) \end{bmatrix} \begin{Bmatrix} ST_{xx} \\ ST_{yy} \\ ST_{xy} \end{Bmatrix} = \begin{Bmatrix} 2.25\ (2.43 \times 10^{-4})\ (3.112 \times 10^6) \\ 2.25\ (4.86 \times 10^{-4})\ (3.112 \times 10^6) \\ 2.25\ (1.39 \times 10^{-4})\ (3.112 \times 10^6) \end{Bmatrix}$$

or

$$\begin{bmatrix} 0 & 8649 & -186 \\ 12100 & 900 & -6600 \\ 289 & 3969 & -2142 \end{bmatrix} \begin{Bmatrix} ST_{xx} \\ ST_{yy} \\ ST_{xy} \end{Bmatrix} = \begin{Bmatrix} 1701 \\ 3403 \\ 973 \end{Bmatrix}$$

which gives

$$ST_{xx} = 0.234 \text{ ft}^2$$
$$ST_{yy} = 0.195 \text{ ft}^2$$
$$ST_{xy} = -0.061 \text{ ft}^2$$

Substituting into Equation 16.10 yields

$$S = \left[\frac{(0.234)\ (0.195) - (-0.061)^2}{3.112 \times 10^6}\right]^{1/2}$$

$$= 1.16 \times 10^{-4}$$

which gives

$$T_{xx}\ \frac{0.234\ \frac{\text{ft}^2}{\text{day}}}{1.16 \times 10^{-4}} = 2016\ \frac{\text{ft}^2}{\text{day}}$$

$$T_{yy}\ \frac{0.195\ \frac{\text{ft}^2}{\text{day}}}{1.16 \times 10^{-4}} = 1680\ \frac{\text{ft}^2}{\text{day}}$$

$$T_{xy}\ \frac{-0.061\ \frac{\text{ft}^2}{\text{day}}}{1.16 \times 10^{-4}} = -526\ \frac{\text{ft}^2}{\text{day}}$$

The principal transmissivities can be computed using Equations 16.11 and 16.12

$$T_{\varepsilon\varepsilon} = \frac{1}{2}\left[(2016 + 1680) + \sqrt{(2016 - 1680)^2 + 4(-526)^2}\right]$$
$$= 2400\ \frac{\text{ft}^2}{\text{day}}$$

$$T_{\eta\eta} = \frac{1}{2}\left[(2016 + 1680) - \sqrt{(2016 - 1680)^2 + 4(-526)^2}\right]$$
$$= 1296\ \frac{\text{ft}^2}{\text{day}}$$

and the axis of the maximum transmissivity is oriented at an angle θ from the x axis

$$\theta = \tan^{-1}\left[\frac{2400 - 2016}{-526}\right] = -36°$$

Chapter 17

MODEL 12: TRANSIENT, UNDERGOING CONVERSION FROM CONFINED TO UNCONFINED

17.1 CONCEPTUAL MODEL

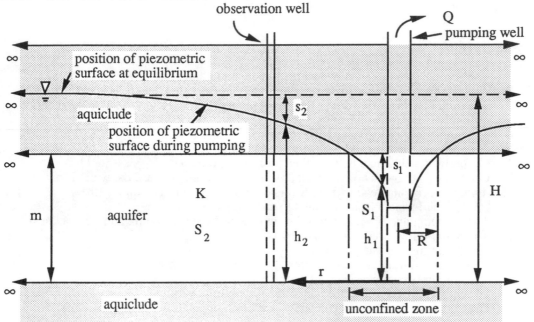

Definition of Terms

h_1 = head in the unconfined zone ($r < R$), L

h_2 = head in the confined zone ($r > R$), L

H = initial piezometric head before pumping, L

K = aquifer hydraulic conductivity, LT^{-1}

m = aquifer thickness, L

Q = constant pumping rate, L^3T^{-1}

r = radial distance from pumping well to a point on the cone of depression (all distances are measured from the center of wells), L

R   = radial distance from the pumping well to the edge of the confined zone, L

s_1 = drawdown in the unconfined zone $(r < R)$, L
          $= m - h_1$

s_2 = drawdown in the confined zone $(r > R)$, L
          $= H - h_2$

S_1 = unconfined zone storativity, dimensionless

S_2 = confined zone storativity, dimensionless

Assumptions

1. The aquifer is bounded above and below by aquicludes.
2. All layers are horizontal and extend infinitely in the radial direction.
3. The initial piezometric surface (before pumping begins) is horizontal and extends infinitely in the radial direction.
4. The aquifer is homogeneous and isotropic.
5. Groundwater density and viscosity are constant.
6. Groundwater flow can be described by Darcy's Law.
7. Groundwater flow is horizontal and is directed radially toward the well. Flow remains horizontal, even after conversion of a portion of the aquifer (the "unconfined zone") from confined to unconfined.
8. The pumping and observation wells are screened over the entire aquifer thickness.
9. The pumping well has an infinitesimal diameter.
10. The pumping rate is constant.
11. Head losses through the well screen and pump intake are negligible.
12. The aquifer is compressible and completely elastic.
13. Groundwater flow above the water table is negligible.
14. Drawdown in the unconfined zone is small compared to the saturated aquifer thickness.
15. As the water level drops below the confining layer, there is immediate conversion of a portion of the aquifer from confined to unconfined.

17.2 MATHEMATICAL MODEL

Governing Equation

Two governing equations are required, one for the unconfined zone and one for the confined zone. Both are derived by combining Darcy's Law with the principle of conservation of mass in a radial coordinate system (Moench and Prickett, 1972)

• For the unconfined zone

$$\frac{\partial^2 h_1}{\partial r^2} + \frac{1}{r}\frac{\partial h_1}{\partial r} = \frac{S_1}{Km}\frac{\partial h_1}{\partial t}, \qquad 0 \leq r \leq R \tag{17.1}$$

• For the confined zone

$$\frac{\partial^2 h_2}{\partial r^2} + \frac{1}{r}\frac{\partial h_2}{\partial r} = \frac{S_2}{Km}\frac{\partial h_2}{\partial t}, \qquad r \geq R \qquad (17.2)$$

where h_1 and h_2 are the heads in the unconfined and confined ones, respectively, r is the radial distance from the pumping well, S_1 and S_2 are the storativities of the unconfined zone and confined zone, respectively, K and m are the hydraulic conductivity and saturated thickness of the aquifer, and t is time.

Initial Conditions

• Before pumping begins drawdown is zero everywhere and the head at each point is equal to the initial piezometric head H

$$h_2(r, t = 0) = H \qquad 0 \leq r \leq \infty \qquad (17.3)$$

Boundary Conditions

• At the radius of water table conversion, the head equals the thickness of the aquifer and the derivative of head is continuous

$$h_2(r = R, t) = h_1(r = R, t) = m \qquad t > 0 \qquad (17.4)$$

$$\frac{\partial h_1 (r = R, t)}{\partial r} = \frac{\partial h_2 (r = R, t)}{\partial r} \qquad t > 0 \qquad (17.5)$$

• Groundwater flow to the pumping well is constant and uniform over the aquifer thickness (which is a result of the assumption of horizontal groundwater flow in the unconfined zone)

$$\lim_{r \to 0} 2\pi r T \frac{\partial h_1(r = 0, t)}{\partial r} = Q \quad t > 0 \qquad (17.6)$$

• At an infinite distance from the pumping well, drawdown is zero and the head is equal to the initial piezometric head H

$$h_2(r = \infty, t) = H \qquad t \geq 0 \qquad (17.7)$$

17.3 ANALYTICAL SOLUTION

17.3.1 General Solution

The solution for heads in the unconfined and confined zones are (Moench and Prickett, 1972)

$$h_1 = m - \frac{Q}{4\pi T}[W(u_1) - W(\upsilon)], \qquad r < R \qquad (17.8)$$

$$h_2 = H - \frac{(H - m)\,W(u_2)}{W(\upsilon S_2/S_1)}, \qquad r > R \qquad (17.9)$$

where $T = Km$, $W(u_1)$, $W(\upsilon)$, $W(u_2)$, and $W(\upsilon S_2/S_1)$ are the Theis well function $W(u)$ (Chapter 8) with u replaced by u_1, υ, u_2, or $\upsilon S_2/S_1$, and

$$u_1 = \frac{r^2 S_1}{4Tt} \qquad (17.10)$$

$$u_2 = \frac{r^2 S_2}{4Tt} \qquad (17.11)$$

$$\upsilon = \frac{R^2 S_1}{4Tt} \qquad (17.12)$$

It is often easier to analyze pumping test results in terms of drawdown, s

$$s_1 = m - h_1 = \frac{Q}{4\pi T} W(u_1, \upsilon) \qquad (17.13)$$

$$s_2 = H - h_2 = \frac{Q}{4\pi T} W(u_2)\exp[\upsilon(S_2/S_1 - 1)] \qquad (17.14)$$

where

$$W(u_1, \upsilon) = W(u_1) - W(\upsilon) = \int_{u_1}^{\upsilon} \frac{e^{-z}}{z} dz \qquad (17.15)$$

Note that s_1 is defined with respect to the aquifer top while s_2 is defined with respect to the initial piezometric surface.

Selected values of the function $W(u_1, \upsilon)$ are in Table 17.1. Values of $W(u_1, \upsilon)$ can also be computed with the computer program TYPE12; drawdown can be computed with the computer program DRAW12 (Appendix B). Selected values of the function $W(u)$ are in Table 8.1. Values of $W(u)$ can also be computed with the computer program TYPE3 (Appendix B).

17.3.2   Special Case Solutions

Early-Time Solution

When $t \le \frac{R^2 S_1}{12T}$, the effects of dewatering in the unconfined zone can be neglected and the Theis solution (Model 3) applies (Moench and Prickett, 1972).

Late-Time Solution

When $t > \frac{5R^2 S_1}{T}$, unconfined drawdown, s_1, can be described by the Theis solution (Model 3) by setting $S = S_y + S_s m$.

Table 17.1   Values of the function $W(u_1, \upsilon)$ for the unconfined zone of an isotropic, homogeneous, nonleaky aquifer undergoing water table conversion (after Moench and Prickett, 1972, p. 497).

$1/u_1$	∞	2	1	.5	.3	.2	.1	.05	.03	.02	.01	.005	.003	.002	.001
7.692×10^3	8.3708	8.3219	8.1515	7.8111	7.4651	7.1482	6.5479	5.9029	5.4117	5.0161	4.3329	3.6447	3.1359	2.7315	2.0393
5.917×10^3	8.1085	8.0596	7.8891	7.5487	7.2028	6.8859	6.2856	5.6406	5.1494	4.7538	4.0706	3.2824	2.8736	2.4691	1.7770
4.552×10^3	7.8462	7.7973	7.6268	7.2864	6.9405	6.6236	6.0233	5.3783	4.8871	4.4915	3.8083	3.1201	2.6113	2.2068	1.5147
3.501×10^3	7.5839	7.5350	7.3645	7.0241	6.6782	6.3613	5.7610	5.1160	4.6248	4.2292	3.5460	2.8578	2.3490	1.9445	1.2524
2.693×10^3	7.3216	7.2727	7.1022	6.7619	6.4160	6.0990	5.4987	4.8537	4.3625	3.9669	3.2837	2.5955	2.0867	1.6822	0.9901
2.072×10^3	7.0594	7.0105	6.8400	6.4996	6.1537	5.8367	5.2365	4.5915	4.1003	3.7047	3.0214	2.3333	1.8245	1.4200	0.7278
1.594×10^3	6.7972	6.7483	6.5778	6.2374	5.8915	5.5745	4.9742	4.3293	3.8380	3.4425	2.7592	2.0711	1.5622	1.1578	0.4656
1.226×10^3	6.5350	6.4861	6.3156	5.9752	5.6293	5.3123	4.7121	4.0671	3.5759	3.1803	2.4971	1.8089	1.3001	0.8956	0.2034
9.430×10^2	6.2729	6.2240	6.0535	5.7131	5.3672	5.0502	4.4499	3.8050	3.3137	2.9182	2.2349	1.5468	1.0379	0.6335	
7.254×10^2	6.0108	5.9619	5.7914	5.4510	5.1051	4.7882	4.1879	3.5429	3.0517	2.6561	1.9729	1.2847	0.7759	0.3714	
5.580×10^2	5.7489	5.7000	5.5295	5.1891	4.8432	4.5262	3.9259	3.2810	2.7897	2.3942	1.7109	1.0228	0.5139	0.1095	
4.292×10^2	5.4870	5.4381	5.2677	4.9273	4.5814	4.2644	3.6641	3.0191	2.5279	2.1323	1.4491	0.7609	0.2521		
3.302×10^2	5.2254	5.1765	5.0060	4.6656	4.3197	4.0027	3.4025	2.7575	2.2663	1.8707	1.1874	0.4993			
2.540×10^2	4.9639	4.9150	4.7445	4.4041	4.0582	3.7413	3.1410	2.4960	2.0048	1.6092	0.9260	0.2378			
1.954×10^2	4.7027	4.6538	4.4833	4.1430	3.7971	3.4801	2.8798	2.2348	1.7436	1.3480	0.6648				
1.503×10^2	4.4419	4.3930	4.2225	3.8821	3.5362	3.2193	2.6190	1.9740	1.4828	1.0872	0.4040				
1.156×10^2	4.1815	4.1326	3.9621	3.6218	3.2759	2.9589	2.3586	1.7136	1.2224	0.8268	0.1436				

Table 17.1   (Continued).

$1/u_1$	∞	2	1	.5	.3	.2	.1	.05	.03	.02	.01	.005	.003	.002	.001
8.893×10^1	3.9217	3.8728	3.7024	3.3620	3.0161	2.6991	2.0988	1.4538	0.9626	0.5670					
6.840×10^1	3.6627	3.6138	3.4433	3.1030	2.7571	2.4401	1.8398	1.1948	0.7036	0.3080					
5.262×10^1	3.4047	3.3558	3.1853	2.8449	2.4990	2.1821	1.5818	0.9368	0.4456	0.0500					
4.048×10^1	3.1480	3.0991	2.9286	2.5882	2.2423	1.9253	1.3251	0.6801	0.1889						
3.114×10^1	2.8929	2.8440	2.6736	2.3332	1.9873	1.6703	1.0700	0.4250							
2.395×10^1	2.6400	2.5911	2.4207	2.0803	1.7344	1.4174	0.8171	0.1721							
1.842×10^1	2.3899	2.3410	2.1705	1.8301	1.4842	1.1673	0.5670								
1.417×10^1	2.1433	2.0944	1.9239	1.5835	1.2376	0.9207	0.3204								
1.090×10^1	1.9013	1.8524	1.6819	1.3415	0.9956	0.6786	0.0784								
8.386×10^0	1.6650	1.6161	1.4457	1.1053	0.7594	0.4424									
6.451×10^0	1.4361	1.3872	1.2167	0.8763	0.5304	0.2135									
4.962×10^0	1.2164	1.1674	0.9970	0.6566	0.3107										
3.817×10^0	1.0080	0.9591	0.7886	0.4482	0.1023										
2.936×10^0	0.8135	0.7646	0.5941	0.2537											
2.259×10^0	0.6356	0.5867	0.4163	0.0759											
1.737×10^0	0.4774	0.4285	0.2580												
1.336×10^0	0.3414	0.2925	0.1220												
1.028×10^0	0.2296	0.1807	0.0103												
7.908×10^{-1}	0.1431	0.0942													

υ

17.4 METHODS OF ANALYSIS

17.4.1 General Solution

Drawdown data can be analyzed by two methods. The first method can be used to interpret time-drawdown data for observation wells within the unconfined zone (i.e., observation wells located near enough to the pumping well that the radius of the unconfined zone R passes the observation well during the test). The second method can be used to interpret time-drawdown data for observation wells that remain within the confined zone for the duration of the test.

17.4.2 Observation Wells Within Unconfined Zone

i   Plot s vs. log t for each observation well. The slope of the data should change abruptly (increasing rate of water level decline) when the conversion from a confined to an unconfined aquifer occurs at the observation well. Use this plot to divide the drawdown data into two sets: those before conversion and those after conversion. Compute values of s_1 using the drawdown data after conversion and the elevation of the aquifer top. The elevation of the aquifer top can be determined from the water elevation in the observation well at the time of conversion or from drilling records.

ii   Prepare a plot of $W(u_1, \upsilon)$ vs. $1/u_1$ on logarithmic paper. This plot is called the type curve (Figure 17.1). The type curve can be prepared using the values of $W(u_1, \upsilon)$ in Table 17.1 or values computed by the computer program TYPE12 (Appendix B).

iii   Plot s_1 vs. t using the same logarithmic scales used to prepare the type curve. This plot is called the data curve.

iv   Overlay the data curve on the type curves. Shift the plots relative to each other, keeping respective axes parallel, until a position of best fit is found between the data curve and one of the type curves.

v   From the best fit portion of the curves select a match-point and record match-point values $W(u_1, \upsilon)^*$, u_1^*, s_1^* and υ^*, and t^*.

vi   Substitute s_1^* and $W(u_1, \upsilon)^*$ into Equation 17.13 and solve for T.

vii   Substitute the computed value for T, u_1^*, and t^* into Equation 17.10 and solve for S_1. Check on the computed values of T and S_1 by substituting υ^* into Equation 17.12.

viii Prepare a plot of $W(u_2)$ vs. $1/u_2$ on logarithmic paper using the values of $W(u)$ in Table 8.1 or values computed by the computer program TYPE3 (Appendix B).

ix   Plot s_2 vs. t using the same logarithmic scales used to prepare the type-curve. Following the match-point method described in Chapter 8, select a match-point and record match-point values $1/u_2^*$ and t^*.

x   Substitute the computed value of T and u_2^* and t^* into Equation 17.11 and solve for S_2.

17.4.3   Observation Wells Within Confined Zone

i   Plot s vs. log t for the <u>pumping</u> well (which has undergone conversion to an unconfined aquifer). Draw a straight line through the data and compute the true aquifer transmissivity by the Cooper and Jacob (1946) method

$$T = -\frac{2.303Q}{2\pi\Delta s} \tag{17.16}$$

where Δs = change in drawdown over one log cycle.

ii Plot s vs. log t for an <u>observation</u> well in the confined zone. Compute the apparent aquifer transmissivity by the Cooper and Jacob method described in step i. The error between the apparent transmissivity and the true transmissivity is proportional to the factor, $\exp[\upsilon(S_2/S_1 - 1)]$. Since S_2/S_1 is often very small, the error is approximately equal to $e^{-\upsilon}$

$$T_{true} = (T_{apparent})\, e^{-\upsilon} \tag{17.17}$$

Solve for υ, substitute into Equation 17.12 and compute S_1. The value of t to use in Equation 17.12 is obtained by projecting a straight line fitted to the semilogarithmic plot of drawdown data for the observation well data and determining the time at which conversion is expected to occur at the observation well (i.e., when R = distance between the pumping well and the observation well).

iii Prepare a $W(u_2)$ vs. $1/u_2$ type curve using the values in Table 8.1 or values computed with the computer program TYPE3 (Appendix B).

iv Prepare a plot of s vs. t (the data curve) for the observation well using the same logarithmic scales used to prepare the type curve.

v   Using the match-point method described in Chapter 8, overlay the two curves, select a match-point and record match-point values $1/u_2^*$ and t^*.

vi Substitute the computed value of T and $1/u_2^*$ and t^* into Equation 17.11 and solve for S_2.

The assumption that the pumping well has an infinitesimal diameter implies that conversion from confined to unconfined conditions near the pumping well occurs immediately after pumping begins. Actually this will take a finite amount of time because of well storage. Test conditions that increase the time required for aquifer conversion near the pumping well are small Q, large T, and large (H - m). This implies a small value for υ. Drawdown will initially follow the Theis solution (Model 3) until conversion occurs at the pumping well; drawdown will then follow the solution in Equation 17.8.

Example 17.1

A pumping test was conducted in a confined aquifer. The pumping rate was 100 gal/min and the initial piezometric surface (before pumping began) was located 1.1 ft above the top of the aquifer. Drawdown data for an observation well located 500 ft from the pumping well are in Table 17.2. Compute T and S_1

Table 17.2   Drawdown data for Example 17.1.

time (min)	drawdown (ft)	time (min)	drawdown (ft)	s_1 (ft)
1	0.10	420	1.05	-
2	0.17	480	1.11	-
4	0.31	540	1.19	0.09
6	0.38	600	1.25	0.15
8	0.44	660	1.30	0.20
10	0.48	720	1.35	0.25
20	0.59	1440	1.84	0.74
30	0.66	2160	2.20	1.10
40	0.70	2880	2.60	1.50
50	0.74	3600	2.65	1.55
60	0.77	5040	3.00	1.90
120	0.86	6480	3.20	2.10
180	0.91	7920	3.41	2.31
240	0.95	9360	3.60	2.50
300	0.98	10800	3.75	2.65
360	1.00			

A plot of s vs. log(t) indicates that conversion from confined to unconfined conditions occurred at the observation well about 470 minutes after pumping began (Figure 17.1). Values for drawdown s_1 are computed by subtracting 1.1 ft from the measured drawdown values for t > 540 minutes (Table 17.2).

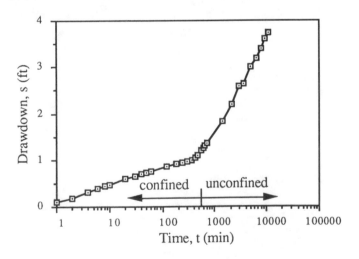

Figure 17.1   Drawdown data for Example 17.1.

Application of the match-point method to the data after conversion results in the following match-point values (Figure 17.2):

t^* = 1440 min
s_1^* = 0.74 ft

$$1/u_1{}^* \quad = 26, \qquad\qquad u_1{}^* = 0.038$$
$$W(u_1, \upsilon)^* = 0.88$$
$$\upsilon^* \qquad = 0.1$$

Substituting into Equation 17.13

$$T = \frac{Q}{4\pi s^*}\, W(u_1, \upsilon)^* = \frac{\left(\dfrac{100\ \text{gal}}{\text{min}}\right)\left(\dfrac{1440\ \text{min}}{\text{day}}\right)(0.88)}{4\ (\pi)\ (0.74\ \text{ft})} = 13627\, \frac{\text{gal}}{\text{day}\cdot\text{ft}}$$

Substituting into Equation 17.10

$$S_1 = \frac{u_1{}^*\ 4Tt^*}{r^2} = \frac{(0.038)\ (4)\left(13627\, \dfrac{\text{gal}}{\text{day}\cdot\text{ft}}\right)\left(\dfrac{\text{ft}}{7.48\ \text{gal}}\right)(1440\ \text{min})\left(\dfrac{\text{day}}{1440\ \text{min}}\right)}{(500\ \text{ft})^2}$$

$$= 0.0011$$

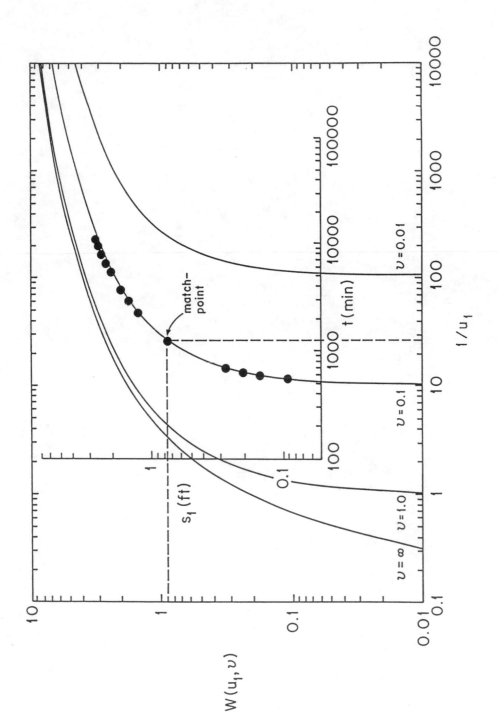

Figure 17.2 Application of match-point method for Example 17.1.

Chapter 18

MODEL 13: TRANSIENT, UNCONFINED, ANISOTROPIC

18.1 CONCEPTUAL MODEL

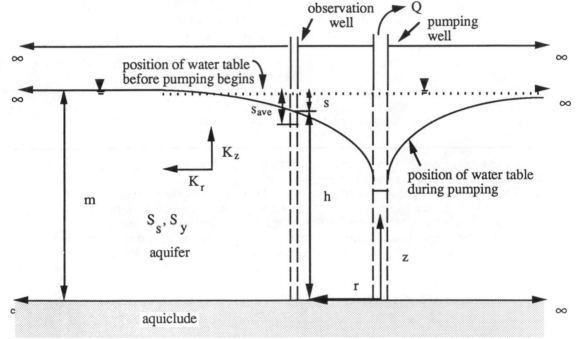

Definition of Terms

K_r = aquifer horizontal hydraulic conductivity, LT^{-1}

K_z = aquifer vertical hydraulic conductivity, LT^{-1}

m = saturated thickness before pumping begins, L

Q = constant pumping rate, L^3T^{-1}

r = radial distance from pumping well to a point on the cone of depression (all distances are measured from the center of wells), L

s = drawdown of water table during pumping, L

s_{ave} = average drawdown of water table for an observation well during pumping, L

S_s = aquifer specific storage, L^{-1}

S_y = aquifer specific yield, dimensionless

z   = vertical coordinate with origin at aquifer base, L

Assumptions

1. The aquifer is bounded below by an aquiclude.
2. All layers are horizontal and extend infinitely in the radial direction.
3. The initial water table (before pumping begins) is horizontal and extends infinitely in the radial direction.
4. The aquifer is homogeneous and anisotropic. The horizontal and vertical hydraulic conductivities of the aquifer may be different.
5. Groundwater density and viscosity are constant.
6. Groundwater flow can be described by Darcy's Law.
7. Pumping and observation wells are screened (or uncased) over the entire saturated thickness.
8. The pumping rate is constant.
9. Head losses through the well screen and pump intake are negligible.
10. The pumping well has an infinitesimal diameter.
11. The aquifer is compressible and completely elastic; pumping instantaneously releases water from storage by expansion of the pore water or compression of the soil skeleton. This water is accounted for in the specific storage, S_s, term.
     As the water table drops, additional water is released from storage due to gravity drainage of the effective pore space. This water is accounted for in the specific yield term, S_y.
12. Groundwater flow above the water table is negligible.
13. Drawdown is small compared to the saturated aquifer thickness.

18.2 MATHEMATICAL MODEL

Governing Equation

The governing equation is derived by combining Darcy's Law with the principle of conservation of mass in a radial coordinate system (Neuman, 1972, 1973a, 1973b)

$$\frac{\partial^2 s}{\partial r^2} + \left(\frac{1}{r}\right)\frac{\partial s}{\partial r} + \frac{K_z}{K_r}\frac{\partial^2 s}{\partial z^2} = \frac{S_s}{K_r}\frac{\partial s}{\partial t} \qquad (18.1)$$

where s is drawdown, r is radial distance from the pumping well, K_z and K_r are the vertical and horizontal hydraulic conductivities of the aquifer, S_s is specific storage, and t is time.

Initial Conditions

• Before pumping begins drawdown is zero everywhere

$$s(r, z, t = 0) = 0 \qquad (18.2)$$

Boundary Conditions

- At an infinite distance from the pumping well, drawdown is zero

$$s(r = \infty, z, t) = 0 \tag{18.3}$$

- At the aquifer base, the change in drawdown with depth is zero

$$\frac{\partial s(r, z = 0, t)}{\partial z} = 0 \tag{18.4}$$

- The water table is a moving boundary. As it drops, water from the effective pore space flows across the water table and into the aquifer. This boundary condition is defined by writing Darcy's Law for flow across the water table

$$\frac{\partial s(r, z = m, t)}{\partial z} = -\frac{S_y}{K_z} \frac{\partial s(r, z = m, t)}{\partial t} \tag{18.5}$$

- Groundwater flow to the pumping well is constant and uniform over the aquifer thickness

$$\lim_{r \to 0} \int_0^m r\frac{\partial s}{\partial r} \, dz = -\frac{Q}{2\pi K_r} \tag{18.6}$$

18.3 ANALYTICAL SOLUTION

18.3.1 General Solution

The solution for average drawdown, s_{ave}, in a fully penetrating, fully screened observation well is (Neuman, 1972; 1973a, 1973b)

$$s_{ave} = \frac{Q}{4\pi T} W(t_s, \sigma, \beta) \tag{18.7}$$

$W(t_s, \sigma, \beta)$ is a well function

$$W(t_s, \sigma, \beta) = \int_0^\infty 4yJ_o (y\beta^{1/2}) \left[u_o(y) + \sum_{n=1}^{\infty} u_n(y) \right] dy \tag{18.8}$$

where

$$u_o(y) = \frac{\{1 - \exp[-t_s\beta(y^2 - \gamma_o^2)]\} \tanh(\gamma_o)}{\{y^2 + (1 + \sigma) \gamma_o^2 - (y^2 - \gamma_o^2)^2 / \sigma\} \gamma_o} \tag{18.9}$$

$$u_n(y) = \frac{\{1 - \exp[-t_s\beta(y^2 + \gamma_n^2)]\} \tan(\gamma_n)}{\{y^2 - (1 + \sigma) \gamma_n^2 - (y^2 + \gamma_n^2)^2 / \sigma\} \gamma_n} \tag{18.10}$$

J_o is the zero order Bessel function of the first kind, and γ_o and γ_n are the roots of the equations

$$\sigma\gamma_o \sinh (\gamma_o) - (y^2 - \gamma_o^2) \cosh (\gamma_o) = 0, \, \gamma_o^2 < y^2 \tag{18.11}$$

$$\sigma\gamma_n \sin (\gamma_n) + (y^2 + \gamma_n^2) \cos (\gamma_n) = 0,$$
$$(2n - 1) (\pi/2) < \gamma_n < n\pi, \quad n \geq 1 \tag{18.12}$$

The dimensionless parameters are

$$t_s = \frac{Tt}{S_s mr^2} \tag{18.13}$$

$$\sigma = \frac{S_s m}{S_y} \tag{18.14}$$

$$\beta = \frac{r^2}{m^2} \frac{K_z}{K_r} \tag{18.15}$$

Values of the function $W(t_s, \sigma, \beta)$ can be computed with the computer program TYPE13; average drawdown can be computed with the computer program DRAW13 (Appendix B).

18.3.2   Special Case Solutions

Late-Time Solution

After relatively long periods of pumping, the effects of delayed yield are negligible and drawdown may be computed using the Theis solution (Model 3). Boulton (1963) developed a curve that can be used to predict the time, t_{wt}, when the effects of delayed yield become negligible (Figure 18.1). The figure gives αt_{wt} as a function of r/D, where

α = an empirical coefficient, T^{-1}
r = radial distance from the pumping well, L
D = $[T/(\alpha S_y)]^{1/2}$, L

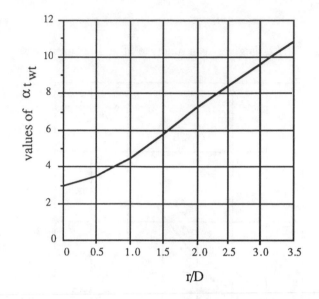

Figure 18.1 Time after which the effects of delayed yield cease to influence drawdown (from Prickett, 1965).

The values of α and D are determined by application of the match-point method (see Section 18.4.1). There have been many attempts to correlate α with physical aquifer properties. If estimates of these properties are available, estimates of α can be obtained by application of the methods listed below:

Gambolati (1976), Neuman (1979)

$$\alpha = \frac{\varepsilon K_z}{S_y m} \tag{18.16}$$

$$\text{where } \varepsilon = 2.4 + \frac{0.384}{\left[\left(\dfrac{K_z}{K_r}\right)^{1/2} \dfrac{r}{m}\right]^{0.886}}$$

$$\text{or, for } \left(\frac{K_z}{K_r}\right)^{1/2} \frac{r}{m} \geq z, \quad \alpha = \frac{2.4 \, K_z}{S_y m}$$

Neuman (1975)

$$\alpha = \frac{K_z}{S_y m}\left[3.063 - 0.567 \log\left(\frac{K_z r^2}{K_r m^2}\right)\right] \tag{18.17}$$

Streltsova (1972)

$$\alpha = \frac{3K_z}{S_y m} \tag{18.18}$$

Prickett (1965)

The "delay index", $1/\alpha$, is given as a function of aquifer grain size (Figure 18.2).

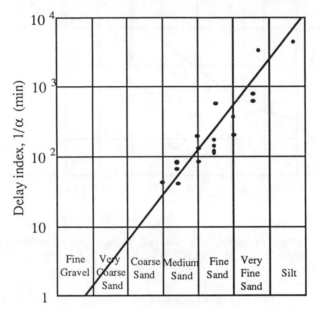

Materials through which gravity drainage takes place

Figure 18.2   Relationship between $1/\alpha$ and aquifer grain size (from Prickett, 1965).

18.4 METHODS OF ANALYSIS

18.4.1 General Solution

Match-Point Method

The well function, $W(t_s, \sigma, \beta)$, in Equation 18.7 is a function of three dimensionless parameters. The number of dimensionless parameters can be reduced from three to two if the aquifer specific yield can be assumed to be much larger than the storativity ($S_y \gg S_s m$) so that $\sigma = 0$. For this case the method of analysis proceeds by dividing the drawdown data into two parts for analysis: early data and late data. For the early data, the effects of gravity drainage above the water table on groundwater flow to the pumping well are negligible; for the late data, the effects of elastic storage within the aquifer are negligible . The result is two asymptotic families of type curves, one for early times which is primarily influenced by elastic storage, S_s, and one for late times where storage from gravity drainage, S_y, becomes important. In deference to earlier work by Boulton (1963), Neuman (1975) called these Type A (early) and Type B (late) curves. Equation 18.7 can be written as

$$s_{ave} = \frac{Q}{4\pi T} W(t_s, \beta), \text{ for early times} \tag{18.19a}$$

$$s_{ave} = \frac{Q}{4\pi T} W(t_y, \beta), \text{ for late times} \tag{18.19b}$$

where $W(t_s, \beta)$ is the well function in Equation 18.8 evaluated with $\sigma = 0$ and $W(t_y, \beta)$ is the well function in Equation 18.8 evaluated with $\sigma = 0$ and t_s replaced by t_y, where

$$t_y = \frac{Tt}{S_y r^2} \tag{18.20}$$

Values for the well functions $W(t_s, \beta)$ and $W(t_y, \beta)$ are in Tables 18.1 and 18.2, respectively. The following match-point method was developed based on this concept:

i   Prepare plots of $W(t_s, \beta)$ vs. t_s and $W(t_y, \beta)$ vs. t_y on logarithmic paper. These plots are called type curves. The curve $W(t_s, \beta)$ vs. t_s (the Type A curve) will be used to interpret early drawdown data and the curve $W(t_y, \beta)$ vs. t_y (the Type B curve) will be used to interpret late drawdown data. Many published type curves combine both plots (e.g., Figure 18.3). The type curves can be plotted using values in Tables 18.1 and 18.2 or values computed by the computer program TYPE13 (Appendix B).

ii Compute the corrected drawdown, $s_{corrected}$ using

$$s_{corrected} = s - (s^2)/2m \tag{18.21}$$

Note that the corrected drawdown data will only be used to match the Type B curves.

iii Plot $s_{corrected}$ vs. t using the same logarithmic scales used to prepare the type curves. This plot is called the data curve. Note that distance-drawdown data ($s_{corrected}$ vs. t/r^2) should not be used with this method (Neuman, 1975).

iv Overlay the corrected data curve on the Type B curve. Shift the plots relative to each other, keeping respective axes parallel, until a position of best fit is found between the data curve and one of the type curves. Use as much late data as possible.

v   From the best fit portion of the curves, select a match-point and record match-point values $W(t_y, \beta)^*$, t_y^*, s^*, t^*, and β^*.

vi Substitute s^* and $W(t_y, \beta)^*$ into Equation 18.19b and solve for T.

vii Substitute t^* and t_y^* and the computed value of T into Equation 18.20 and solve for S_y.

viii Repeat the match-point procedure using the uncorrected drawdown data and the Type A type curve. Use as much early data as possible.

ix From the best fit portion of the curves, select a match-point and record match-point values $W(t_s, \beta)^*$, t_s^*, s^*, t^*, and β^*.

x   Substitute s^* and $W(t_s, b)^*$ into Equation 18.19a and solve for T.

Table 18.1   Values for the function $W(t_s, \beta)$ for an unconfined aquifer at early times with fully penetrating, infinitesimally small wells (after Neuman, 1975, pp. 332-333).

t_s	$\beta = 0.001$	$\beta = 0.004$	$\beta = 0.01$	$\beta = 0.03$	$\beta = 0.06$	$\beta = 0.1$	$\beta = 0.2$	$\beta = 0.4$	$\beta = 0.6$
1×10^{-1}	2.48×10^{-2}	2.43×10^{-2}	2.41×10^{-2}	2.35×10^{-2}	2.30×10^{-2}	2.24×10^{-2}	2.14×10^{-2}	1.99×10^{-2}	1.88×10^{-2}
2×10^{-1}	1.45×10^{-1}	1.42×10^{-1}	1.40×10^{-1}	1.36×10^{-1}	1.31×10^{-1}	1.27×10^{-1}	1.19×10^{-1}	1.08×10^{-1}	9.88×10^{-2}
3.5×10^{-1}	3.58×10^{-1}	3.52×10^{-1}	3.45×10^{-1}	3.31×10^{-1}	3.18×10^{-1}	3.04×10^{-1}	2.79×10^{-1}	2.44×10^{-1}	2.17×10^{-1}
6×10^{-1}	6.62×10^{-1}	6.48×10^{-1}	6.33×10^{-1}	6.01×10^{-1}	5.70×10^{-1}	5.40×10^{-1}	4.83×10^{-1}	4.03×10^{-1}	3.43×10^{-1}
1×10^{0}	1.02×10^{0}	9.92×10^{-1}	9.63×10^{-1}	9.05×10^{-1}	8.49×10^{-1}	7.92×10^{-1}	6.88×10^{-1}	5.42×10^{-1}	4.38×10^{-1}
2×10^{0}	1.57×10^{0}	1.52×10^{0}	1.46×10^{0}	1.35×10^{0}	1.23×10^{0}	1.12×10^{0}	9.18×10^{-1}	6.59×10^{-1}	4.97×10^{-1}
3.5×10^{0}	2.05×10^{0}	1.97×10^{0}	1.88×10^{0}	1.70×10^{0}	1.51×10^{0}	1.34×10^{0}	1.03×10^{0}	6.90×10^{-1}	5.07×10^{-1}
6×10^{0}	2.52×10^{0}	2.41×10^{0}	2.27×10^{0}	1.99×10^{0}	1.73×10^{0}	1.47×10^{0}	1.07×10^{0}	6.96×10^{-1}	
1×10^{1}	2.97×10^{0}	2.80×10^{0}	2.61×10^{0}	2.22×10^{0}	1.85×10^{0}	1.53×10^{0}	1.08×10^{0}		
2×10^{1}	3.56×10^{0}	3.30×10^{0}	3.00×10^{0}	2.41×10^{0}	1.92×10^{0}	1.55×10^{0}			
3.5×10^{1}	4.01×10^{0}	3.65×10^{0}	3.23×10^{0}	2.48×10^{0}	1.93×10^{0}				
6×10^{1}	4.42×10^{0}	3.93×10^{0}	3.37×10^{0}	2.49×10^{0}	1.94×10^{0}				
1×10^{2}	4.77×10^{0}	4.12×10^{0}	3.43×10^{0}	2.50×10^{0}					
2×10^{2}	5.16×10^{0}	4.26×10^{0}	3.45×10^{0}						
3.5×10^{2}	5.40×10^{0}	4.29×10^{0}	3.46×10^{0}						
6×10^{2}	5.54×10^{0}	4.30×10^{0}							
1×10^{3}	5.59×10^{0}								
2×10^{3}	5.62×10^{0}								
3.5×10^{3}	5.62×10^{0}	4.30×10^{0}	3.46×10^{0}	2.50×10^{0}	1.94×10^{0}	1.55×10^{0}	1.08×10^{0}	6.96×10^{-1}	5.07×10^{-1}

Values were obtained from Equation 18.8 with $\sigma = 10^{-9}$

Table 18.1 (Continued).

t_s	$\beta = 0.8$	$\beta = 1.0$	$\beta = 1.5$	$\beta = 2.0$	$\beta = 2.5$	$\beta = 3.0$	$\beta = 4.0$	$\beta = 5.0$	$\beta = 6.0$	$\beta = 7.0$
1×10^{-1}	1.79×10^{-2}	1.70×10^{-2}	1.53×10^{-2}	1.38×10^{-2}	1.25×10^{-2}	1.13×10^{-2}	9.33×10^{-3}	7.72×10^{-3}	6.39×10^{-3}	5.30×10^{-3}
2×10^{-1}	9.15×10^{-2}	8.49×10^{-2}	7.13×10^{-2}	6.03×10^{-2}	5.11×10^{-2}	4.35×10^{-2}	3.17×10^{-2}	2.34×10^{-2}	1.74×10^{-2}	1.31×10^{-2}
3.5×10^{-1}	1.94×10^{-1}	1.75×10^{-1}	1.36×10^{-1}	1.07×10^{-1}	8.46×10^{-2}	6.78×10^{-2}	4.45×10^{-2}	3.02×10^{-2}	2.10×10^{-2}	1.51×10^{-2}
6×10^{-1}	2.96×10^{-1}	2.56×10^{-1}	1.82×10^{-1}	1.33×10^{-1}	1.01×10^{-1}	7.67×10^{-2}	4.76×10^{-2}	3.13×10^{-2}	2.14×10^{-2}	1.52×10^{-2}
1×10^{0}	3.60×10^{-1}	3.00×10^{-1}	1.99×10^{-1}	1.40×10^{-1}	1.03×10^{-1}	7.79×10^{-2}	4.78×10^{-2}		2.15×10^{-2}	
2×10^{0}	3.91×10^{-1}	3.17×10^{-1}	2.03×10^{-1}	1.41×10^{-1}						
3.5×10^{0}	3.94×10^{-1}									
6×10^{0}										
1×10^{1}										
2×10^{1}										
3.5×10^{1}										
6×10^{1}										
1×10^{2}										
2×10^{2}										
3.5×10^{2}										
6×10^{2}										
1×10^{3}										
2×10^{3}										
3.5×10^{3}	3.94×10^{-1}	3.17×10^{-1}	2.03×10^{-1}	1.41×10^{-1}	1.03×10^{-1}	7.79×10^{-2}	4.78×10^{-2}	3.13×10^{-2}	2.15×10^{-2}	1.52×10^{-2}

Values were obtained from Equation 18.8 with $\sigma = 10^{-9}$

Table 18.2   Values for the function $W(t_y, \beta)$ for an unconfined aquifer at late times with fully penetrating, infinitesimally small wells (after Neuman, 1975, pp. 332-333).

t_y	$\beta = 0.001$	$\beta = 0.004$	$\beta = 0.01$	$\beta = 0.03$	$\beta = 0.06$	$\beta = 0.1$	$\beta = 0.2$	$\beta = 0.4$	$\beta = 0.6$
1×10^{-4}	5.62×10^{0}	4.30×10^{0}	3.46×10^{0}	2.50×10^{0}	1.94×10^{0}	1.56×10^{0}	1.09×10^{0}	6.97×10^{-1}	5.08×10^{-1}
2×10^{-4}									
3.5×10^{-4}									
6×10^{-4}									
1×10^{-3}								6.97×10^{-1}	5.08×10^{-1}
2×10^{-3}								6.97×10^{-1}	5.09×10^{-1}
3.5×10^{-3}								6.98×10^{-1}	5.10×10^{-1}
6×10^{-3}								7.00×10^{-1}	5.12×10^{-1}
1×10^{-2}								7.03×10^{-1}	5.16×10^{-1}
2×10^{-2}							1.09×10^{0}	7.10×10^{-1}	5.24×10^{-1}
3.5×10^{-2}						1.56×10^{0}	1.10×10^{0}	7.20×10^{-1}	5.37×10^{-1}
6×10^{-2}					1.95×10^{0}	1.57×10^{0}	1.11×10^{0}	7.37×10^{-1}	5.57×10^{-1}
1×10^{-1}				2.51×10^{0}	1.96×10^{0}	1.58×10^{0}	1.13×10^{0}	7.63×10^{-1}	5.89×10^{-1}
2×10^{-1}	5.62×10^{0}	4.30×10^{0}		2.52×10^{0}	1.98×10^{0}	1.61×10^{0}	1.18×10^{0}	8.29×10^{-1}	6.67×10^{-1}
3.5×10^{-1}	5.63×10^{0}	4.31×10^{0}	3.47×10^{0}	2.54×10^{0}	2.01×10^{0}	1.66×10^{0}	1.24×10^{0}	9.22×10^{-1}	7.80×10^{-1}
6×10^{-1}	5.63×10^{0}	4.31×10^{0}	3.49×10^{0}	2.57×10^{0}	2.06×10^{0}	1.73×10^{0}	1.35×10^{0}	1.07×10^{0}	9.54×10^{-1}
1×10^{0}	5.63×10^{0}	4.32×10^{0}	3.51×10^{0}	2.62×10^{0}	2.13×10^{0}	1.83×10^{0}	1.50×10^{0}	1.29×10^{0}	1.20×10^{0}
2×10^{0}	5.64×10^{0}	4.35×10^{0}	3.56×10^{0}	2.73×10^{0}	2.31×10^{0}	2.07×10^{0}	1.85×10^{0}	1.72×10^{0}	1.68×10^{0}
3.5×10^{0}	5.65×10^{0}	4.38×10^{0}	3.63×10^{0}	2.88×10^{0}	2.55×10^{0}	2.37×10^{0}	2.23×10^{0}	2.17×10^{0}	2.15×10^{0}
6×10^{0}	5.67×10^{0}	4.44×10^{0}	3.74×10^{0}	3.11×10^{0}	2.86×10^{0}	2.75×10^{0}	2.68×10^{0}	2.66×10^{0}	2.65×10^{0}
1×10^{1}	5.70×10^{0}	4.52×10^{0}	3.90×10^{0}	3.40×10^{0}	3.24×10^{0}	3.18×10^{0}	3.15×10^{0}	3.14×10^{0}	3.14×10^{0}
2×10^{1}	5.76×10^{0}	4.71×10^{0}	4.22×10^{0}	3.92×10^{0}	3.85×10^{0}	3.83×10^{0}	3.82×10^{0}	3.82×10^{0}	3.82×10^{0}
3.5×10^{1}	5.85×10^{0}	4.94×10^{0}	4.58×10^{0}	4.40×10^{0}	4.38×10^{0}	4.38×10^{0}	4.37×10^{0}	4.37×10^{0}	4.37×10^{0}
6×10^{1}	5.99×10^{0}	5.23×10^{0}	5.00×10^{0}	4.92×10^{0}	4.91×10^{0}	4.91×10^{0}	4.91×10^{0}	4.91×10^{0}	4.91×10^{0}
1×10^{2}	6.16×10^{0}	5.59×10^{0}	5.46×10^{0}	5.42×10^{0}	5.42×10^{0}	5.42×10^{0}	5.42×10^{0}	5.42×10^{0}	5.42×10^{0}

Values were obtained from Equation 18.8 with $\sigma = 10^{-9}$

Table 18.2   (Continued).

t_y	$\beta = 0.8$	$\beta = 1.0$	$\beta = 1.5$	$\beta = 2.0$	$\beta = 2.5$	$\beta = 3.0$	$\beta = 4.0$	$\beta = 5.0$	$\beta = 6.0$	$\beta = 7.0$
1×10^{-4}	3.95×10^{-1}	3.18×10^{-1}	2.04×10^{-1}	1.42×10^{-1}	1.03×10^{-1}	7.80×10^{-2}	4.79×10^{-2}	3.14×10^{-2}	2.15×10^{-2}	1.53×10^{-2}
2×10^{-4}						7.81×10^{-2}	4.80×10^{-2}	3.15×10^{-2}	2.16×10^{-2}	1.53×10^{-2}
3.5×10^{-4}					1.03×10^{-1}	7.83×10^{-2}	4.81×10^{-2}	3.16×10^{-2}	2.17×10^{-2}	1.54×10^{-2}
6×10^{-4}					1.04×10^{-1}	7.85×10^{-2}	4.84×10^{-2}	3.18×10^{-2}	2.19×10^{-2}	1.56×10^{-2}
1×10^{-3}	3.95×10^{-1}	3.18×10^{-1}	2.04×10^{-1}	1.42×10^{-1}	1.04×10^{-1}	7.89×10^{-2}	4.78×10^{-2}	3.21×10^{-2}	2.21×10^{-2}	1.58×10^{-2}
2×10^{-3}	3.96×10^{-1}	3.19×10^{-1}	2.05×10^{-1}	1.43×10^{-1}	1.05×10^{-1}	7.99×10^{-2}	4.96×10^{-2}	3.29×10^{-2}	2.28×10^{-2}	1.64×10^{-2}
3.5×10^{-3}	3.97×10^{-1}	3.21×10^{-1}	2.07×10^{-1}	1.45×10^{-1}	1.07×10^{-1}	8.14×10^{-2}	5.09×10^{-2}	3.41×10^{-2}	2.39×10^{-2}	1.73×10^{-2}
6×10^{-3}	3.99×10^{-1}	3.23×10^{-1}	2.09×10^{-1}	1.47×10^{-1}	1.09×10^{-1}	8.38×10^{-2}	5.32×10^{-2}	3.61×10^{-2}	2.57×10^{-2}	1.89×10^{-2}
1×10^{-2}	4.03×10^{-1}	3.27×10^{-1}	2.13×10^{-1}	1.52×10^{-1}	1.13×10^{-1}	8.79×10^{-2}	5.68×10^{-2}	3.93×10^{-2}	2.86×10^{-2}	2.15×10^{-2}
2×10^{-2}	4.12×10^{-1}	3.37×10^{-1}	2.24×10^{-1}	1.62×10^{-1}	1.24×10^{-1}	9.80×10^{-2}	6.61×10^{-2}	4.78×10^{-2}	3.62×10^{-2}	2.84×10^{-2}
3.5×10^{-2}	4.25×10^{-1}	3.50×10^{-1}	2.39×10^{-1}	1.78×10^{-1}	1.39×10^{-1}	1.13×10^{-1}	8.06×10^{-2}	6.12×10^{-2}	4.86×10^{-2}	3.98×10^{-2}
6×10^{-2}	4.47×10^{-1}	3.74×10^{-1}	2.65×10^{-1}	2.05×10^{-1}	1.66×10^{-1}	1.40×10^{-1}	1.06×10^{-1}	8.53×10^{-2}	7.14×10^{-2}	6.14×10^{-2}
1×10^{-1}	4.83×10^{-1}	4.12×10^{-1}	3.07×10^{-1}	2.48×10^{-1}	2.10×10^{-1}	1.84×10^{-1}	1.49×10^{-1}	1.28×10^{-1}	1.13×10^{-1}	1.02×10^{-1}
2×10^{-1}	5.71×10^{-1}	5.06×10^{-1}	4.10×10^{-1}	3.57×10^{-1}	3.23×10^{-1}	2.98×10^{-1}	2.66×10^{-1}	2.45×10^{-1}	2.31×10^{-1}	2.20×10^{-1}
3.5×10^{-1}	6.97×10^{-1}	6.42×10^{-1}	5.62×10^{-1}	5.17×10^{-1}	4.89×10^{-1}	4.70×10^{-1}	4.45×10^{-1}	4.30×10^{-1}	4.19×10^{-1}	4.11×10^{-1}
6×10^{-1}	8.89×10^{-1}	8.50×10^{-1}	7.92×10^{-1}	7.63×10^{-1}	7.45×10^{-1}	7.33×10^{-1}	7.18×10^{-1}	7.09×10^{-1}	7.03×10^{-1}	6.99×10^{-1}
1×10^{0}	1.16×10^{0}	1.13×10^{0}	1.10×10^{0}	1.08×10^{0}	1.07×10^{0}	1.07×10^{0}	1.06×10^{0}	1.06×10^{0}	1.05×10^{0}	1.05×10^{0}
2×10^{0}	1.66×10^{0}	1.65×10^{0}	1.64×10^{0}	1.63×10^{0}	1.63×10^{0}	1.63×10^{0}	1.63×10^{0}	1.63×10^{0}	1.63×10^{0}	1.63×10^{0}
3.5×10^{0}	2.15×10^{0}	2.14×10^{0}	2.14×10^{0}	2.14×10^{0}	2.14×10^{0}	2.14×10^{0}	2.14×10^{0}	2.14×10^{0}	2.14×10^{0}	2.14×10^{0}
6×10^{0}	2.65×10^{0}	2.65×10^{0}	2.65×10^{0}	2.64×10^{0}	2.64×10^{0}	2.64×10^{0}	2.64×10^{0}	2.64×10^{0}	2.64×10^{0}	2.64×10^{0}
1×10^{1}	3.14×10^{0}	3.14×10^{0}	3.14×10^{0}	3.14×10^{0}	3.14×10^{0}	3.14×10^{0}	3.14×10^{0}	3.14×10^{0}	3.14×10^{0}	3.14×10^{0}
2×10^{1}	3.82×10^{0}	3.82×10^{0}	3.82×10^{0}	3.82×10^{0}	3.82×10^{0}	3.82×10^{0}	3.82×10^{0}	3.82×10^{0}	3.82×10^{0}	3.82×10^{0}
3.5×10^{1}	4.37×10^{0}	4.37×10^{0}	4.37×10^{0}	4.37×10^{0}	4.37×10^{0}	4.37×10^{0}	4.37×10^{0}	4.37×10^{0}	4.37×10^{0}	4.37×10^{0}
6×10^{1}	4.91×10^{0}	4.91×10^{0}	4.91×10^{0}	4.91×10^{0}	4.91×10^{0}	4.91×10^{0}	4.91×10^{0}	4.91×10^{0}	4.91×10^{0}	4.91×10^{0}
1×10^{2}	5.42×10^{0}	5.42×10^{0}	5.42×10^{0}	5.42×10^{0}	5.42×10^{0}	5.42×10^{0}	5.42×10^{0}	5.42×10^{0}	5.42×10^{0}	5.42×10^{0}

Values were obtained from Equation 18.8 with $\sigma = 10^{-9}$

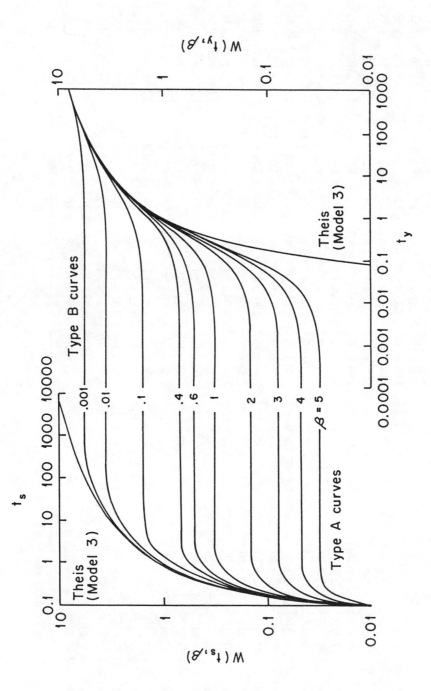

Figure 18.3   Type A and B Type curves for use with early and late data, respectively.

xi   Substitute the computed value of T, m, t_s*, and t* into Equation 18.13 and solve for S_s.

xii   Compute the horizontal hydraulic conductivity, K_r, using

$$K_r = \frac{T}{m} \qquad (18.22)$$

xiii   Substitute the computed value of K_r and β* into Equation 18.15 and solve for K_z.

xiv   If desired compute D and determine t_{wt} from Figure 18.1.

Example 18.1

A pumping test was conducted in an unconfined sand and gravel aquifer. The initial saturated thickness was 100 ft and the pumping rate was 50 gal/min. Drawdown data are listed below for an observation well located 75 ft from the pumping well. Compute T, K_r, K_z, S_s, and S_y.

t (min)	s (ft)	$s_{corrected}$ (ft)	t (min)	s (ft)	$s_{corrected}$ (ft)
1	0.3		480	18.8	17.0
2	1.6		960	18.7	17.0
4	4.5		1440	19.1	17.3
6	7.0		1920	20.0	18.0
8	8.2		2400	20.1	18.1
10	9.0		2880	21.2	19.0
20	12.4		3360	21.8	19.4
30	14.0		3840	22.2	19.7
60	17.0	15.6	4320	22.8	20.2
120	18.0	16.4	4800	23.3	20.6
240	18.5	16.8	7200	24.8	21.7
360	18.6	16.9	9600	26.2	22.8

Corrected drawdown was computed for a portion of the data using Equation 18.21. Application of the match-point method for these data using the Type B curve is shown in Figure 18.4. The match-point values are:

$$
\begin{aligned}
s^* &= 14 \text{ ft} \\
t^* &= 3000 \text{ min} \\
\beta^* &= 0.03 \\
W(t_y, \beta)^* &= 2.7 \\
t_y^* &= 2.9
\end{aligned}
$$

Substituting these values into Equation 18.19b gives

$$T = \frac{Q}{4\pi s^*} W(t_y, \beta)^* = \frac{\left(50\frac{\text{gal}}{\text{min}}\right)\left(\frac{\text{ft}^3}{7.48 \text{ gal}}\right)\left(\frac{1440 \text{ min}}{\text{d}}\right)(2.7)}{4\pi (14 \text{ ft}))} = 148 \frac{\text{ft}^2}{\text{d}}$$

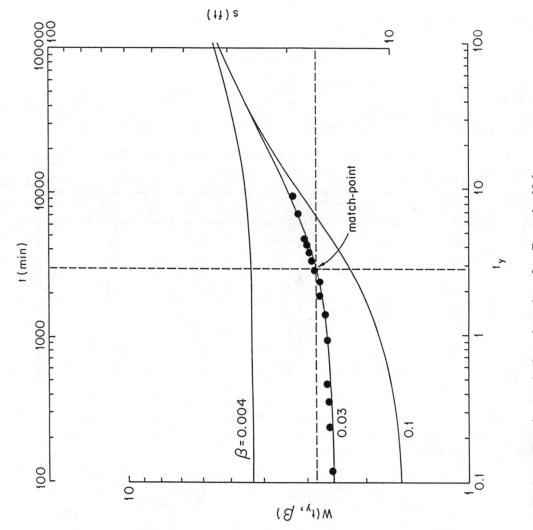

Figure 18.4 Application of match-point method to late data for Example 18.1.

From Equation 18.20

$$S_y = \frac{Tt^*}{t_y^* r^2} = \frac{\left(148\frac{ft^2}{d}\right)(3000 \text{ min})\left(\frac{d}{1440 \text{ min}}\right)}{(2.9)(75 \text{ ft})^2} = 0.02$$

Application of the match-point method using the <u>uncorrected</u> data and the Type A curve is shown in Figure 18.5. The match-point values are:

$$
\begin{aligned}
s^* &= 9 \text{ ft} \\
t^* &= 10 \text{ min} \\
\beta^* &= 0.03 \\
W(t_s, \beta)^* &= 1.7 \\
t_s^* &= 2
\end{aligned}
$$

Substituting into Equation 18.19a gives

$$T = \frac{Q}{4\pi s^*} W(t_s, \beta) = \frac{\left(50\frac{gal}{min}\right)\left(\frac{ft^3}{7.48 \text{ gal}}\right)\left(1440\frac{min}{d}\right)(1.7)}{4\pi(9ft)} = 145 \frac{ft^2}{d}$$

The two values of T are within 2%. The average value is $(148 + 145)/2 = 147$ ft²/d. From Equation 18.13

$$S_s = \frac{Tt^*}{t_s^* mr^2} = \frac{\left(147\frac{ft^2}{d}\right)\left(\frac{d}{1440 \text{ min}}\right)(10 \text{ min})}{(2)(100 \text{ ft})(75 \text{ ft})^2} = 1 \times 10^{-6} \text{ ft}^{-1}$$

From Equation 18.22, $K_r = T/m = 1.47$ ft/d, and from Equation 18.15

$$K_z = \frac{\beta^* K_r m^2}{r^2} = \frac{(0.03)\left(1.47\frac{ft}{d}\right)(100 \text{ ft}^2)}{(75 \text{ ft})^2} = 0.08 \frac{ft}{d}$$

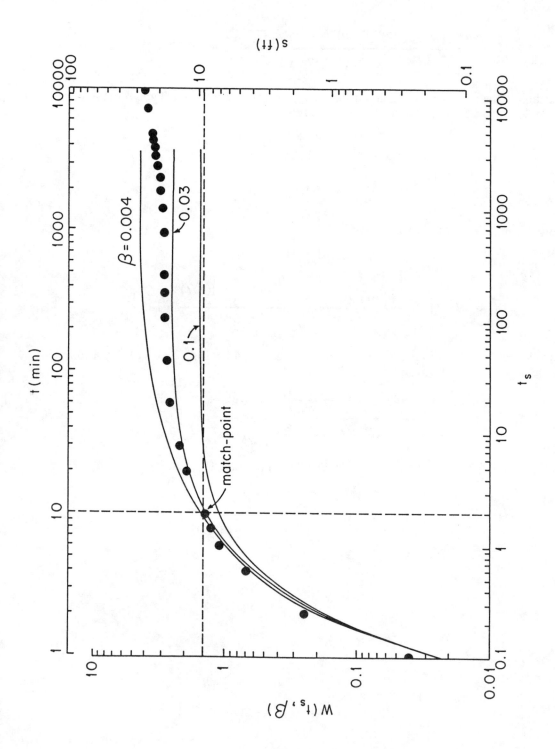

Figure 18.5 Application of match-point method to early data for Example 18.1.

Straight-Line Method

Neuman (1975) developed a procedure based on methodology outlined by Berkaloff (1963) which makes use of the three distinct portions of time-drawdown data when plotted on a semilogarithmic scale. Figure 18.6 shows values of the well function from Tables 18.1 and Table 18.2 plotted to this scale.

It can be seen that early data fall near the line

$$\frac{s4\pi T}{Q} = W(t_s, \beta) = 2.303 \log(2.246 t_s) \tag{18.23}$$

while late data fall near the line

$$\frac{s4\pi T}{Q} = W(t_y, \beta) = 2.303 \log(2.246 t_y) \tag{18.24}$$

where Equations 18.23 and 18.24 are the Cooper and Jacob (1946) approximations for the Theis well function for small u (large t_s or t_y) values. (See the discussion of Model 3 and Appendix A for an explanation of the Cooper and Jacob approximation.) The intermediate time data tend to fall on a horizontal line.

If, for any value of β, the horizontal line of the intermediate range is extended, it intersects the line described by Equation 18.24 at a point labeled $t_{y\beta}$. In Figure 18.6, the horizontal line for $\beta = 0.03$ is extended, resulting in $t_{y\beta} = 2.1$. When values of $1/\beta$ vs. $t_{y\beta}$ are plotted on a logarithmic scale, the result is the curve shown in Figure 18.7. This curve is linear within the range $4.0 \leq t_{y\beta} \leq 100.0$ and can be approximated within this range as

$$\beta = \frac{0.195}{(t_{y\beta})^{1.1053}} \tag{18.25}$$

Based on the above, the aquifer properties K_r, S_y, K_z, and often S can be obtained from a semilogarithmic plot of s vs. t. The procedure is:

i Plot s vs. log(t).

ii Fit a straight line to the late portion of the data. The intersection of this line with the horizontal axis where s = 0 is denoted by t_{oL}. The slope of this line is the change in drawdown over one log cycle, denoted by Δs_L.

iii From Equation 18.24,

$$T = \frac{2.303\,Q}{4\pi}(\Delta s_L) \tag{18.26}$$

Use this equation to solve for transmissivity, then compute horizontal hydraulic conductivity as $K_r = T/m$.

iv The specific yield can be computed using

$$S_y = \frac{2.246\,T t_{oL}}{r^2} \tag{18.27}$$

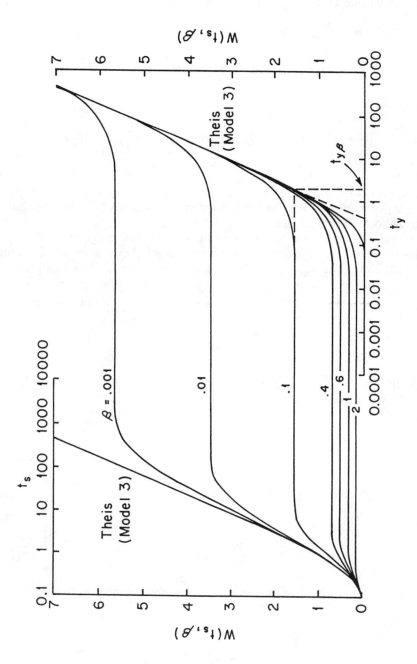

Figure 18.6 Semilogarithmic relationship between the well function and dimensionless time.

v Fit a horizontal line to the intermediate-time data. Record as $t_{y\beta}$ the value of time where this line intersects the straight line through the late data.

vi Using the computed values of T and S_y, solve for dimensionless time, $t_{y\beta}$, from the equation

$$t_{y\beta} = \frac{T\,t_\beta}{S_y r^2} \tag{18.28}$$

Equation 18.25 or Figure 18.7 can be used to find β from $t_{y\beta}$. Equation 18.15 can be used to solve for K_z as

$$K_z = \frac{\beta K_r m^2}{r^2}$$

vii Draw a straight line through the early data. If the slope is significantly different from the line through the late data, this method cannot be used to find S_s. If the two lines are roughly parallel, extend the line through the horizontal axis where s = 0. Call this point t_{oE}. Record the change in drawdown over one log cycle as Δs_E.

viii Recheck the transmissivity calculated in step iii by substituting Δs_E into Equation 18.23

$$T = \frac{2.303\,Q}{4\pi}(\Delta s_E) \tag{18.29}$$

ix Compute the specific storage using

$$S_s = \frac{2.246\,T t_{oE}}{m r^2} \tag{18.30}$$

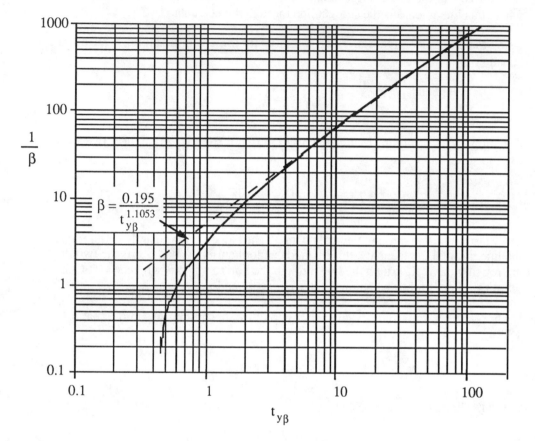

Figure 18.7 Logarithmic plot of $1/\beta$ vs. $t_{y\beta}$.

Chapter 19

MODEL 14: TRANSIENT, UNCONFINED, PARTIAL PENETRATION, ANISOTROPIC

19.1 CONCEPTUAL MODEL

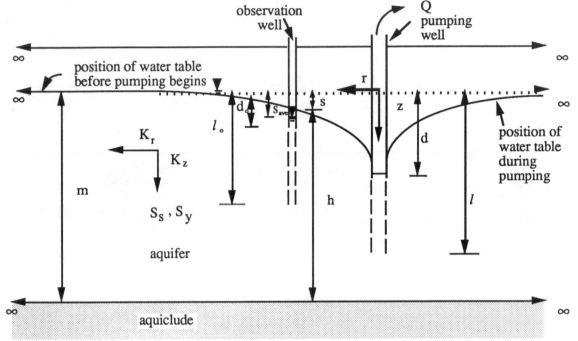

Definition of Terms

d   = distance from the static water table (the water table before pumping begins) to the top of the pumping well screen, L

d_o = distance from the static water table to the top of the observation well screen, L

K_r = aquifer horizontal hydraulic conductivity, LT^{-1}

K_z = aquifer vertical hydraulic conductivity, LT^{-1}

l = distance from the static water table to the bottom of the pumping well screen, L

l_o = distance from the static water table to the bottom of the observation well screen, L

m   = saturated thickness before pumping begins, L

Q   = constant pumping rate, L^3T^{-1}

r   = radial distance from the pumping well to a point on the cone of depression
           (all distances are measured from the center of wells), L

s   = drawdown of water table during pumping, L

s_{ave} = average drawdown for an observation well screened between the depths d_o and
           l_o, L

S_s = aquifer specific storage, L^{-1}

S_y = aquifer specific yield, dimensionless

z   = vertical coordinate with origin at the static water table, L

Assumptions

1. The aquifer is bounded below by an aquiclude.
2. All layers are horizontal and extend infinitely in the radial direction.
3. The initial water table (before pumping begins) is horizontal and extends infinitely in the radial direction.
4. The aquifer is homogeneous and anisotropic. The horizontal and vertical hydraulic conductivities of the aquifer may be different.
5. Groundwater density and viscosity are constant.
6. Groundwater flow can be described by Darcy's Law.
7. The pumping rate is constant.
8. Head losses through the well screen and pump intake are negligible.
9. The pumping well has an infinitesimal diameter.
10. The aquifer is compressible and completely elastic; pumping instantaneously releases water from storage by expansion of the pore water or compression of the soil skeleton. This water is accounted for in the specific storage, S_s, term.
    As the water table drops, additional water is released from storage due to gravity drainage of the effective pore space. This water is accounted for in the specific yield term, S_y.
11. Groundwater flow above the water table is negligible.
12. Drawdown is small compared to the saturated aquifer thickness.

19.2 MATHEMATICAL MODEL

Governing Equation

The governing equation is derived by combining Darcy's Law with the principle of conservation of mass in a radial coordinate system (Neuman, 1974)

$$\frac{\partial^2 s}{\partial r^2} + \left(\frac{1}{r}\right)\frac{\partial s}{\partial r} + \frac{K_z}{K_r}\frac{\partial^2 s}{\partial z^2} = \frac{S_s}{K_r}\frac{\partial s}{\partial t} \tag{19.1}$$

where s is drawdown, r is radial distance from the pumping well, K_z and K_r are the vertical and horizontal hydraulic conductivities of the aquifer, S_s is specific storage, and t is time.

Initial Conditions

- Before pumping begins drawdown is zero everywhere

$$s(r, z, t = 0) = 0 \qquad (19.2)$$

Boundary Conditions

- At an infinite distance from the pumping well, drawdown is zero

$$s(r = \infty, z, t) = 0 \qquad (19.3)$$

- At the aquifer base, the change in drawdown with depth is zero (i.e., the aquifer is impermeable)

$$\frac{\partial s(r, z = m, t)}{\partial z} = 0 \qquad (19.4)$$

- The free surface is a moving boundary. As it drops, water from the effective pore space becomes available to flow through the boundary and into the aquifer. This boundary condition is simply a statement of Darcy's Law for flow through the upper surface

$$\frac{\partial s(r, z = 0, t)}{\partial z} = -\frac{S_y}{K_z} \frac{\partial s(r, z = 0, t)}{\partial t} \qquad (19.5)$$

- Groundwater flow to the well is constant and uniform along the screened length

$$\lim_{r \to 0} r\frac{\partial s}{\partial r} = -\frac{Q}{2\pi K_r (l - d)} \qquad d < z < l \qquad (19.6)$$

- Change in drawdown is zero at the well center outside of the screened length

$$\frac{\partial s}{\partial z} (r = 0, z, t) = 0 \quad 0 \le z \le d \text{ and } l \le z \le m \qquad (19.7)$$

19.3 ANALYTICAL SOLUTION

19.3.1   General Solution

The solution for average drawdown, s_{ave}, in an observation well screened between the depths d_o and l_o is (Neuman, 1974)

$$s_{ave} = \frac{Q}{4\pi K_r m} W(t_s, t_y, \beta, \sigma, d/m, l/m, d_o/m, l_o/m) \qquad (19.8)$$

where W is a well function

$$W(t_s, t_y, \beta, \sigma, d/m, l/m, d_o/m, l_o/m) =$$

$$\int_0^\infty 4yJ_o\,(y\beta^{1/2})\left[u_o(y) + \sum_{n=1}^\infty u_n(y)\right] dy \qquad (19.9)$$

J_o is the zero order Bessel function of the first kind,

$u_o(y) =$

$$\frac{\{1-\exp[-t_s\beta(y^2-\gamma_o)]\}\{\sinh[\gamma_o(1-(d_o/m))]-\sinh[\gamma_o(1-(l_o/m))]\}\{\sinh[\gamma_o(1-(d/m))]-\sinh[\gamma_o(1-(l/m))]\}}{\left[y^2+(1+\sigma)\gamma_o^2-\dfrac{(y^2-\gamma_o^2)^2}{\sigma}\right]\cosh(\gamma_o)\cdot[(l_o/m)-(d_o/m)]\gamma_o[(l/m)-(d/m)]\sinh(\gamma_o)} \qquad (19.10)$$

$u_n(y) =$

$$\frac{\{1-\exp[-t_s\beta(y^2-\gamma_n)]\}\{\sinh[\gamma_n(1-(d_o/m))]-\sinh[\gamma_n(1-(l_o/m))]\}\{\sinh[\gamma_n(1-(d/m))]-\sinh[\gamma_n(1-(l/m))]\}}{\left[y^2+(1+\sigma)\gamma_n^2-\dfrac{(y^2-\gamma_n^2)^2}{\sigma}\right]\cosh(\gamma_n)\cdot[(l_o/m)-(d_o/m)]\gamma_n[(l/m)-(d/m)]\sinh(\gamma_o)} \qquad (19.11)$$

γ_o and γ_n are the roots of the equations

$$\sigma\gamma_o\,\sinh\,(\gamma_o) - (y^2 - \gamma_o^2)\,\cosh\,(\gamma_o) = 0,\text{ for } \gamma_o^2 < y^2 \qquad (19.12)$$

$$\sigma\gamma_n\,\sinh\,(\gamma_n) + (y^2 + \gamma_n^2)\,\cos\,(\gamma_n) = 0,$$
$$\text{for } (2n-1)\,(\pi/2) < \gamma_n < n\pi, n \geq 1 \qquad (19.13)$$

and

$$t_s = \frac{Tt}{S_s mr^2} \qquad (19.14)$$

$$t_y = \frac{Tt}{S_y r^2} \qquad (19.15)$$

$$\beta = \frac{r^2}{m^2}\frac{K_z}{K_r} \qquad (19.16)$$

$$\sigma = \frac{S_s m}{S_y} \qquad (19.17)$$

19.3.2   Special Case Solutions

Late-Time and Large-Distance Solutions

As distance from the pumping well and time increase, the effects of partial penetration decrease. According to Neuman (1974), at distances where

$$r > m \sqrt{\frac{K_r}{K_z}} \quad \text{and when } t > 10 \, S_y r^2 / K_r m$$

the effects of partial penetration disappear and drawdown can be computed from Model 3 with $S = S_y$.

Fully Penetrating Observation Well Solution

If an observation well is screened over the entire saturated thickness, the drawdown at

$$r > m \sqrt{\frac{K_r}{K_z}} \quad \text{and } t > S_y r^2 / K_r m$$

will follow the Theis solution (Model 3) with $S = S_y$ (Neuman, 1974). At distances where

$$r < 0.03 m \sqrt{\frac{K_r}{K_z}} \quad \text{and } t < S_s r^2 / K_r m$$

the drawdown will follow the early Theis curve with $S = S_s m$.

19.4 METHODS OF ANALYSIS

19.4.1   General Solution

Match-Point Method

The well function defined in Equation 19.9 is a function of eight dimensionless parameters. If the aquifer thickness and screen locations of all wells are known, four parameters are unknown: t_s, t_y, β, and σ. The number of dimensionless parameters can be reduced from three to two if the aquifer specific yield can be assumed to be much larger than the storativity ($S_y \gg S_s m$) so that $\sigma = 0$. For this case the method of analysis proceeds by dividing the drawdown data into two parts for analysis: early data and late data. For the early data, the effects of gravity drainage above the water table on groundwater flow to the pumping well are negligible; for the late data, the effects of elastic storage within the aquifer are negligible . The result is two asymptotic families of type curves, one for early times (the Type A curves) which is primarily influenced by elastic storage, S_s, and one for late times (the Type B curves) where storage from gravity drainage, S_y, becomes important. Equation 19.8 can be written as

$$s_{ave} = \frac{Q}{4\pi T} W(t_s, \beta), \text{ for early times} \tag{19.18a}$$

$$s_{ave} = \frac{Q}{4\pi T} W(t_y, \beta), \text{ for late times} \tag{19.18b}$$

where $W(t_s, \beta)$ is the well function in Equation 19.9 evaluated with $\sigma = 0$ and $W(t_y, \beta)$ is the well function in Equation 19.9 evaluated with $\sigma = 0$ and t_s replaced by t_y, where

$$t_y = \frac{Tt}{S_y r^2} \tag{19.19}$$

The following match-point method was developed based on this concept:

i   Prepare plots of $W(t_s, \beta)$ vs. t_s and $W(t_y, \beta)$ vs. t_y on logarithmic paper. These plots are called type curves. The curve $W(t_s, \beta)$ vs. t_s (the Type A curve) will be used to interpret early drawdown data and the curve $W(t_y, \beta)$ vs. t_y (the Type B curve) will be used to interpret late drawdown data. A separate set of Type A and Type B curves must be prepared for each well with different combinations of d_o and l_o. The type curves can be plotted using values computed by the computer program TYPE14 (Appendix B).

ii   Compute the corrected drawdown, $s_{corrected}$, using

$$s_{corrected} = s - (s^2)/2m \tag{19.20}$$

Note that the corrected drawdown data will only be used to match the Type B curves.

iii   Plot $s_{corrected}$ vs. t using the same logarithmic scales used to prepare the type curves. This plot is called the data curve. Note that distance-drawdown data ($s_{corrected}$ vs. t/r^2) should not be used with this method (Neuman, 1975).

iv   Overlay the corrected data curve on the Type B curve. Shift the plots relative to each other, keeping respective axes parallel, until a position of best fit is found between the data curve and one of the type curves. Use as much late data as possible.

v   From the best fit portion of the curves, select a match-point and record match-point values $W(t_y, \beta)^*$, t_y^*, s^*, t^*, and β^*.

vi   Substitute s^* and $W(t_y, \beta)^*$ into Equation 19.18b and solve for T.

vii   Substitute t^* and t_y^* and the computed value of T into Equation 19.15 and solve for S_y.

viii Repeat the match-point procedure using the uncorrected drawdown data and the Type A type curve. Use as much early data as possible.

ix   From the best fit portion of the curves, select a match-point and record match-point values $W(t_s, \beta)^*$, t_s^*, s^*, t^*, and β^*.

x   Substitute s^* and $W(t_s, b)^*$ into Equation 19.18a and solve for T.

xi   Substitute t_s^* and t^* into Equation 19.14 and solve for S_s.

xii   Compute the horizontal hydraulic conductivity, K_r, using

$$K_r = \frac{T}{m} \tag{19.21}$$

xiii Substitute the computed value of K_r and β^* in Equation 19.16 and solve for K_z.

Chapter 20

MODEL 15: TRANSIENT, UNCONFINED, PARTIAL PENETRATION, ANISOTROPIC, WELL STORAGE

20.1 CONCEPTUAL MODEL

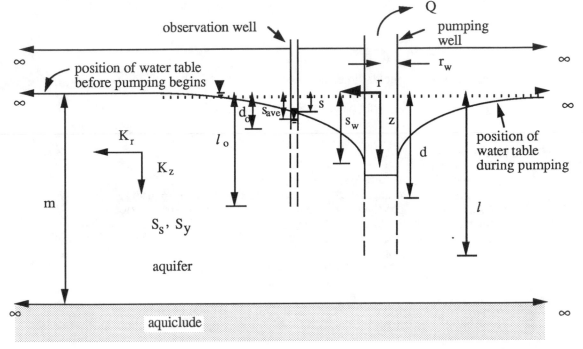

Definition of Terms

d = distance from the static water table (the water table before pumping begins) to the top of the pumping well screen, L

d_o = distance from the static water table to the top of the observation well screen, L

K_r = aquifer horizontal hydraulic conductivity, LT^{-1}

K_z = aquifer vertical hydraulic conductivity, LT^{-1}

l = distance from the static water table to bottom of pumping well screen, L

l_o = distance from the static water table to bottom of observation well screen, L

m = saturated thickness before pumping begins, L

Q   = constant pumping rate, L^3T^{-1}

r   = radial distance from the pumping well to a point on the cone of depression
           (all distances are measured from the center of wells), L

r_w = effective pumping well radius, L

s   = drawdown of water table during pumping, L

s_{ave} = average drawdown for an observation well screened between the depths d_o and
           l_o, L

S_s = aquifer specific storage, L^{-1}

S_y = aquifer specific yield, dimensionless

z   = vertical coordinate with origin at the static water table, L

Assumptions

1. The aquifer is bounded below by an aquiclude.
2. All layers are horizontal and extend infinitely in the radial direction.
3. The initial water table (before pumping begins) is horizontal and extends infinitely in the radial direction.
4. The aquifer is homogeneous and anisotropic. The horizontal and vertical hydraulic conductivities of the aquifer may be different.
5. Groundwater density and viscosity are constant.
6. Groundwater flow can be described by Darcy's Law.
7. The pumping rate is constant.
8. Head losses through the well screen and pump intake are negligible.
9. The aquifer is compressible and completely elastic; pumping instantaneously releases water from storage by expansion of the pore water or compression of the soil skeleton. This water is accounted for in the specific storage, S_s, term.
   As the water table drops, additional water is released from storage due to gravity drainage of the effective pore space. This water is accounted for in the specific yield term S_y.
10. Groundwater flow above the water table is negligible.
11. Drawdown is small compared to the saturated aquifer thickness.

20.2 MATHEMATICAL MODEL

Governing Equation

The following equations and solution for drawdown are developed for early times <u>only</u> (Neuman, 1974). The solution of Model 14 (Chapter 19) can be used for late times because the effect of well storage on late drawdown is negligible. The governing equation is derived by combining Darcy's Law with the principle of conservation of mass in a radial coordinate system

$$\frac{\partial^2 s}{\partial r^2} + \left(\frac{1}{r}\right)\frac{\partial s}{\partial r} + \frac{K_z}{K_r}\frac{\partial^2 s}{\partial z^2} = \frac{S_s}{K_r}\frac{\partial s}{\partial t} \tag{20.1}$$

where s is drawdown, r is radial distance from the pumping well, K_z and K_r are the vertical and horizontal hydraulic conductivities of the aquifer, S_s is specific storage, and t is time.

Initial Conditions

• Before pumping begins drawdown is zero everywhere

$$s(r, z, t = 0) = 0, \quad r \geq r_w \tag{20.2}$$

Boundary Conditions

• At an infinite distance from the pumping well, drawdown is zero

$$s(r = \infty, z, t) = 0 \tag{20.3}$$

• At the aquifer base, the change in drawdown with depth is zero

$$\frac{\partial s(r, z = m, t)}{\partial z} = 0 \tag{20.4}$$

• Elastic storage effects are important only for relatively small values of time. For small values of time, drawdown at the free surface is very small compared to drawdown at some finite depth z. This statement becomes more accurate as the ratio of $S_y/(S_s m)$ increases

$$s(r, z = 0, t) = 0 \tag{20.5}$$

• Constant flow, Q, out of the well is derived from inflow through the screen and the change in well storage. There is a constant flux over the length of the screen

$$Q = \pi r_w^2 \frac{\partial s(r = r_w, z, t)}{\partial t} - 2\,\pi r_w\,(l - d)\,K_r \frac{\partial s(r = r_w, z, t)}{\partial t} \tag{20.6}$$

$$d < z < l$$

• So that a solution which is independent of depth can be developed, it is assumed that the drawdown in the pumping well, s_w, is the average value of $s(r_w, z, t)$ over the screened portion of the well

$$s_w = \frac{1}{l - d} \int_d^l \frac{\partial s(r = r_w, z, t)}{\partial t}\, dz \tag{20.7}$$

• By integrating both sides of Equation 20.7 and substituting the relation of Equation 20.6, the boundary condition along the well screen becomes

$$Q = \pi r_w^2 \frac{\partial s_w}{\partial t} - 2\pi r_w\,(l - d)\,K_r \int_d^l \frac{\partial s(r = r_w, z, t)}{\partial t}\, dz \tag{20.8}$$

$$\text{for} \quad d \le z \le l$$

- No groundwater flows into the pumping well along the unscreened portion

$$0 = \pi r_w^2 \frac{\partial s_w}{\partial t} - 2\pi r_w \ (l-d) \ K_r \int_d^l \frac{\partial s(r = r_w, z, t)}{\partial t} \ dz \tag{20.9}$$

$$\text{for} \quad 0 < z < d \text{ and } l < z < m$$

20.3 ANALYTICAL SOLUTION

20.3.1   General Solution

The solution is (Boulton and Streltsova, 1976)

$$s = \frac{Q}{4\pi T} \ W(t_s, \ \beta, \ S_s m, \ \rho, \ l/m, \ d/m, \ z/m) \tag{20.10}$$

where $W(t_s, \beta, S_s m, \rho, l/m, d/m, z/m)$ is a well function

$$W(t_s, \beta, S_s/m, \rho, l/m, d/m, z/m) =$$

$$\frac{Q}{4\pi T} \sum_{n=1,3,5,\dots}^{\infty} G_n \ \sin\left(\frac{n\pi z}{2m}\right) \Bigg\{ \frac{\pi K_o(\gamma_n \ \rho) \ [1 - \exp \ (-\phi_n t_{sw}/4)]}{K_1(\gamma_n)[4S_s(l-d)\gamma_n(1-S_s(l-d))+\phi_n^2/\gamma_n]}$$

$$+ \int_0^\infty \frac{[P_2 J_o(y\rho) - P_1 Y_o(y\rho)] \ \{1 - \exp[-t_{sw}/4(y^2 + c_n^2)]\}}{(P_1^2 + P_2^2) \ (y^2 + c_n^2)} \ y \ dy \Bigg\} \tag{20.11}$$

where

$$G_n = \frac{32 S_s m}{\pi^2} \left[\frac{1}{n} \left(\cos \frac{n\pi d}{2m} - \cos \frac{n\pi l}{2m} \right) \right] \tag{20.12}$$

$$P_1 = (y^2 + c_n^2) \ J_o(y) - 2(l - d) \ S_s y J_1(y) \tag{20.13}$$

$$P_2 = (y^2 + c_n^2) \ Y_o(y) - 2(l - d) \ S_s y Y_1(y) \tag{20.14}$$

$$c_n = \frac{n\pi r_w}{2m} \sqrt{\frac{K_z}{K_r}} \tag{20.15}$$

$$t_{sw} = \frac{4Tt}{S_s m r_w^2} \tag{20.16}$$

γ_n is the positive root of

$$c_n^2 - \gamma_n^2 = 2S_s(l - d)\frac{\gamma_n K_1(\gamma_n)}{K_o(\gamma_n)} \tag{20.17}$$

$$\phi_n = c_n^2 - \gamma_n^2 \tag{20.18}$$

J_o = zero order Bessel function of the first kind
J_1 = first order Bessel function of the first kind

K_o = zero order modified Bessel function of the second kind

Y_o = zero order Bessel function of the second kind
Y_1 = first order Bessel function of the second kind

and

$$t_s = \frac{4\pi t}{S_s m r^2} \tag{20.19}$$

$$\beta = \frac{K_z r^2}{K_r m^2} \tag{20.20}$$

$$\rho = \frac{r}{r_w} \tag{20.21}$$

To obtain the solution for average drawdown in an observation well screened between the depths d_o and l_o, Equation 20.10 is integrated with respect to z between the limits d_o and l_o and the result divided by the screen length (l_o - d_o). The integrated form of Equation 20.10 can be written

$$s_{ave} = \frac{Q}{4\pi T} W(t_s, \beta, S_s m, \rho, l/m, d/m, d_o/m, l_o/m) \tag{20.22}$$

Values of this integrated well function can be computed with the computer program TYPE15; average drawdown s_{ave} can be computed with the computer program DRAW15 (Appendix B).

20.3.2   Special Case Solutions

The effects of well storage on drawdown can be neglected if either

$t > \dfrac{2.5 \times 10^3 \ r_c^2}{T}$ or r > $300r_w$ (Papadopulos and Cooper, 1967). For this case the solution of Model 14 (Chapter 19) can be used.

20.4 METHODS OF ANALYSIS

20.4.1 General Solution

Match-Point Method

The well function defined in Equation 20.22 is a function of six dimensionless parameters (t_s, β, $S_s m$, ρ, l/m, d/m, d_o/m, l_o/m). If the aquifer thickness and screen locations of all wells are known, three parameters are unknown: t_s, β, and $S_s m$. To proceed it is necessary to assume the value of either S_s or the ratio K_r/K_z. This will allow the calculation of $S_s m$ or β; the remaining two parameters can be determined by application of the match-point method. The following method of analysis is presented with the assumption that S_s can be estimated; the method can be easily modified for the case of known K_r/K_z:

i   Estimate the aquifer specific storage S_s. Prepare a plot of $W(t_s$, β, $S_s m$, ρ, l/m, d/m, d_o/m, $l_o/m)$ vs. t_s on logarithmic paper. This plot is called the type curve. The type curve can be plotted using the program TYPE15 (Appendix B). A separate type curve will be required for each observation well with different values of r, l_o, or d_o.

ii   Plot s vs. t for each observation well using the same logarithmic scales used to prepare the type curve(s). This plot is called the data curve. Each observation well will have a separate data curve.

iii   Overlay the data curve for an observation well on the type curves developed for that well. Shift the plots relative to each other, keeping the respective axes parallel, until a position of best fit is found between the data curve and one of the type curves.

iv   From the best fit portion of the curves, select a match-point and record match-point values $W(t_s$, β, $S_s m$, ρ, l/m, d/m, d_o/m, $l_o/m)^*$, t_s^*, β^*, s_{ave}^*, and t^*.

v   Substitute $W(t_s$, β, $S_s m$, ρ, l/m, d/m, d_o/m, $l_o/m)^*$ and s_{ave}^* into Equation 20.22 and solve for T. Compute $K_r = T/m$.

vi   Substitute the computed value of K_r and β^* and t^* into Equation 20.20 and solve for K_z.

vii   Substitute t_s^* into Equation 20.19 and solve for S_s. This value should agree with the initial estimate for S_s used to prepare the type curve(s). However, the computed values of K_r and K_z are not very sensitive to the choice of S_s used to prepare the type curves.

viii   Repeat steps iii through vii for each observation well. Average the values of K_r and K_z from all observation wells to determine the effective properties of the aquifer.

Chapter 21

MODEL 16: TRANSIENT, UNCONFINED, PARTIAL PENETRATION, AQUITARD-AQUIFER

21.1 CONCEPTUAL MODEL

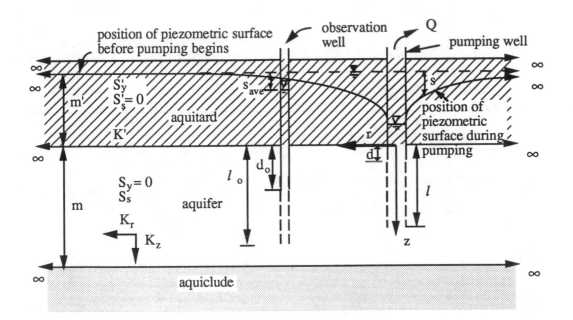

Definition of Terms

d = distance from the static water table (the water table before pumping begins) to the top of the pumping well screen, L

d_o = distance from the static water table to the top of the observation well screen, L

K_r = aquifer horizontal hydraulic conductivity, LT^{-1}

K_z = aquifer vertical hydraulic conductivity, LT^{-1}

K' = aquitard vertical hydraulic conductivity, LT^{-1}

l = distance from the static water table to the bottom of the pumping well screen, L

l_o = distance from the static water table to the bottom of the observation well screen, L m   = aquifer thickness, L

m'   = aquitard saturated thickness, L

Q   = constant pumping rate, L^3T^{-1}

r   = radial distance from the pumping well to a point on the cone of depression (all distances are measured from the center of wells), L

s   = drawdown of piezometric surface during pumping, L

s_{ave} = average drawdown over the saturated portion of the aquifer at a given distance r from the pumping well, L

S_s = aquifer specific storage, L^{-1}

S_s' = aquitard specific storage, L^{-1}
       = 0

S_y = aquifer specific yield, dimensionless
       = 0

S_y' = aquitard specific yield, dimensionless

z   = vertical coordinate with origin at the static water table, L

Assumptions

This model can represent a wide range of conditions. If the hydraulic conductivity of the aquitard is negligible, the system acts as a confined aquifer. If the hydraulic conductivity of the aquitard is less than the hydraulic conductivity of the aquifer the system responds as a heterogeneous (two layer) aquifer. If the hydraulic conductivity of the aquitard is the same as the aquifer, the system responds as a single unconfined aquifer.

The system is different than the multiple leaky aquifer system of Neuman and Witherspoon (1972) or the Hantush (1964) models for leaky aquifers because those models assume the presence of an unconfined aquifer (the "source bed") above the aquitard which is unaffected by pumping. The assumptions of Model 16 are:

1. The aquifer is bounded below by an aquiclude and bounded above by an aquitard.
2. All layers are horizontal and extend infinitely in the radial direction.
3. The initial piezometric surface (before pumping begins) is horizontal and extends infinitely in the radial direction.
4. The aquifer is homogeneous and anisotropic. The horizontal and vertical hydraulic conductivities of the aquifer may be different. The aquitard is homogeneous and isotropic.
5. Groundwater density and viscosity are constant.
6. Groundwater flow can be described by Darcy's Law.
7. Groundwater flow in the aquitard is vertical. Groundwater flow in the aquifer is horizontal and directed radially toward the well. This assumption is valid when $K_r/K' > 100$ (Neuman and Witherspoon, 1969a and 1969b).

8. The pumping rate is constant.
9. The pumping well has an infinitesimal diameter.
10. Head losses through the well screen and pump intake are negligible.
11. The aquifer is compressible and completely elastic. The aquitard is incompressible (i.e., no water is released from storage during pumping).

21.2 MATHEMATICAL MODEL

Governing Equation

The governing equation is derived by combining Darcy's Law with the principle of conservation of mass in a radial coordinate system (Neuman, 1974)

$$\frac{1}{r}\frac{\partial s}{\partial r}\left(r\,\frac{\partial s'}{\partial r}\right)+\left(\frac{K_z}{K_r}\right)\frac{\partial^2 s}{\partial z^2}=\frac{S_s}{K_r}\frac{\partial s}{\partial t} \tag{21.1}$$

where s and s' are the drawdown in the aquifer and aquitard, respectively, r is radial distance from the pumping well, K_z and K_r are the vertical and horizontal hydraulic conductivities of the aquifer, S_s is specific storage, and t is time.

Initial Conditions

• Before pumping begins drawdown is zero everywhere

$$s(r, z, t = 0) = 0 \tag{21.2}$$

Boundary Conditions

• The free surface is a moving boundary. As it drops, water from the effective pore space becomes available to flow through the boundary and into the aquifer. This boundary condition is simply a statement of Darcy's Law for flow through the upper surface

$$\frac{\partial s(r,\ z = \text{-}m,\ t)}{\partial z}=-\frac{S_y'}{K'}\frac{\partial s(r,\ z = \text{-}m,\ t)}{\partial t} \tag{21.3}$$

• Groundwater flow to the well is constant and uniform along the screened length

$$\lim_{r \to 0}\ r\frac{\partial s}{\partial r}=-\frac{Q}{2\pi K_r\,(l-d)} \qquad d < z < l \tag{21.4}$$

$$=0 \qquad\qquad 0 < z < d \tag{21.5}$$

$$=0 \qquad\qquad l < z < l \tag{21.6}$$

• At the bottom of the aquifer, the change in drawdown with depth is zero (the lower boundary is impermeable)

$$\frac{\partial s\ (r,\ z = m,\ t)}{\partial z}=0 \tag{21.7}$$

21.3 ANALYTICAL SOLUTION

21.3.1   General Solution

The average drawdown, s_{ave}, in an observation well screened between the depths d_o and l_o, is (Boulton and Streltsova, 1975)

$$s_{ave} = \frac{Q}{4\pi T} \, W(t_s, \beta, c, \sigma, d/m, l/m, d_o/m, l_o/m) \qquad (21.8)$$

where $W(t_s, \beta, c, \sigma, d/m, l/m, d_o/m, l_o/m)$ is a well function

$W(t_s, \beta, c, \sigma, d/m, l/m, d_o/m, l_o/m) =$

$$\frac{1}{(l_o - d_o)(1 - d)} \int_0^\infty 4y J_o \, (y\beta^{1/2}) \left[u_o(y) + \sum_{n=1}^\infty u_n(y) \right] dy \quad (21.9)$$

where

$u_o(y) =$

$$\frac{\{\sinh[\gamma_o(1 - d/m)] - \sinh[\gamma_n(1 - l/m)]\} \{\sinh[\gamma_o(1 - d_o)] - \sinh[\gamma_o(1 - l_o)]\} \{1 - \exp[-(y^2 + \gamma_n^2) \, t_s\beta]\}}{(y^2 - \gamma_n^2) \, x_o \cos (\gamma_n)}$$

$$(21.10)$$

$u_n(y) =$

$$\frac{\{\sinh[\gamma_n(1 - d_o/m)] - \sinh[\gamma_n(1 - l/m)]\} \{\sinh[\gamma_o(1 - d_o)] - \sinh[\gamma_o(1 - l_o)]\} \{1 - \exp[-(y^2 + \gamma_n^2) \, t_s\beta]\}}{(y^2 - \gamma_n^2) \, x_n \cos (\gamma_n)}$$

$$(21.11)$$

and

$$x_o = \left[1 + \frac{c \, (y^2 - \gamma_o^2)}{y^2 - \gamma_o^2 - c/\upsilon} \right] \sinh(\gamma_o) + \left[1 + \frac{2c\sigma}{(y^2 - \gamma_o^2 - c\sigma)^2} \right] \gamma_o \cosh(\gamma_o) \qquad (21.12)$$

$$x_m = \left[1 + \frac{c \, (y^2 - \gamma_n^2)}{y^2 + \gamma_o^2 - c\sigma} \right] \sin(\gamma_n) + \left[1 + \frac{2c^2\sigma}{(y^2 + \gamma_n^2 - c\sigma)^2} \right] \gamma_n \cos(\gamma_n) \qquad (21.13)$$

while γ_o is the positive root of

$$(y^2 - \gamma_o^2 - c\sigma)\gamma_o \sinh(\gamma_o) + c(y^2 - \gamma_o^2) \cosh(\gamma_o) = 0, \quad \gamma_o^2 < y^2 \qquad (21.14)$$

and γ_n is the nth positive root of

$$(y^2 - \gamma_n^2 - c\sigma)\gamma_n \sin(\gamma_n) + c(y^2 - \gamma_n^2) \cos(\gamma_n) = 0 \qquad (21.15)$$

and

$$t_s = \frac{Tt}{S_s m r^2} \qquad (21.16)$$

$$\beta = \frac{K_z r}{K_r m} \qquad (21.17)$$

$$c = \frac{K'm}{K_z m'} \qquad (21.18)$$

$$\sigma = \frac{S_s m}{S_y} \qquad (21.19)$$

and J_o is the zero order Bessel function of the first kind. Values of $W(t_s, \beta, c, \sigma, d/m, l/m, d_o/m, l_o/m)$ can be computed using the computer program TYPE16; average drawdown can be computed using the computer program DRAW16 (Appendix B).

21.4 METHODS OF ANALYSIS

21.4.1   General Solution

Match-Point Method

The function $W(t_s, \beta, c, \sigma, d/m, l/m, d_o/m, l_o/m)$ defined in Equation 21.8 is a function of seven dimensionless parameters. If the aquifer thickness and screen locations of all wells are known, the unknown parameters are t_s, β, c, and σ. To proceed it is necessary to assume the value of two of the following ratios: S_s/S_y (so that σ can be computed using Equation 21.19), K'/K_z (so that c can be computed using Equation 21.18), or K_z/K_r (so that β can be computed using Equation 21.17). The remaining dimensionless parameters can be determined by application of the match-point method. The following method of analysis is based on the assumption that estimates of S_s/S_y and K'/K_z are available. The method can be easily modified for the case where estimates are available for other combinations of aquifer and aquitard properties.

The theoretical basis for this method of analysis is described in Chapter 4. The specific steps are:

i   Estimate S_s/S_y and K'/K_z. Compute c using Equation 21.18 and σ using Equation 21.19. Using the computed values of c and σ prepare a plot of $W(t_s, \beta, c, \sigma, d/m, l/m, d_o/m, l_o/m)$ vs. t_s on logarithmic paper. This plot is called the type curve. The type curve can be plotted using the program TYPE16 (Appendix B). A separate type curve will be required for each observation well with different values of r, l_o, or d_o.

ii Plot s vs. t for each observation well using the same logarithmic scales used to prepare the type curve(s). This plot is called the data curve. Each observation well will have a separate data curve.

iii Overlay the data curve for an observation well on the type curves developed for that well. Shift the plots relative to each other, keeping the respective axes parallel, until a position of best fit is found between the data curve and one of the type curves.

iv From the best fit portion of the curves, select a match-point and record match-point values $W(t_s, \beta, c, \sigma, d/m, l/m, d_o/m, l_o/m)^*$, t_s^*, β^*, s_{ave}^*, and t^*.

v Substitute $W(t_s, \beta, c, \sigma, d/m, l/m, d_o/m, l_o/m)^*$ and s_{ave}^* into Equation 21.8 and solve for T. Compute $K_r = T/m$.

vi Substitute the computed value of K_r and β^* into Equation 21.17 and solve for K_z.

vii Substitute the computed value of K_z into Equation 21.18 and solve for K'.

viii Substitute t_s^* and the computed value of T into Equation 21.16 and solve for S_s.

ix Compute S_y using the Equation 21.19.

Chapter 22

MODEL 17 : SLUG TEST - CONFINED, INFINITE OR SEMI-INFINITE DEPTH, INCOMPRESSIBLE AQUIFER, ANISOTROPIC

22.1 CONCEPTUAL MODEL

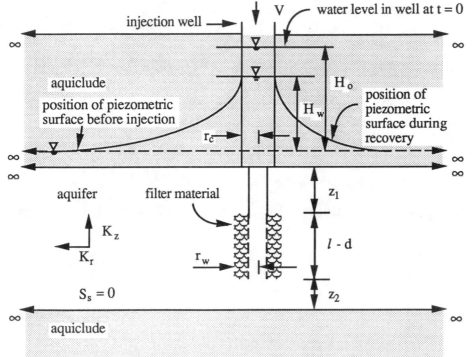

Definition of Terms

H_o = initial change in head in the well casing due to an injection of volume V at time t = 0, L

H_w = height of water in the well above the equilibrium level at time t > 0, L

K_r = aquifer horizontal hydraulic conductivity, LT^{-1}

K_z = aquifer vertical hydraulic conductivity, LT^{-1}

l - d   = length of screen, filter pack, or open hole over which water leaves or enters the well, L

n   = porosity of filter material or developed zone, dimensionless

r_c = effective radius of the well casing over which the water level in the well changes, L

          = r_w if the water level is always above the well screen

$$= \sqrt{r_i^2(1-n) + nr_o^2}$$

         if the water level is falling within the screened length of the well <u>and</u> the hydraulic conductivity of the filter material or developed zone is much larger than the hydraulic conductivity of the aquifer

r_w = effective radius of the well bore or open hole, L

         = borehole radius if the filter is much more permeable than the aquifer

         = screen radius if no filter is used or if the filter has a hydraulic conductivity similar to that of the aquifer

r_i = inside radius of well screen, L

r_o = outside radius of filter material or developed zone, L

S_s = aquifer specific (elastic) storage, L^{-1}

         = 0

V = volume of water injected into the well at time $t = 0$, L^3

z_1 = vertical distance from the top of the screen, filter pack, or open hole to the top of the aquifer, L

z_2 = vertical distance from the bottom of the screen, filter pack, or open hole to the bottom of the aquifer, L

Assumptions

1. The aquifer is bounded above and below by aquicludes.
2. All layers are horizontal and extend infinitely in the radial direction.
3. The initial piezometric surface (before injection) is horizontal and extends infinitely in the radial direction.
4. The aquifer is homogeneous and anisotropic.
5. Groundwater density and viscosity are constant.
6. Groundwater flow can be described by Darcy's Law.
7. Groundwater flow is horizontal and is directed laterally away from the injection well.
8. A volume of water, V, is injected instantaneously at time $t = 0$.
9. The injection well is considered to be a slot (line source) with an infinitesimal width.
10. Head losses through the well screen and filter (if present) are negligible.
11. The aquifer is incompressible.

22.2 MATHEMATICAL MODEL

The model applies to the instantaneous injection or withdrawal of a volume of water from the well casing. If water is injected into the well casing, V is positive, the initial head is above the equilibrium level, and H_w is referred to as buildup. If water is withdrawn from the well casing, V is negative, the initial head is below the equilibrium level, and H_w is referred to as drawdown.

Governing Equation

The governing equation is derived by writing Darcy's Law for the flow of injected water from the well (Hvorslev, 1951)

$$Q = K_r H_w F \qquad (22.1)$$

where Q $(L^3 T^{-1})$ is the flow rate, K_r is horizontal hydraulic conductivity, and F (L) is a shape factor which depends on the geometry of the well intake and its location within the aquifer.

Hvorslev (1951) assumed that the volume of water that flows out of the well is equal to the change in volume of water stored in the well casing

$$\frac{dH_w}{dt} = -\frac{Q}{\pi r_c^2} \qquad (22.2)$$

where r_c is the effective casing radius. Substituting Equation 22.1 into Equation 22.2 and rearranging yields

$$\frac{dH_w}{H_w} = \frac{K_r F}{\pi r_c^2} \, dt \qquad (22.3)$$

The time lag, t_L, is defined as the time required for the buildup due to injection to dissipate if the initial flow rate is maintained (see Section 22.4)

$$t_L = \frac{V}{Q} = \frac{\pi r_c^2 H_o}{K_r F H_o} = \frac{\pi r_c^2}{K_r F} \qquad (22.4)$$

The time lag is substituted into Equation 22.3 to give the final form of the governing differential equation

$$\frac{dH_w}{H_w} = \frac{dt}{t_L} \qquad (22.5)$$

Initial Condition

- A volume ("slug") of water is instantaneously introduced into the well at time t = 0, causing an initial rise in the water level in the well, H_o

$$H_w(t = 0) = H_o \qquad (22.6)$$

22.3 ANALYTICAL SOLUTION

Integrating Equation 22.5 yields

$$t_L = \frac{t}{\ln \left(\frac{H_w}{H_o}\right)} \tag{22.7}$$

Substituting Equation 22.7 into Equation 22.4 yields

$$K_r = \frac{\pi r_c^2}{F t_L} \tag{22.8}$$

Hvorslev (1951) presents a compilation of empirical equations for computing the shape factor, F. Three of these equations are applicable to most situations. In the following discussion, a correction factor, a_k, has been added to the shape factors to account for the influence of differences between horizontal and vertical hydraulic conductivity

$$a_k = \sqrt{\frac{K_r}{K_z}} \tag{22.9}$$

Table 22.1 contains the equations for the shape factors and a description of the conditions for each case.

In Case A, the injection well partially penetrates a very deep aquifer (Figure 22.1). In Case B, the injection well screen is located in the center of a very thick aquifer. Case C applies to a fully penetrating injection well.

Note that to calculate F for Case C requires an estimate of the radius of influence, R. Although empirical equations have been developed for this purpose (e.g., see Chapter 6) their use is not recommended. Instead slug test data conforming to Case C conditions should be interpreted using Model 18, where estimates for R have been developed from electrical resistance analogs.

Table 22.1   Selected equations for the shape factor, F.

Case	Shape Factor	Conditions of Development
A	$$F = \dfrac{2\pi\,(l-d)}{\ln\left[\dfrac{a_k\,(l-d)}{r_w} + \sqrt{1 + \left(\dfrac{a_k\,(l-d)}{r_w}\right)^2}\,\right]}$$ $$F = \dfrac{2\pi\,(l-d)}{\ln\left[\dfrac{2a_k\,(l-d)}{r_w}\right]}, \quad \text{for}\ \dfrac{a_k\,(l-d)}{r_w} > 4$$	$z_1 = 0,\ z_2 = \infty$ Flow emanates from a line source for which the equipotential surfaces are semi-ellipsoids.
B	$$F = \dfrac{2\pi\,(l-d)}{\ln\left[\dfrac{a_k\,(l-d)}{2r_w} + \sqrt{1 + \left(\dfrac{a_k\,(l-d)}{2r_w}\right)^2}\,\right]}$$ $$F = \dfrac{2\pi\,(l-d)}{\ln\left[\dfrac{a_k\,(l-d)}{r_w}\right]}, \quad \text{for}\ \dfrac{a_k\,(l-d)}{2r_w} > 4$$	$z_1 = \infty,\ z_2 = \infty$ Flow emanates from a line source. Flow lines are symmetrical with respect to a horizontal plane through the center of the intake.
C	$$F = \dfrac{2\pi\,(l-d)}{\ln\,(R/r_w)}$$	$z_1 = 0,\ z_2 = 0$ Flow emanates from a line source and moves radially outward.

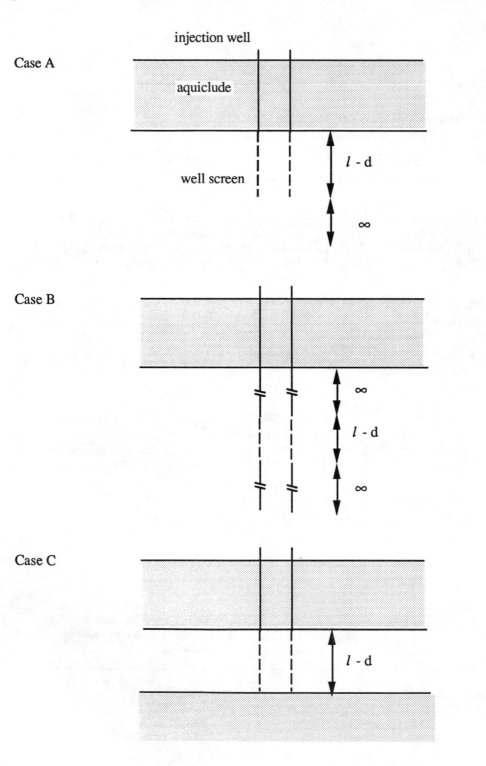

Figure 22.1 Aquifer conditions for Cases A, B, and C.

22.4 METHODS OF ANALYSIS

The analytical solution described in Section 22.3 can be used to interpret slug test data for either injection or withdrawal of a volume of water from the well casing. To compute the hydraulic conductivity, one should determine the value of the shape factor for the case that most accurately represents test conditions, determine the time lag, and substitute these values into Equation 22.8. The value of a_k must be estimated from the geology of the site. The procedure for determining the time lag is as follows:

i   Plot $\log(H_w/H_o)$ vs. t as shown in the example of Figure 22.2.

ii   Theoretically, the data should fall along a straight line originating at $H_w/H_o = 1$ and $t = 0$ from which t_L can be computed. In practice, the data tend to form a slight, concave upward curve because the aquifer is somewhat compressible. Therefore, t_L should be computed as the inverse of the slope of a straight line fitted to the data (i.e.,
$$\frac{1}{t_L} = \frac{\ln (H_w/H_o)_1 - \ln (H_w/H_o)_2}{t_1 - t_2}$$ where the subscripts 1 and 2 refer to
any two points on the fitted line).

The procedure for finding t_L is commonly simplified by drawing a second line, parallel to the fitted line, and intersecting the line defined by $H_w/H_o = 1.0$. For this second line,

$$\ln (H_w/H_o)_1 = 0$$

$$t_1 = 0$$

and t_L is the time, t_2, where $1/\ln (H_w/H_o)_2 = 1$ (i.e., where $(H_w/H_o)_2 = 0.37$).

Slug test data often plot as a curved line on this plot (e.g., see Figure 22.2). In part, this is because real aquifers are compressible. When a slug of water is added to the well, the total vertical stresses on the soil remain essentially constant, but the pore water pressure is increased such that the effective vertical stress decreases. This decrease in effective stress causes an initial swelling of the soil around the well screen and an increase in the rate of flow of water from the well; thus, the $\ln \dfrac{H_w}{H_o}$ vs. t curve (recovery curve) decreases sharply. As the water level in the well drops, the rate of excess flow decreases and the slope of the recovery curve becomes less steep. At this point, the pore pressure in the voids is the same as that indicated by the piezometer. As the water level in the well decreases further, causing decreased pore pressures, the soil begins to reconsolidate and the rate of flow of water to the soil decreases. The decrease in flow rate causes the slope of the curve to decrease further.

A similar event occurs when the water level is instantaneously lowered and allowed to rise. An initial decrease in pore pressure causes consolidation followed by pressure equalization and, finally, swelling as the water level in the well rises. The shape of the recovery curve is the same as that when a slug of water is added, providing that the aquifer is perfectly elastic.

Example 22.1

A slug test was performed in a confined, fine sand aquifer. The aquifer is 30 meters thick and a piezometer extends 4 meters below the upper confining layer. The lower 2.5 meters of the piezometer are screened. A sealed end pipe was lowered into the well at time t = 0, causing the water level in the well to rise 3.0 meters. At all times $t \geq 0$, the water level in the well was above the top of the sealed end pipe such that r_c remained constant at 5 cm. Values of H_w/H_o for the test are plotted in Figure 22.2. Assume that the ratio of horizontal to vertical hydraulic conductivity for the aquifer is 50. Compute K.

From the given information:

$$
\begin{aligned}
l\text{-}d &= 250 \text{ cm} \\
r_w &= 8 \text{ cm} \\
r_c &= 5 \text{ cm} \\
z_1 &= 150 \text{ cm} \\
z_2 &= 2600 \text{ cm} \\
a_k &= \sqrt{50}
\end{aligned}
$$

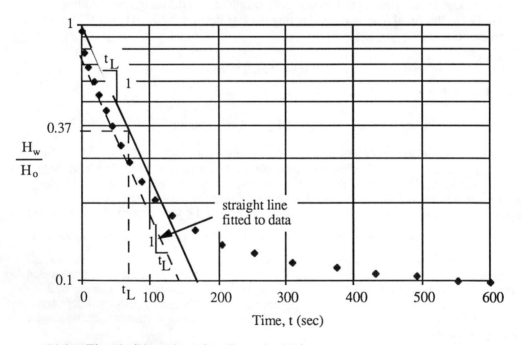

Figure 22.2   Time-buildup data for Example 22.1.

To solve for hydraulic conductivity, the time lag, t_L, and an appropriate shape factor, F, must be determined. From Figure 22.2, the time lag can be determined from the slope of the fitted (dashed) line

$$
\frac{1}{t_L} = \frac{\ln (H_w/H_o)_1 - \ln (H_w/H_o)_2}{t_1 - t_2}
$$

where the subscripts 1 and 2 refer to any two points along the straight line. In this case,

$$\frac{1}{t_L} = \frac{\ln(0.75) - \ln(0.1)}{0 - 145 \text{ sec}}$$

$$t_L = \frac{1}{0.0139 \text{ sec}^{-1}} = 72 \text{ sec}$$

Alternatively, t_L could be determined by the shortcut method where a line of the same slope is extended through the point where $H_w/H_o = 1.0$ and $t = 0$. With this arrangement, t_L is the time where $(H_w/H_o)_2 = 0.37$. The time lag determined by this shortcut method is shown in Figure 22.2. A second (solid) line was drawn parallel to the fitted (dashed) line and shifted to the right so that it intersects the line $H_w/H_o = 1.0$ and the point $t = 0$.

In this example the shape factor for Case A applies. Though the example well screen does not begin precisely at the top of the aquifer, this is better than assuming the aquifer thickness is infinite (Case B). From Table 22.1 with $\frac{\sqrt{50}\ (250)}{8} > 4$

$$F = \frac{2\pi(l - d)}{\ln[2a_k(l - d)/r_w]}$$

Substituting F into Equation 22.8 yields

$$K_r = \frac{\pi r_c^2 \ln[2a_k(l - d)/r_w]}{2\pi(l - d)\ t_L}$$

$$= \frac{(5\text{cm})^2 \ln[2\ \sqrt{50}\ (250\ \text{cm})/8\text{cm}]}{2\ (250\ \text{cm})\ (72\text{sec})}$$

$$= 4.2 \times 10^{-3} \text{ cm/sec}$$

Because of the low water volumes involved, a slug test "samples" a much smaller aquifer volume than does a pumping test. Though the distance $z_1 = 150$ cm is very small compared to $z_2 = 2600$ cm, it is not so small when compared to $l - d = 250$ cm. Thus, the use of the Case B equation for F may be appropriate for this case. If the Case B shape factor is used, K_r would be 3.7×10^{-3} cm/sec.

Chapter 23

MODEL 18 : SLUG TEST - UNCONFINED OR LEAKY CONFINED, INCOMPRESSIBLE AQUIFER, PARTIAL PENETRATION

23.1 CONCEPTUAL MODEL

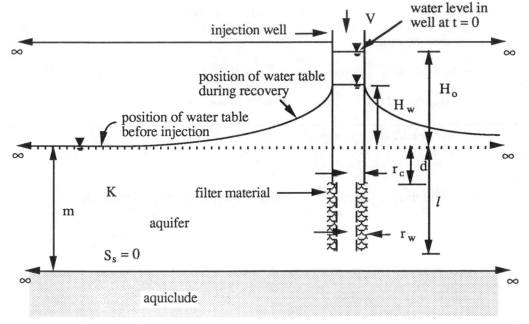

Definition of Terms

d   = distance from the static water table (the water table before injection) to the top of the well screen or open hole (d ≥ 0), L

H_o = instantaneous change in head in the well casing due to an injection of a volume of water V at time t = 0, L

H_w = height of water in the well above the static water table at time t > 0, L

K   = aquifer hydraulic conductivity, LT^{-1}

l   = vertical distance from the static water table to the bottom of the well screen or open hole, L

m   = aquifer saturated thickness, L

n   = porosity of filter material or developed zone, dimensionless

r_c = effective radius of the well casing over which the water level in the well changes, L

     $= r_w$ if the water level is always above the well screen

     $= \sqrt{r_i^2(1 - n) + nr_o^2}$
     if the water level is falling within the screened length of the well <u>and</u> the hydraulic conductivity of the filter material or developed zone is much larger than the hydraulic conductivity of the aquifer

r_w = effective radius of the well bore or open hole, L
     = borehole radius if the filter is much more permeable than the aquifer
     = screen radius if no filter is used or if the filter has a hydraulic conductivity similar to that of the aquifer

r_i = inside radius of well screen, L

r_o = outside radius of filter material or developed zone, L

S_s = aquifer specific (elastic) storage, L^{-1}

t = time since injection, T

V = volume of water injected into the well (sometimes used to calculate H_o), L^3

Assumptions

  1. The aquifer is bounded below by an aquiclude.
  2. All layers are horizontal and extend infinitely in the radial direction.
  3. The initial water table (before injection) is horizontal and extends infinitely in the radial direction.
  4. The aquifer is homogeneous and isotropic.
  5. Groundwater density and viscosity are constant.
  6. Groundwater flow can be described by Darcy's Law.
  7. A volume of water, V, is injected instantaneously at time $t = 0$.
  8. Head losses through the well screen, filter material, and developed zone (if present) are negligible.
  9. The aquifer is incompressible.
  10. Buildup of the water table is small compared to the aquifer saturated thickness.

23.2 MATHEMATICAL MODEL

     The model applies to the instantaneous injection or withdrawal of a volume of water from the well casing. If water is injected into the well casing, V is positive, the initial head is above the equilibrium level, and H_w is referred to as buildup. If water is withdrawn from the well casing, V is negative, the initial head is below the equilibrium level, and H_w is referred to as drawdown.

Governing Equation

The equations describing flow are based on a modified form of the Thiem equation derived in Chapter 6. The rate of groundwater flow, Q (L^3T^{-1}), from a well screened between the depths d and l, for a specified water level in the well, H_w, is

$$Q = \frac{2\pi K\,(l - d)\,H_w}{\ln\,(R/r_w)} \tag{23.1}$$

where K is hydraulic conductivity and R is the radius of influence of the injection well. The rate of fall of the water level in the well is equal to the flow rate divided by the effective cross-sectional area of the well casing

$$\frac{dH_w}{dt} = -\frac{Q}{\pi r_c^2} \tag{23.2}$$

Combining Equations 23.1 and 23.2 yields

$$\frac{dH_w}{H_w} = -\frac{2K(l - d)}{r_c^2\,\ln(R/r_w)}\,dt \tag{23.3}$$

Initial Condition

- A volume ("slug") of water is instantaneously introduced into the well at time $t = 0$, causing an initial rise in the water level in the well, H_o

$$H_w(t = 0) = H_o \tag{23.4}$$

Boundary Condition

- At a distance R from the well, buildup is zero

$$H_w(r = R) = 0 \tag{23.5}$$

23.3 ANALYTICAL SOLUTION

23.3.1 General Solution

Integrating Equation 23.3 gives

$$K = \frac{r_c^2\,\ln(R/r_w)\,\ln(H_o/H_w)}{2(l - d)\,t} \tag{23.6}$$

$$= \frac{r_c^2\,\ln(R/r_w)}{2(l - d)\,t_L} \tag{23.7}$$

where

$$t_L = \frac{t}{\ln(H_w/H_o)} \quad \text{is called the time lag} \tag{23.8}$$

Bouwer and Rice (1976) determined the radius of influence, R, for different values of r_w, $(l - d)$, H_w, and m using measurements made with an electrical resistance analog model. In their electrical resistance model, the top of the aquifer was set as a boundary of constant potential (head). Therefore, their solution can be applied to both unconfined aquifers and leaky aquifers with a continuous source of leakage. From their experiments, the following empirical equation was developed for estimating R:

$$\ln(R/r_w) = \left\{ \frac{1.1}{\ln(l/r_w)} + \frac{A + B \ln[(m - l)/r_w]}{(l - d)/r_w} \right\}^{-1} \tag{23.9}$$

where A and B are dimensionless coefficients which are functions of $(l - d)/r_w$ as shown in Figure 23.1. The results of the analog experiments indicated that the effect of partial penetration reaches a maximum when $\ln[(m - l)/r_w] = 6$ and this is the largest value that should be substituted for that expression (Bouwer and Rice, 1976).

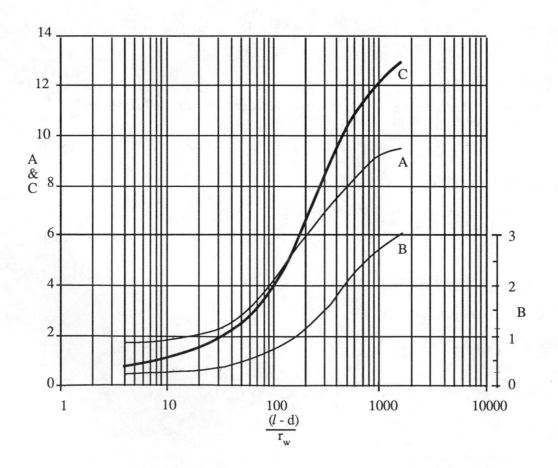

Figure 23.1   Values of the coefficients A, B, and C for use in estimating the radius of influence, R (from Bouwer and Rice, 1976, p. 426).

23.3.2   Special Case Solution

Well Screen Extending to Bottom of Aquifer

If the injection well fully-penetrates the aquifer (i.e., $l = m$), Equation 23.9 cannot be used and the appropriate expression for $\ln(R/r_w)$ is

$$\ln(R/r_w) = \left\{ \frac{1.1}{\ln(l/r_w)} + \frac{C}{(l - d)/r_w} \right\}^{-1} \qquad (23.10)$$

where C is a dimensionless coefficient shown in Figure 23.1. Equations 23.9 and 23.10 are accurate to within about 10 % of the results obtained in the analog model if $(l\text{-d}) > 0.4\ l$, and within about 25 % if $(l\text{-d}) < 0.1\ l$ (Bouwer and Rice, 1976).

23.4 METHODS OF ANALYSIS

The analytical solution described in Section 23.3 can be used to interpret slug test data for either injection or removal of a volume of water from the well casing. The method of analysis is based on Equation 23.8 which indicates that a plot of $\ln (H_w/H_o)$ vs. t should yield a straight line with a slope of $1/t_L$. The slope of the line can be determined directly from the plot. Because natural logarithms (ln) are related to common logarithms (log) by a constant, i.e., $\ln(x) = 2.303 \log(x)$, the analysis can be conveniently performed using a plot of $\log(H_w/H_o)$ vs. t :

i   Plot $\log(H_w/H_o)$ vs. t as shown in the example in Figure 23.3.

ii   Fit a straight line through the data (giving most weight to the early data) and compute the slope of the fitted line, $1/t_L$. The time lag can be computed using the values of t and H_w/H_o for any two points on the fitted line

$$t_L = \frac{t_1 - t_2}{\ln(H_w/H_o)_1 - \ln(H_w/H_o)_2}$$

or by drawing another line, parallel to the fitted line and intersecting the point where $H_w/H_o = 1.0$. For this point

$$\ln(H_w/H_o)_1 = 0$$

$$t_1 = 0$$

and t_L can be conveniently found as the time, t_2, where $1/\ln(H_w/H_o)_2 = -1$, i.e., where $(H_w/H_o)_2 = 0.37$. An example of this shortcut procedure is shown in Figure 23.3.

iii Compute the radius of influence by first selecting values of the coefficients A and B or C from Figure 23.1. Compute R using Equation 23.9 or 23.10. If $\ln[(m - l)/r_w] > 6$, then Equation 23.9 becomes

$$\ln (R/r_w) = \left\{ \frac{1.1}{\ln(l/r_w)} + \frac{A + 6B}{(l - d)/r_w} \right\}^{-1} \qquad (23.11)$$

iv   Compute K by substituting values for R, r_w, l, d, and r_c into Equation 23.7.

Example 23.1

A slug test was performed on a clayey silt aquifer. The well and aquifer geometry are shown in Figure 23.2.

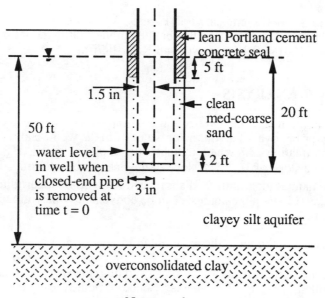

Not to scale

Figure 23.2   Well and aquifer geometry for slug test of Example 23.1.

Prior to the test, a closed-end pipe was inserted into the well; the water had returned to its equilibrium level before the start of the test. When the pipe was removed at the beginning of the test, the water level was lowered by 18 feet. Figure 23.3 shows the results of the test plotted as $\log(H_w/H_o)$ vs. t, with $H_o = 18$ feet.

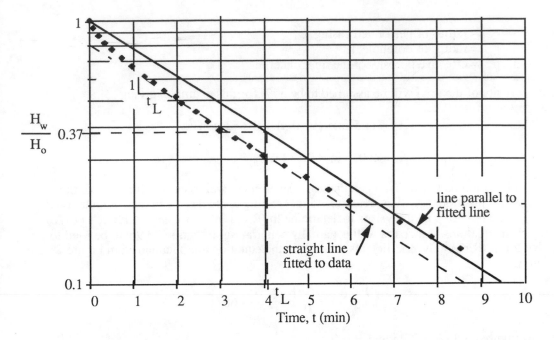

Figure 23.3 Recovery data for a slug test on the aquifer of Example 23.1. The inverse of the slope of a straight line fitted to the data is the time lag, t_L.

The aquifer is unconfined and the overconsolidated clay can be assumed to be impermeable relative to the clayey silt. The first step is to determine values for the aquifer and well geometry.

i From Figure 23.2, the saturated thickness, m = 50 ft.

ii Since the medium-coarse sand filter is much more permeable than the surrounding clayey silt, l is measured to the bottom of the sand filter, i.e., l = 20 ft.

iii The distance to the top of the sand pack is used to select d = 5 ft.

iv The distance to the undisturbed aquifer is the well radius plus the thickness of the filter, since the hydraulic conductivity of the sand is much larger than the clayey silt. Therefore use r_w = 3 in = 0.25 ft.

v The effective radius of the well casing, r_c is used in the computation for the change in volume of water within the well. If, as in this example, the water level in the well is recovering over the screened interval of the well, the effective casing radius is larger than the inside radius of the casing since water can move into the voids of the sand filter with virtually no resistance. The effective radius in this case can be computed using

$$r_c^2 = r_i^2 (1 - n) + nr_o^2 \qquad (23.12)$$

where

r_i = radius of perforated casing, L
r_o = radius of borehole, L
n = effective porosity of the filter material

The porosity of the sand will be assumed to be 30% for this example, giving

$$r_c^2 = (1.5 \text{ in})^2 (1 - 0.3) + 0.3 (3.0 \text{ in})^2$$

$$r_c = 2.076 \text{ in} = 0.172 \text{ ft}$$

It should be noted that when the water level in the well rises to within 5 ft of the equilibrium position, r_c will decrease to 1.5 in, creating a small error in the computations.

The next step in analysis is to determine $\ln (R/r_w)$ from Bouwer and Rice's (1976) empirical equations. Since the well is partially penetrating, Equation 23.9 will be used to solve for $\ln (R/r_w)$. The coefficients A and B in the equation are obtained from Figure 23.1 with

$$\frac{l - d}{r_w} = \frac{20 \text{ ft} - 5 \text{ ft}}{0.25 \text{ ft}} = 60$$

From Figure 23.1, A = 3.25 and B = 0.5.

Before applying Equation 23.9, the value of $\ln \left(\dfrac{m - l}{r_w} \right)$ should be checked.

$$\ln \left(\frac{50 \text{ ft} - 20 \text{ ft}}{0.25 \text{ ft}} \right) = 4.79$$

Since this value is less than 6.0, Equation 23.9 can be applied without alteration.

$$\ln (R/r_w) = \left\{ \frac{1.1}{\ln \left(\dfrac{20 \text{ ft}}{0.25 \text{ ft}} \right)} + \frac{3.25 + 0.5 \ln \left[\dfrac{50 \text{ ft} - 20 \text{ ft}}{0.25 \text{ ft}} \right]}{(20 \text{ ft} - 5 \text{ ft})/0.25 \text{ ft}} \right\}^{-1}$$

$$\ln (R/r_w) = 2.90$$

Next, the time lag, t_L, is determined from Figure 23.3. It can be computed from the fitted line using any two points, e.g.,

$$\frac{1}{t_L} = \frac{\ln (0.8) - \ln (0.1)}{8.5 \text{ min} - 0 \text{ min}} = 0.245 \text{ min}^{-1}$$

$$t_L = \frac{1}{0.245 \text{ min}^{-1}} = 4.1 \text{ min}$$

or from the shortcut procedure where t is read from the parallel line fitted through H_w/H_o = 1.0 at the point where H_w/H_o = 0.37 as shown in Figure 23.3. The hydraulic conductivity of the aquifer is computed using Equation 23.7

.08579

$$K = \frac{r_c^2 \ln(R/r_w)}{2(l - d)t_L} = \frac{(0.172 \text{ ft})^2 (2.90)}{2(20 \text{ ft} - 5 \text{ ft}) (4.1 \text{ min})} = 7.0 \times 10^{-4} \text{ ft/min}$$
$$= 1.0 \text{ ft/day}$$

Chapter 24

MODEL 19: SLUG TEST - CONFINED, COMPRESSIBLE AQUIFER, WELL STORAGE

24.1 CONCEPTUAL MODEL

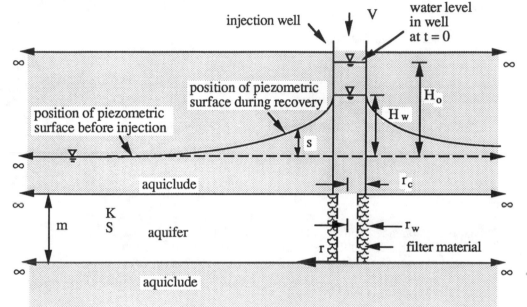

Definition of Terms

H_o = initial change in head in the well casing due to an injection volume V at time $t = 0$, L

H_w = height of water in the well above the equilibrium level at time $t > 0$, L

K   = aquifer hydraulic conductivity, LT^{-1}

m   = aquifer saturated thickness, L

n   = porosity of filter material or developed zone, dimensionless

r   = radial distance from the injection well to a point on the cone of depression (all distances are measured from the center of wells), L

r_c = effective radius of the well casing over which the water level in the well changes, L
        = r_w if the water level is always above the well screen

$$= \sqrt{r_i^2(1 - n) + nr_o^2}$$

if the water level is falling within the screened length of the well <u>and</u> the hydraulic conductivity of the filter material or developed zone is much larger than the hydraulic conductivity of the aquifer

r_w = effective radius of the well bore or open hole, L
          = borehole radius if the filter is much more permeable than the aquifer
          = screen radius if no filter is used or if the filter has a hydraulic conductivity similar to that of the aquifer

r_i = inside radius of well screen, L

r_o = outside radius of filter material or developed zone, L

S   = aquifer storativity, dimensionless

s   = drawdown or buildup, at a distance r from the well at time t > 0, L

V   = volume of water injected into the well at time t = 0, L^3

Assumptions

1. The aquifer is bounded above and below by aquicludes.
2. All layers are horizontal and extend infinitely in the radial direction.
3. The initial piezometric surface (before injection) is horizontal and extends infinitely in the radial direction.
4. The aquifer is homogeneous and isotropic.
5. Groundwater density and viscosity are constant.
6. Groundwater flow can be described by Darcy's Law.
7. Groundwater flow is horizontal and directed radially away from the injection well.
8. The injection well is screened over the entire saturated thickness of the aquifer.
9. A volume of water, V, is injected instantaneously at time t = 0.
10. Head losses through the well screen and filter (if present) are negligible.
11. The aquifer is compressible and completely elastic.

24.2 MATHEMATICAL MODEL

The model applies to the instantaneous injection or withdrawal of a volume of water from the well casing. If water is injected into the well casing, V and s are positive, the initial head is above the equilibrium level, and s is referred to as buildup. If water is withdrawn from the well casing, V and s are negative, the initial head is below the equilibrium level, and s is referred to as drawdown.

Governing Equation

The governing equation is derived by combining Darcy's Law with the principle of conservation of mass in a radial coordinate system (Cooper et al., 1967)

$$\frac{\partial^2 s}{\partial r^2} + \left(\frac{1}{r}\right)\frac{\partial s}{\partial r} = \left(\frac{S}{T}\right)\frac{\partial s}{\partial t}, \ r > r_w \qquad (24.1)$$

where s is buildup, r is radial distance from the injection well, S and T = Km are the storativity and transmissivity of the aquifer, and t is time.

Initial Conditions

• Before pumping begins buildup is zero everywhere in the aquifer and equal to H_o inside the well

$$s(r > r_w, t = 0) = 0 \qquad (24.2)$$

$$H_w(t = 0) = H_o = \frac{V}{\pi\, r_c^2} \qquad (24.3)$$

where H_w is buildup in the well.

Boundary Conditions

• Buildup in the aquifer at the face of the well and in the well casing are equal

$$s(r = r_w, t) = H_w \qquad (24.4)$$

• At an infinite distance from the injection well buildup is zero

$$s(r = \infty, t) = 0 \qquad (24.5)$$

• The rate of flow of water into the aquifer is equal to the rate of decrease in the volume of water stored in the well

$$(2\pi r_w T)\frac{\partial s(r = r_w, t)}{\partial r} = \pi\, r_c^2 \frac{\partial H_w(t)}{\partial t} \qquad (24.6)$$

24.3 ANALYTICAL SOLUTION

24.3.1 General Solution

In the Aquifer

The solution for buildup in the aquifer is (Cooper et al., 1967)

$$s = \frac{2H_o}{\pi} \int_0^\infty \exp\left(-\frac{\beta y^2}{\alpha}\right) \left\{ J_o\left(\frac{yr}{r_w}\right) [yY_o(y) - 2\alpha Y_1(y)] - \right.$$
$$\left. Y_o(yr/r_w)\, [yJ_o(y) - 2\alpha J_1(y)] \right\} \frac{dy}{\Delta(y)} \qquad (24.7)$$

where

$$\alpha = Sr_w{}^2/r_c^2 \qquad (24.8)$$

$$\beta = Tt/r_c^2 \qquad (24.9)$$

$$\Delta(y) = [uJ_o(y) - 2\alpha J_1(y)]^2 + [yY_o(y) - 2\alpha Y_1(y)]^2 \tag{24.10}$$

and

J_o = zero order Bessel function of the first kind
J_1 = first order Bessel function of the first kind
Y_o = zero order Bessel function of the second kind
Y_1 = first order Bessel function of the second kind

In the Injection Well

The solution for the buildup in the injection well is (Cooper et al., 1967)

$$\frac{H_w}{H_o} = F(\alpha, \beta) \tag{24.11}$$

$$F(\alpha, \beta) = \left(\frac{8\alpha}{\pi^2}\right) \int_0^\infty \frac{\exp(-\beta y^2/\alpha)}{y \, \Delta(y)} \, dy \tag{24.12}$$

where $F(\alpha, \beta)$ is a well function. Selected values of $F(\alpha, \beta)$ are in Table 24.1. Values of $F(\alpha, \beta)$ may also be computed using the computer program TYPE19; drawdown can be computed using the computer program DRAW19 (Appendix B).

24.3.2 Special Case Solution

Late-Time Solution

When $t > 100r_c^2/T$, well casing storage becomes unimportant and the well can be treated as a line source or sink without significant error. Under these conditions, the following solutions apply (Ferris et al., 1962)

$$s/H_o = \frac{r_c^2 \exp\left(-\frac{r^2 S}{4Tt}\right)}{4Tt}, \qquad t > 100 \; r_c^2/T \tag{24.13}$$

$$H_w/H_o = r_c^2/4Tt, \qquad t > 100 \; r_c^2/T \tag{24.14}$$

Table 24.1   Values of H_w/H_0 for a slug test in a well of finite diameter in an isotropic, nonleaky aquifer (from Cooper et al., 1967 and Papadopulos et al., 1973).

	$F(\alpha, \beta)$				
β	$\alpha = 10^{-1}$	$\alpha = 10^{-2}$	$\alpha = 10^{-3}$	$\alpha = 10^{-4}$	$\alpha = 10^{-5}$
1.00×10^{-3}	0.9771	0.9920	0.9969	0.9985	0.9992
2.15×10^{-3}	0.9658	0.9876	0.9949	0.9974	0.9985
4.64×10^{-3}	0.9490	0.9807	0.9914	0.9954	0.9970
1.00×10^{-2}	0.9238	0.9693	0.9853	0.9915	0.9942
2.15×10^{-2}	0.8860	0.9505	0.9744	0.9841	0.9888
4.64×10^{-2}	0.8293	0.9187	0.9545	0.9701	0.9781
1.00×10^{-1}	0.7460	0.8655	0.9183	0.9434	0.9572
2.15×10^{-1}	0.6289	0.7782	0.8538	0.8935	0.9167
4.64×10^{-1}	0.4782	0.6436	0.7436	0.8031	0.8410
1.00×10^{0}	0.3117	0.4598	0.5729	0.6520	0.7080
2.15×10^{0}	0.1665	0.2597	0.3543	0.4364	0.5038
4.64×10^{0}	0.07415	0.1086	0.1554	0.2082	0.2620
7.00×10^{0}	0.04625	0.06204	0.08519	0.1161	0.1521
1.00×10^{1}	0.03065	0.03780	0.04821	0.06355	0.08378
1.40×10^{1}	0.02092	0.02414	0.02844	0.03492	0.04426
2.15×10^{1}	0.01297	0.01414	0.01545	0.01723	0.01999
3.00×10^{1}	0.009070	0.009615	0.01016	0.01083	0.01169
4.64×10^{1}	0.005711	0.005919	0.006111	0.006319	0.006554
7.00×10^{1}	0.003722	0.003809	0.003884	0.003962	0.004046
1.00×10^{2}	0.002577	0.002618	0.002653	0.002688	0.002725
2.15×10^{2}	0.001179	0.001187	0.001194	0.001201	0.001208

Table 24.1   (Continued).

β	$\alpha = 10^{-6}$	$\alpha = 10^{-7}$	$F(\alpha, \beta)$ $\alpha = 10^{-8}$	$\alpha = 10^{-9}$	$\alpha = 10^{-10}$
0.001	0.9994	0.9996	0.9996	0.9997	0.9997
0.002	0.9989	0.9992	0.9993	0.9994	0.9995
0.004	0.9980	0.9985	0.9987	0.9989	0.9991
0.006	0.9972	0.9978	0.9982	0.9984	0.9986
0.008	0.9964	0.9971	0.9976	0.9980	0.9982
0.01	0.9956	0.9965	0.9971	0.9975	0.9978
0.02	0.9919	0.9934	0.9944	0.9952	0.9958
0.04	0.9848	0.9875	0.9894	0.9908	0.9919
0.06	0.9782	0.9819	0.9846	0.9866	0.9881
0.08	0.9718	0.9765	0.9799	0.9824	0.9844
0.1	0.9655	0.9712	0.9753	0.9784	0.9807
0.2	0.9361	0.9459	0.9532	0.9587	0.9631
0.4	0.8828	0.8995	0.9122	0.9220	0.9298
0.6	0.8345	0.8569	0.8741	0.8875	0.8984
0.8	0.7901	0.8173	0.8383	0.8550	0.8686
1.0	0.7489	0.7801	0.8045	0.8240	0.8401
2.0	0.5800	0.6235	0.6591	0.6889	0.7139
3.0	0.4554	0.5033	0.5442	0.5792	0.6096
4.0	0.3613	0.4093	0.4517	0.4891	0.5222
5.0	0.2893	0.3351	0.3768	0.4146	0.4487
6.0	0.2337	0.2759	0.3157	0.3525	0.3865
7.0	0.1903	0.2285	0.2655	0.3007	0.3337
8.0	0.1562	0.1903	0.2243	0.2573	0.2888
9.0	0.1292	0.1594	0.1902	0.2208	0.2505
10.0	0.1078	0.1343	0.1620	0.1900	0.2178
20.0	0.02720	0.03343	0.04129	0.05071	0.06149
30.0	0.01286	0.01448	0.01667	0.01956	0.02320
40.0	0.008337	0.008898	0.009637	0.01062	0.01190
50.0	0.006209	0.006470	0.006789	0.007192	0.007709
60.0	0.004961	0.005111	0.005283	0.005487	0.005735
80.0	0.003547	0.003617	0.003691	0.003773	0.003863
100.0	0.002763	0.002803	0.002845	0.002890	0.002938
200.0	0.001313	0.001322	0.001330	0.001339	0.001348

24.4 METHODS OF ANALYSIS

24.4.1   General Solution

Match-Point Method

This method can be used to determine aquifer properties T and S. However, the shapes of the recovery curves are not very sensitive to even large changes in storativity. The authors of this model believe that the best one could expect is to be able to estimate the dimensionless parameter α to within one or two orders of magnitude (Papadopulos et al., 1973). This magnitude of error for values of $\alpha < 10^{-5}$ would produce an error in the computed transmissivity of less than 30 percent.

The theoretical basis for this method of analysis is described in Chapter 4. The specific steps are:

i   Prepare a plot of $F(\beta, \alpha)$ vs. β on semilogarithmic paper. This plot is called the type curve (Figure 24.2). The type curve can be plotted using the values in Table 24.1 or values computed by the computer program TYPE19 (Appendix B).

ii   Plot H_w/H_o vs. log(t) using the same semilogarithmic scales used to prepare the type curve. This plot is called the data curve.

iii   Overlay the data curve on the type curves. Shift the plots relative to each other, keeping respective axes parallel, until a position of best fit is found between the data curve and one of the type curves.

iv   From the overlapping portions of the curves, select a match-point and record match-point values β^*, α^* and t^*.

v   Substitute β^* and t^* into Equation 24.9 and solve for T.

vi   Substitute α^* into Equation 24.8 and solve for S.

Example 24.1

A slug test was performed on a well in the confined aquifer (Figure 24.1). At time $t = 0$, a slug of water was added to the well, increasing the water level by 10 feet. The following measurements were recorded.

Time after injection (min)	H_w/H_o	Time after injection (min)	H_w/H_o
0.15	0.97	50.00	0.19
0.20	0.95	60.00	0.15
0.70	0.92	70.00	0.10
1.50	0.87	80.00	0.05
2.00	0.78	100.00	0.04
4.00	0.71	120.00	0.02
8.00	0.63	240.00	0.001
10.00	0.55	300.00	0.0005
15.00	0.35		
20.00	0.25		

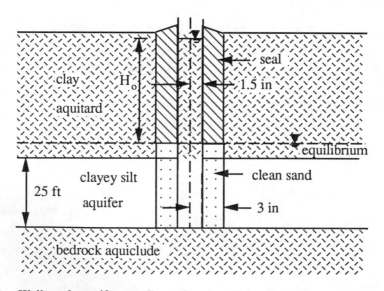

Figure 24.1 Well and aquifer configuration for Example 24.1.

The above data plotted on a semilogarithmic scale and overlaid on a family of type curves are shown in Figure 24.2. The lower and left-hand axes represent $F(\beta, \alpha)$ and β of the type curve, while the upper and right axes represent H_w/H_o and time in the data plot.

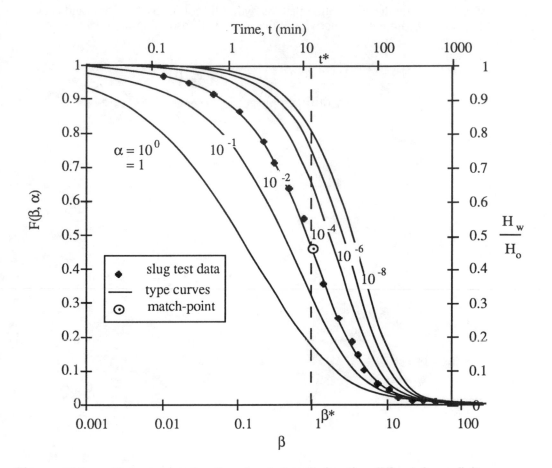

Figure 24.2   Type curves for the slug test well function F(β, α) for a finite
diameter well in a nonleaky confined aquifer overlain by the data curve
from Example 24.1.

The shape of the data curve most closely matches the α = 10^{-2} type curve. With the
curves aligned, the following match-point values were selected (* denotes match-point):

$$\beta^* = 1.0$$
$$t^* = 18 \text{ min}$$
$$\alpha^* = 1 \times 10^{-2}$$

From Figure 24.1, the values for the well and aquifer dimensions are taken as

$$r_w = 3 \text{ in} = 0.25 \text{ ft}$$
$$r_c = 1.5 \text{ in} = 0.125 \text{ ft}$$
$$m = 25 \text{ ft}$$

The choice of r_w is based on the assumption that the hydraulic conductivity of the sand pack is much larger than the hydraulic conductivity of the aquifer. Substituting these values into Equation 24.9 gives

$$T = \frac{\beta^* \, r_c^2}{t^*} = \frac{1.0 \, (0.125 \text{ ft})^2}{18 \text{ min} \left(\dfrac{\text{day}}{1440 \text{ min}}\right)} = 1.25 \, \frac{\text{ft}^2}{\text{day}}$$

and Equation 24.8 yields

$$S = \frac{\alpha^* \, r_c^2}{r_w^2} = \frac{10^{-2} \, (0.125 \text{ ft})^2}{(0.25 \text{ ft})^2} = 0.0025$$

Appendix A

DERIVATION OF COOPER AND JACOB (1946) STRAIGHT-LINE METHOD

In the Theis method (Model 3), drawdown is computed using the following equation

$$s = \frac{Q}{4\pi T} W(u) \qquad (A.1)$$

where s is drawdown, Q is the pumping rate, T is the transmissivity, and W(u) is a well function

$$W(u) = \int_u^\infty \left[\frac{e^{-u}}{u} \right] du$$

$$= -0.577216 - \ln(u) + u - u^2(2 \cdot 2!) + u^3(3 \cdot 3!) - u^4(4 \cdot 4!) + \ldots \qquad (A.2)$$

where u is defined

$$u = \frac{r^2 S}{4Tt} \qquad (A.3)$$

and r is the radial distance from the pumping well and t is time. Neglecting all but the first two terms in the expression for W(u) we have

$$s = \frac{Q}{4\pi T} \left(-0.577216 - \ln\left(\frac{r^2 S}{4Tt}\right) \right)$$

$$= \frac{Q}{4\pi T} \left(-\ln(1.78) + \ln\left(\frac{4Tt}{r^2 S}\right) \right)$$

$$S = \frac{Q}{4\pi T} \ln\left(\frac{4Tt}{1.78 r^2 S}\right)$$

$$= \frac{2.3Q}{4\pi T} \log\left(\frac{4Tt}{1.78 r^2 S}\right)$$

$$= \frac{0.183Q}{T} \log\left(\frac{2.25\ Tt}{r^2 S}\right) \qquad (A.4)$$

The change in drawdown Δs for an interval of time Δt is

$$\Delta s = \frac{0.183Q}{T} \log\left(\frac{2.25\ T\Delta t}{r^2S}\right)$$

$$= \frac{0.183Q}{T}\left[\log\left(\frac{2.25T}{r^2S}\right) + \log(\Delta t)\right]$$

$$= \frac{0.183Q}{T} \log\left(\frac{2.25T}{r^2S}\right) + \frac{0.183Q}{T} \log(\Delta t) \qquad (A.5)$$

Equation A.5 is the equation of a straight line on a plot of drawdown vs log(time). The slope of the line is

$$\frac{\Delta s}{\log(\Delta t)} = \frac{0.183Q}{T} \qquad (A.6)$$

If Δs is measured over one log cycle of time, $\log(\Delta t) = \log(10) = 1$ and we have

$$T = \frac{0.183Q}{\Delta s} \qquad (A.7)$$

When drawdown is zero, Equation A.4 can be written

$$s = 0 = \frac{0.183\ Q}{T} \log\left(\frac{2.25\ Tt_o}{r^2S}\right) \qquad (A.8)$$

or

$$S = \frac{2.25\ Tt_o}{r^2} \qquad (A.9)$$

where t_o is the time when $s = 0$.
Equation A.4 can also be used to analyze distance drawdown data. The change in drawdown Δs for an interval of distance Δr is

$$\Delta s = \frac{0.183\ Q}{T} \log\left(\frac{2.25\ Tt}{(\Delta r)^2S}\right)$$

$$= \frac{0.183\ Q}{T}\left[\log\left(\frac{2.25\ Tt}{S}\right) + \log\left(\frac{1}{(\Delta r)^2}\right)\right]$$

$$= \frac{0.183\ Q}{T} \log\left(\frac{2.25\ Tt}{S}\right) + \frac{0.183\ Q}{T} \log\left(\frac{1}{(\Delta r)^2}\right) \qquad (A.10)$$

Expanding the second term on the right hand side of Equation A.10 we have

$$\frac{0.183\ Q}{T} \log\left(\frac{1}{(\Delta r)^2}\right) = -\frac{(2)\ 0.183\ Q}{T} \log(\Delta r) \qquad (A.11)$$

Equation A.10 is the equation of a straight line on a plot of drawdown vs log (distance).

The slope of the line is

$$\frac{\Delta s}{\log (\Delta r)} = -\frac{0.183\ Q\ (2)}{T} = -\frac{0.366\ Q}{T} \tag{A.12}$$

where the negative sign can be dropped if we define Δs to be positive (decreasing drawdown with increasing distance). If Δs is measured over one log cycle log $(\Delta r) = 1$ we have

$$\Delta s = \frac{0.366\ Q}{T} \tag{A.13}$$

When drawdown is zero, Equation A.4 can be written

$$s = 0 = \frac{0.183\ Q}{T} \log \left(\frac{2.25\ Tt}{r^2 S}\right) \tag{A.14}$$

or

$$S = \frac{2.25\ Tt}{r_o^2} \tag{A.15}$$

where r_o is the distance where $s = 0$.

Appendix B

COMPUTER PROGRAMS FOR COMPUTING TYPE CURVES AND DRAWDOWN

Computer programs (written in Basic) referred to in the text are provided on the attached diskettes. These programs will operate on any IBM-compatible microcomputer. One set of programs computes type curve ordinates for use with the match-point method (TYPE3, TYPE4, etc.). Programs are provided for use with each conceptual model that utilizes type curves to analyze test data (e.g., TYPE7 computes type curve ordinates for Model 7). Another set of programs computes drawdown for use in test design (DRAW3, DRAW4, etc.). Programs are provided for each conceptual model except Models 1, 2, 17, and 18; type curves are not required to analyze test data for these four conceptual models and the calculation of drawdown is trivial and did not warrant the development of computer programs for this purpose. A menu program (MENU.EXE) is also provided to simplify program use. Two files are provided for each type curve or drawdown program: the "source code" file (e.g., DRAW3.BAS, DRAW4.BAS, etc.) for use with a Basic interpreter or compiler and the "executable" file (e.g., DRAW3.EXE, DRAW4.EXE, etc.).

Instructions for using the computer programs are contained on the file README.DOC. To print this file put any diskette into drive A and type "PRINT A: README.DOC"

The following table shows the correspondence between text chapters, conceptual model number, and computer program names:

| | | Computer Programs | |
Chapter	Model	Type curve	Drawdown
8	3	TYPE3	DRAW3
9	4	TYPE4	DRAW4
10	5	TYPE5	DRAW5
11	6	TYPE6	DRAW6
12	7	TYPE7	DRAW7
13	8	TYPE8	DRAW8
14	9	TYPE9	DRAW9
15	10	TYPE10	DRAW10
16	11	TYPE11	DRAW11
17	12	TYPE12	DRAW12
18	13	TYPE13	DRAW13
19	14	TYPE14	DRAW14
20	15	TYPE15	DRAW15
21	16	TYPE16	DRAW16
24	19	TYPE19	DRAW19

References

Azzouz, A.S., R.J. Krizek, and R.B. Corotis. 1976. Regression analysis of soil compressibility. Soils and Foundations, Vol. 16, No. 2, pp. 19-29.

Berkaloff, E. 1963. Essai de puits-Interpretation-Nappe libre avec strate conductrice d'eau privilegiee. Bur. Rech. Geol. Min. Rep. DS 63 A 18, Orleans, France.

Bierschenk, W.H. 1964. Determining well efficiency by multiple step-drawdown tests. International Association of Scientific Hydrology, Publication 64, pp. 493-505.

Boulton, N.S. 1954a. The drawdown of the water table under non-steady conditions near a pumped well in an unconfined formation. Proceedings, Institute of Civil Engineers, Vol 3, Part 3, pp. 564-579.

Boulton, N.S. 1954b. Unsteady radial flow to a pumped well allowing for delayed yield from storage. International Association of Science Hydrology, Publication 37, pp. 472-477.

Boulton, N.S. 1963. Analysis of data from nonequilibrium pumping tests allowing for delayed yield from storage. Proceedings, Institute of Civil Engineers, 26(6693), pp. 469-482.

Boulton, N.S. and T.D. Streltsova. 1975. New equations for determining the formation constants of an aquifer from pumping test data. Water Resources Research, Vol. 11, No. 1, pp. 148-153.

Boulton, N.S. and T.D. Streltsova. 1976. The drawdown near an abstraction well of large diameter under non-steady conditions in an unconfined aquifer. Journal of Hydrology, Vol. 30, pp. 29-46.

Boulton, N.S. and T.D. Streltsova. 1977. Unsteady flow to a pumped well in a fissured water-bearing formation. Journal of Hydrology, Vol. 35, pp. 257-270.

Bouwer, H. and R.C. Rice. 1976. A slug test for determining hydraulic conductivity of unconfined aquifers with completely or partially penetrating wells. Water Resources Research. Vol. 12, No. 3, pp. 423-428.

Bowles, J.E. 1982. Foundation analysis and design. Third Edition. McGraw-Hill, Inc., New York, NY. 816 p.

Bowles, J.E. 1984. Physical and geotechnical properties of soils. McGraw-Hill, Inc., New York, NY. 570 p.

Bruin, J. and H.E. Hudson, Jr. 1955. Selected methods for pumping test analysis. Illinois Water Survey, Dept. of Investigations, No. 25.

Clarke, D.K. 1988. Groundwater Discharge Tests: Simulation and Analysis, Developments in Groundwater Science, 37. Elsevier, Amsterdam Oxford/New York/Tokyo. 375 p.

Cooper, H.H., Jr. and C.E. Jacob. 1946. A generalized graphic method for evaluating formation constants and summarizing well-field history. Transactions, American Geophysical Union, Vol. 27, No. 4, pp. 526-534.

Cooper, H.H., Jr., J.D. Bredehoeft, and I.S. Papadopulos. 1967. Response of a finite-diameter well to an instantaneous charge of water. Water Resource Research, Vol. 3, No. 1, pp. 263-269.

de Marsily, G. 1986. Quantitative Hydrogeology. Academic Press, Inc., Orlando, FL. p. 440.

Delhomme, J.P. 1979. Spatial variability and uncertainty in groundwater flow parameters: a geostatistical approach. Water Resources Research, Vol. 15, No. 2, pp. 269-280.

Driscoll, F.G. 1986. Groundwater and wells. Second Edition. Johnson Division, St. Paul, MN. 1089 p.

Earlougher, Jr., R.C. 1977. Advances in well tests analysis. Society of Petroleum Engineers of AIME Monograph 5. 264 p.

Emsellem, Y. and G. deMarsily. 1971. An automatic solution for the inverse problem. Water Resources Research, Vol. 7, No. 5, pp. 1264-1283.

Ferris, J.G. 1959. Groundwater: in Wisler, C.O. and E.F. Brater, Editors. Hydrology, Chapter 7. John Wiley and Sons, Inc., New York, NY.

Ferris, J.G., D.B. Knowles, R.H. Brown, and R.W. Stallman. 1962. Theory of aquifer tests. U.S. Geological Survey. Water-Supply Paper 1536-E. 174 p.

Freeze, R.A. 1971. Three-dimensional, transient, saturated-unsaturated flow in a groundwater basin. Water Resources Research, Vol. 7, No. 2, pp. 347-366.

Freeze, R.A. 1972. Regionalization of hydrogeologic parameters for use in mathematical models of groundwater flow. In Gill, J.E. (editor). Hydrogeology. Harpell's. Gardenvale, Quebec.

Freeze, R.A. 1975. A stochastic-conceptual analysis of one-dimensional groundwater flow in nonuniform homogeneous media. Water Resources Research, Vol. 11, No. 45, pp. 725-741.

Freeze, R.A. and J.A. Cherry. 1979. Groundwater. Prentice-Hall, Inc., Englewood Cliffs, NJ. 604 p.

Gambolati, G. 1976. Transient free surface flow to a well: An analysis of theoretical solutions. Water Resources Research, Vol. 12, pp. 27-39.

Gringarten, A.C. and H.J. Ramey. 1974. Unsteady state pressure distributions created by a well with a single horizontal fracture, partial penetration, or restricted entry. Society of Petroleum Engineers Journal, pp. 413-426.

Guitjens, J.G. and J. N. Luthin. 1971. Effect of soil moisture hysteresis on the water table profile around a gravity well. Water Resources Research, Vol. 7, No. 2, pp. 334-346.

Hantush, M.S. 1956. Analysis of data from pumping tests in leaky aquifers. Transactions, American Geophysical Union, Vol. 37, No. 6, pp. 702-714.

Hantush, M.S. 1960. Modification of the theory of leaky aquifers. Journal of Geophysical Research, Vol. 65, No. 11, pp. 3713-3725.

Hantush, M.S. 1961. Aquifer tests on partially penetrating wells. Journal of Hydraulics Div., Proceedings American Society of Civil Engineers., HY5, pp. 171-195.

Hantush, M.S. 1964. Hydraulics of Wells. In Chow, V.T. Ed., Advances in Hydroscience. Vol. 1. Academic Press. New York/London. pp. 281-442.

Hantush, M.S. 1967. Flow of groundwater in relatively thick leaky aquifers. Water Resources Research, Vol. 3, No. 2, pp. 583-590.

Hantush, M.S. and C.E. Jacob. 1955. Non-steady radial flow in an infinite leaky aquifer. Transactions, American Geophysical Union, Vol. 36, No. 1, pp. 95-100.

Holtz, R.D. and W.D. Kovacs. 1981. An introduction to geotechnical engineering. Prentice-Hall, Inc. Englewood Cliffs, NJ. 733 p.

Hvorslev, M.J. 1951. Time lag and soil permeability in groundwater observations. U.S. Army Corps of Engineers, Waterways Experiment Station Bulletin 36. Vicksburg, MS. 50 p.

Hunt, B.E. 1983. Mathematical analysis of groundwater resources. Butterworth & Co.. 271 p.

Jacob, C.E. 1940. On the flow of water in an elastic artesian aquifer. Transactions, American Geophysical Union, pp. 574-586.

Jacob, C.E. 1944. Notes on determining permeability by pumping tests under watertable conditions. U.S. Geological Survey, Open File Report.

Jacob, C.E. 1946. Radial flow in a leaky artesian aquifer. Transactions, American Geophysical Union, Vol. 27, No II, pp. 198-208.

Journel, A.G. and CH.J. Huijbregts. 1978. Mining geostatistics. Academic Press Inc., London. 600 p.

Kirkham, D. 1946. Proposed method for field measurement of permeability of soil below the water table. Soil Science Society Proceedings, Vol. 28, pp. 58-68.

Kitanidis, P.K. and E.G. Vomvoris. 1983. A geostatistical approach to the inverse problem in groundwater modeling (steady state) and one-dimensional simulations. Water Resources Research, Vol. 19, No. 3, pp. 677-690.

Kroszynski, V.I. and G. Dagan. 1975. Well pumping in unconfined aquifers: The influence of the unsaturated zone. Water Resources Research, Vol. 11, No. 3, pp. 479-490.

Lai, R.Y.S. and C.W. Su. 1974. Nonsteady flow to a large well in a leaky aquifer. Journal of Hydrology, Vol. 22, pp. 333-345.

Leonards, G.A. 1976. Estimation consolidation settlements of shallow foundations on overconsolidated clays. Special Report 163, Transportation Research Board, pp. 13-16.

Mansur, C.I. and R.I. Kaufman. 1962. Dewatering: in Leonards, G.A. Editor. Foundation Engineering, Chapter 3. McGraw-Hill Inc., New York, NY.

Maslia, M.L. and R.B. Randolph. 1987. Methods and computer program documentation for determining anisotropic transmissivity tensor components of two-dimensional ground-water flow. U.S.G.S. Water-Supply Paper 2308. 46 p.

Moench, A.F. and T.A. Prickett. 1972. Radial flow in an infinite aquifer undergoing conversion from artesian to water table conditions. Water Resources Research, Vol 8, No. 2, pp. 494-499.

Muskat, M. 1937. The flow of homogeneous fluids through porous media. McGraw-Hill Book Company, Inc. New York, NY. 763 p.

Neuman, S.P. 1972. Theory of flow in unconfined aquifers considering delayed response of the water table. Water Resources Research, Vol. 8, No. 4, pp. 1031-1045.

Neuman, S.P. 1973a. Supplementary comments on 'Theory of flow in unconfined aquifers considering delayed response of the water table'. Water Resources Research, Vol. 9, No. 4, pp. 1102-1103.

Neuman, S.P. 1973b. Calibration of distributed parameter groundwater flow models viewed as a multiple-objective decision process under uncertainty. Water Resources Research, Vol. 9, No. 4, pp. 1006-1021.

Neuman, S.P. 1974. Effect of partial penetration on flow in unconfined aquifers considering delayed gravity response. Water Resources Research, Vol. 10, No. 2, pp. 303-312.

Neuman, S.P. 1975. Analysis of pumping test data from anisotropic unconfined aquifers considering delayed gravity response. Water Resources Research, Vol. 11, No. 2, pp. 329-342.

Neuman, S.P. 1979. Perspective on 'delayed yield'. Water Resources Research, Vol. 15, No. 4, pp. 899-908.

Neuman, S.P. and P.A. Witherspoon. 1969a. Theory of flow in a confined two aquifer system. Water Resources Research, Vol. 5, No. 4, pp. 803-816.

Neuman, S.P. and P.A. Witherspoon. 1969b. Applicability of current theories of flow in leaky aquifers. Water Resources Research, Vol. 5, No. 4, pp. 817-829.

Neuman, S.P. and P.A. Witherspoon. 1972. Field determination of the hydraulic properties of leaky multiple aquifer systems. Water Resources Research, Vol. 8, No. 5, pp. 1284-1298.

Neuman, S.P. and D.A. Gardner. 1989. Determination of aquitard/aquiclude hydraulic properties from arbitrary water-level fluctuations by deconvolation. Groundwater, Vol. 27, No. 1, pp. 66-76.

Nguyen, V. and G.F. Pinder. 1984. Direct calculation of aquifer parameters in slug test analysis. IN: Groundwater hydraulics, Rosenshein, J. and G.D. Bennett, Editors. American Geophysical Union, Water Resources Monograph 9, pp. 22-240.

Papadopulos, I.S. 1965. Nonsteady flow to a well in an infinite anisotropic aquifer. International Association of Scientific Hydrology Symposium, Dubrovnik, Oct., Vol. 1, No. 73, pp. 21-31.

Papadopulos, I.S. 1967. Drawdown distributor around a large-diameter well. Proceedings of the National Symposium on Ground-Water Hydrology, San Francisco. Nov. 1967. Am. Water Resources Assoc. Proc., No. 4, pp. 157-168.

Papadopulos, I.S. and H.H. Cooper, Jr., 1967. Drawdown in a well of large diameter. Water Resource Research, Vol. 3, No. 1, pp. 241-244.

Papadopulos, I.S. and J.D. Bredehoeft, and H.H. Cooper, Jr. 1973. On the analysis of 'slug test' data. Water Resource Research, Vol. 9, No. 4, pp. 1087-1089.

Peck, A., S. Gorelick, G. deMarsily, S. Foster, and V. Kovalevsky. 1988. Consequences of spatial variability in aquifer properties and data limitations for groundwater modeling practice. International Association of Hydrological Sciences Publication No. 175. IAHS Press. Institute of Hydrology, Wallingford, Oxfordshire, U.K. 272 p.

Peck, R.B., W.E. Hanson, and T.H. Thornburn. 1974. Foundation engineering. John Wiley & Sons, New York, NY. 514 p.

Prickett, T.A. 1965. Type-curve solution to aquifer tests under water-table conditions. Groundwater, Vol. 3, No. 3, pp. 5-14.

Reed, J.E. 1980. Type curves for selected problems of flow to wells in confined aquifers. Techniques of water resources investigations of the U.S. Geological Survey, Book 3, Chapter B3, pp. 106.

Rorabaugh, M.I. 1953. Graphical and theoretical analysis of step-drawdown test of artesian well. Proceedings, American Society of Civil Engineers, Vol. 79.

Sammel, E.A. 1974. Aquifer tests in large-diameter wells in India. Groundwater, Vol. 12, No. 5, pp. 265-272.

Stallman, R.W. 1956. Numerical analysis of regional water levels to define aquifer hydrology. Transactions, American Geophysical Union, Vol. 37, No. 4, pp. 451-460.

Streeter, V.L. and E.B. Wylie. 1979. Fluid Mechanics. McGraw-Hill, Inc., New York, NY. 562 p.

Streltsova, T.D. 1972. Unsteady radial flow in an unconfined aquifer. Water Resources Research, Vol. 8, No. 4, pp. 1059-1066.

Streltsova, T.D. 1974. Drawdown in compressible unconfined aquifer. Journal of the Hydraulics Division, American Society of Civil Engineers, HY11, pp. 1601-1616.

Taylor, G.S. and J.N. Luthin. 1969. Computer methods for transient analysis of water-table aquifers. Water Resources Research, Vol. 5, No. 1, pp. 144-152.

Terzaghi, K. and R.B. Peck. 1967. Soil mechanics in engineering practice. John Wiley & Sons, New York, NY. 729 p.

Theis, C.V. 1935. The relation between the lowering of the piezometric surface and the rate and duration of discharge of a well using ground-water storage. Transactions, American Geophysical Union, Vol. 16, pp. 519-524.

Thiem, G. 1906. Hydrologische methoden (Hydrologic methods). J.M. Gebhardt. Leipzig. 56 p.

U.S. Army Corps of Engineers. 1956. Investigation of underseepage and its control, Lower Mississippi River Levees. Waterways Experiment Station. TM 3-424, Vicksburg, MS.

Walton, W.C. 1960. Application and limitation of methods used to analyze pumping test data. Water Well Journal. Feb-March.

Walton, W.C. 1970. Groundwater Resource Evaluation. McGraw-Hill Kogakusha, Ltd. Tokyo. 664 p.

Walton W.C. 1984. Practical aspects of groundwater modeling. National Water Well Association. 566 p.

Walton, W.C. 1988. Groundwater Pumping Tests. Lewis Publishers, Inc., Chelsea, MI. 201 p.

Warren, J.E. and P.J. Root. 1963. The behavior of naturally fractured reservoirs. Society of Petroleum Engineers Journal, Vol. 9, pp. 245-255.

Witherspoon, P.A. and S.P. Neuman. 1972. Hydrodynamics of fluid injection. American Association of Petroleum Geologists, Memoir No. 18, pp. 258-272.

Yeh, W.W.-G., Y.S. Yoon, and K.S. Lee. 1983. Aquifer parameter identification with kriging and optimum parameterization. Water Resources Research, Vol. 19, No. 1, pp. 225-233.

Youngs, E.G. 1968. Shape factors for Kirkham's piezometer method for determining the hydraulic conductivity of soil in situ for soils overlying an impermeable floor or infinitely permeable stratum. Soil Science, Vol. 106, No. 3, pp. 235-237.

Index